Lecture Notes in Physics

Volume 1025

The series Lecture Notes in Physics (LNP), founded in 1969, reports new developments in physics research and teaching - quickly and informally, but with a high quality and the explicit aim to summarize and communicate current knowledge in an accessible way. Books published in this series are conceived as bridging material between advanced graduate textbooks and the forefront of research and to serve three purposes:

- to be a compact and modern up-to-date source of reference on a well-defined topic;
- to serve as an accessible introduction to the field to postgraduate students and non-specialist researchers from related areas;
- to be a source of advanced teaching material for specialized seminars, courses and schools.

Both monographs and multi-author volumes will be considered for publication. Edited volumes should however consist of a very limited number of contributions only. Proceedings will not be considered for LNP.
Volumes published in LNP are disseminated both in print and in electronic formats, the electronic archive being available at springerlink.com. The series content is indexed, abstracted and referenced by many abstracting and information services, bibliographic networks, subscription agencies, library networks, and consortia.
Proposals should be sent to a member of the Editorial Board, or directly to the responsible editor at Springer:

Dr Lisa Scalone
lisa.scalone@springernature.com

Ramón Aguado • Roberta Citro •
Maciej Lewenstein • Michael Stern
Editors

New Trends and Platforms for Quantum Technologies

Editors

Ramón Aguado
Institute of Materials Science Madrid
Spanish National Research Council
Madrid, Spain

Roberta Citro
Department of Physics "E.R.Caianiello"
University of Salerno
Fisciano, Italy

Maciej Lewenstein
Institute of Photonic Sciences
Castelldefels (Barcelona), Spain

Michael Stern
Physics Department
Bar-Ilan University
Ramat Gan, Israel

ISSN 0075-8450 ISSN 1616-6361 (electronic)
Lecture Notes in Physics
ISBN 978-3-031-55656-2 ISBN 978-3-031-55657-9 (eBook)
https://doi.org/10.1007/978-3-031-55657-9

This Springer imprint is published by the registered company Springer Nature Switzerland AG
The registered company address is: Gewerbestrasse 11, 6330 Cham, Switzerland

Preface

This book is a comprehensive introduction to quantum technologies (QT) and quantum computing platforms, inspired by recent experimental advancements in detecting and manipulating single quantum objects. It covers recent progresses on QT from theoretical and experimental point of view which would be beneficial to researchers approaching the field. The readers will also strongly benefit from the different overviews on the topic.

The organization of the book is the following: Chap. 1 covers the Josephson phenomenon and the Josephson junction devices with application in quantum computing; Chap. 2 describes the superconducting quantum circuits with a focus on the superconducting qubits and their use as quantum gates via the coupling to resonators; Chap. 3 is a comprehensive overview of semiconductor-superconductor devices where Andreev qubits can be realized; Chap. 4 is an introduction to quantum entanglement and presents different applications in many-body systems.

Madrid, Spain
Fisciano, Italy
Castelldefels (Barcelona), Spain
Ramat Gan, Israel
May 2024

Ramón Aguado
Roberta Citro
Maciej Lewenstein
Michael Stern

Contents

1 Josephson Junctions, Superconducting Circuits, and Qubit for Quantum Technologies 1
Roberta Citro, Claudio Guarcello, and Sergio Pagano
1.1 The Josephson Effect 1
1.1.1 First Josephson Equation 2
1.1.2 Second Josephson Equation 4
1.1.3 Estimation of the Maximum Josephson Current Density 5
1.1.4 Anomalous Josephson Effect 7
1.1.5 Short JJ: The Resistively and Capacitively Shunted Junction (RCSJ) Model 10
1.1.6 Long JJ: The Sine-Gordon Model 18
1.1.7 The Superconducting Quantum Interference Device (SQUID) 24
1.2 Josephson Devices 30
1.2.1 JJ Fabrication Technology 31
1.2.2 Voltage Standard 35
1.2.3 SQUIDs 37
1.2.4 Classical Digital Circuits 40
1.3 Towards Quantum Computing with Superconducting Qubits 42
1.3.1 Bloch Sphere Representation 42
1.3.2 Qubit Control 43
1.3.3 Boolean Logic Gates in Classical Computers 44
1.3.4 Quantum Logic Gates Used in Quantum Computers 46
1.3.5 Classical Versus Quantum Gates 49
1.3.6 Further Developments for Superconducting Qubits 51
References 52

2 Introduction to Superconducting Quantum Circuits 61
Michael Stern
2.1 Quantization of an Electrical Circuit 62
2.1.1 The Lumped-Element Circuit Model 62
2.1.2 Quantization of a Lumped Element Circuit 64
2.1.3 Transmission Lines 67
2.1.4 Quantization of a Transmission Line 70

2.1.5 Transmission Line Resonators ... 74
2.1.6 Quantization of Transmission Line Resonators ... 79
2.2 Superconducting Qubits ... 81
2.2.1 Using the Non-linearity of Josephson Junctions ... 81
2.2.2 The Charge Qubit ... 82
2.2.3 The Superconducting Flux Qubit ... 87
2.3 Coupling Qubits and Resonators ... 94
2.3.1 Coupling a Qubit with a Resonator ... 94
2.3.2 Single Qubit Gates ... 98
2.3.3 Two-Qubit Gates ... 99
2.4 Relaxation and Decoherence ... 105
2.4.1 Relaxation ... 106
2.4.2 Pure Dephasing ... 114
2.4.3 Decoherence Under Driven Evolution ... 121
2.5 Appendix ... 122
2.5.1 Master Equation Formalism ... 122
2.5.2 Schrieffer Wolff Transformation ... 128
References ... 130

3 Subgap States in Semiconductor-Superconductor Devices for Quantum Technologies: Andreev Qubits and Minimal Majorana Chains ... 133
Rubén Seoane Souto and Ramón Aguado
3.1 Subgap States in Superconductors ... 136
3.1.1 Bogoliubov de Gennes Quasiparticles in Superconductors ... 137
3.1.2 Andreev Bound States ... 140
3.1.3 Subgap Levels in Quantum Dot Junctions ... 141
3.2 Kitaev Chain: Basics ... 148
3.2.1 Majorana Zero Modes ... 148
3.2.2 The Kitaev Model ... 149
3.2.3 Majorana Non-Abelian Properties ... 154
3.3 Top Down Approach Toward Topological Superconductivity ... 156
3.3.1 The Semiconductor-Superconductor Platform ... 157
3.4 Bottom-Up Approach: Minimal Kitaev Chains ... 161
3.4.1 Majorana States in Double Quantum Dots ... 162
3.5 Hybrid Semiconductor-Based Superconducting Qubits ... 181
3.5.1 Qubits Based on a Single Andreev State ... 182
3.5.2 Beyond Single Andreev States ... 189
3.6 Coherent Experiments with Majorana States ... 192
3.6.1 Topological Qubits ... 193
3.6.2 Majorana Fusion Rules ... 196
3.6.3 Majorana Braiding ... 197
3.6.4 Coherent Experiments with PMMs ... 200
3.6.5 Non-abelian Experiments with PMMs ... 205
References ... 209

4 Introduction to Quantum Entanglement in Many-Body Systems 225
Anubhav Kumar Srivastava, Guillem Müller-Rigat, Maciej Lewenstein, and Grzegorz Rajchel-Mieldzioć
4.1 Introduction ... 226
4.2 Mathematical Foundations and Bipartite Systems ... 227
4.2.1 Bipartite Systems ... 230
4.2.2 Detection of Entanglement ... 233
4.2.3 Bell Nonlocality ... 240
4.2.4 Applications of Entanglement ... 241
4.3 Entanglement in Multipartite Systems ... 242
4.3.1 To Be *Separable* or Not to Be *Separable* ... 243
4.3.2 Multipartite Entanglement as a Resource ... 245
4.4 Use and Detection of Many-Body Quantum Entanglement ... 254
4.4.1 Useful (and Useless) Many-Body Entanglement ... 257
4.4.2 Scalable Entanglement Certification ... 266
4.5 Open Problems ... 272
References ... 274

Contributors

Ramón Aguado Instituto de Ciencia de Materiales de Madrid (ICMM), Consejo Superior de Investigaciones Científicas (CSIC), Madrid, Spain

Roberta Citro Department of Physics "E.R. Caianello", University of Salerno, Fisciano, Italy
INFN, Sezione di Napoli Gruppo Collegato di Salerno, Complesso Universitario di Monte S. Angelo, Napoli, Italy
CNR-SPIN c/o Universitá degli Studi di Salerno, Fisciano (Sa), Italy

Claudio Guarcello Department of Physics "E.R. Caianello", University of Salerno, Fisciano, Italy
INFN, Sezione di Napoli Gruppo Collegato di Salerno, Complesso Universitario di Monte S. Angelo, Napoli, Italy

Maciej Lewenstein ICFO - Institut de Ciencies Fotoniques, The Barcelona Institute of Science and Technology, Castelldefels (Barcelona), Spain
ICREA, Barcelona, Spain

Guillem Müller-Rigat ICFO - Institut de Ciencies Fotoniques, The Barcelona Institute of Science and Technology, Castelldefels (Barcelona), Spain

Sergio Pagano Department of Physics "E.R. Caianello", University of Salerno, Fisciano, Italy
INFN, Sezione di Napoli Gruppo Collegato di Salerno, Complesso Universitario di Monte S. Angelo, Napoli, Italy

Grzegorz Rajchel-Mieldzioć ICFO - Institut de Ciencies Fotoniques, The Barcelona Institute of Science and Technology, Castelldefels (Barcelona), Spain
NASK National Research Institute, Warszawa, Poland

Rubén Seoane Souto Instituto de Ciencia de Materiales de Madrid (ICMM), Consejo Superior de Investigaciones Científicas (CSIC), Madrid, Spain

Anubhav Kumar Srivastava ICFO - Institut de Ciencies Fotoniques, The Barcelona Institute of Science and Technology, Castelldefels (Barcelona), Spain

Michael Stern Department of Physics, Quantum Nanoelectronics Laboratory, Bar Ilan University, Ramat Gan, Israel

Josephson Junctions, Superconducting Circuits, and Qubit for Quantum Technologies

1

Roberta Citro, Claudio Guarcello, and Sergio Pagano

Abstract

In the realm of physics, a pivotal moment occurred six decades ago when Brian Josephson made a groundbreaking prediction, setting in motion a series of events that would eventually earn him the prestigious Nobel Prize 11 years later. This prediction centered around what is now known as the Josephson effect, a phenomenon with far-reaching implications. At the heart of this effect lies the Josephson junction (JJ), a device that has become a linchpin in various scientific applications. This chapter delves into the foundational principles of the Josephson effect and the remarkable properties of JJs. From their role in metrology to their application in radiation detectors, these junctions have ushered in a new era of electronics. Exploiting the unique features of superconductive devices, such as high speed, low dissipation, and dispersion, JJs find today practical implementation in the development of superconductive qubits and nanotechnology applications.

1.1 The Josephson Effect

The Josephson effect is a quantum phenomenon observed in superconducting systems. In its simplest form, the Josephson effect involves the flow of supercurrent (a current without resistance) between two superconductors separated by a thin insulating barrier, typically a JJ. The key feature of the Josephson effect is the phase coherence of the wave functions of the superconducting electrons on either side of the barrier, as we will see in this Chapter. There are two main types of Josephson

R. Citro (✉) · C. Guarcello · S. Pagano
Department of Physics "E.R. Caianiello", University of Salerno, Fisciano, Italy
e-mail: rocitro@unisa.it; cguarcello@unisa.it; spagano@unisa.it

R. Aguado et al. (eds.), *New Trends and Platforms for Quantum Technologies*, Lecture Notes in Physics 1025, https://doi.org/10.1007/978-3-031-55657-9_1

effects: (a) *dc Josephson effect* when a constant voltage is applied across the JJ, resulting in a steady supercurrent flow through the junction. The current is directly proportional to the sine of the phase difference between the wave functions of the superconducting electrons on either side of the barrier; (b) *ac Josephson effect* when an alternating voltage is applied across the JJ, leading to an alternating supercurrent that oscillates at the same frequency as the applied voltage. This phenomenon is used in various applications such as superconducting quantum interference devices (SQUIDs), Josephson voltage standards, and superconducting qubits for quantum computing.

The Josephson effect has significant practical applications in areas such as metrology, quantum computing, and high-speed digital electronics. It also provides insights into the fundamental properties of superconductors and quantum mechanics.

1.1.1 First Josephson Equation

When thinking about the derivation of the Josephson equations, one usually refers to Feynman's formulation [1], which is based on a "two-level system" picture. The interested reader can find a comprehensive overview of this method in the Barone and Paternó's book [2]. Here, we follow a different line of thinking, proposing an approach due to Landau and Lifschitz [3] and later discussed in Ref. [4].

Let's start with some considerations. First, it is reasonable to assume that the supercurrent depends on the density of Cooper pairs in the superconductors forming the junction, $|\Psi_1|^2 = n^\star_{s,1}$ and $|\Psi_2|^2 = n^\star_{s,2}$. Furthermore, since the coupling between the two superconductors is "weak", we can also assume that the supercurrent density between the two superconducting electrodes does not change $|\Psi|^2$. On the other side, it is reasonable to expect that the supercurrent density depends on the phases of the wave functions. In a bulk superconductor, the supercurrent density is proportional to the gauge-invariant phase gradient $\gamma(\mathbf{r}, t)$ according to

$$J_s(\mathbf{r}, t) = \frac{q^\star n^\star_s \hbar}{m^\star} \gamma(\mathbf{r}, t) \qquad \text{where} \qquad \gamma(\mathbf{r}, t) = \nabla\theta - \frac{2\pi}{\Phi_0}\mathbf{A}(\mathbf{r}, t). \tag{1.1}$$

Here, $\Phi_0 = h/(2e)$ is the flux quantum, $\hbar = h/(2\pi)$ is the reduced Planck constant, $\mathbf{A}$ is the potential vector, $m^\star$ and $q^\star$ are the charge and the mass of superelectrons, respectively.

We simplify our discussion by introducing two key assumptions. Firstly, we posit that the current density is uniformly distributed, namely, the junction area is sufficiently small. Secondly, we consider two weakly connected superconductors and that the phase gradient in the superconducting electrodes undergoes negligible variation: this condition holds when the Cooper pairs density in the electrodes is greater than in the coupling region. Since the supercurrent density remains constant in the electrodes (owing to current conservation), the only relevant phase gradient is that in the interlayer region, in accordance with Eq. (1.1). Thus, we can focus on the *gauge-invariant phase difference*, given by

$$\varphi(\mathbf{r},t) = \int_1^2 \gamma(\mathbf{r},t) = \int_1^2 \left(\nabla\theta(\mathbf{r},t) - \frac{2\pi}{\Phi_0}\mathbf{A}(\mathbf{r},t)\right) d\mathbf{l}$$
$$= \theta_2(\mathbf{r},t) - \theta_1(\mathbf{r},t) - \frac{2\pi}{\Phi_0}\int_1^2 \mathbf{A}(\mathbf{r},t)\, d\mathbf{l}, \tag{1.2}$$

with the integration path along the flow direction of the current. One can reasonably assume that the supercurrent density is proportional to the phase difference, i.e, $J_s = J_s(\varphi)$. Moreover, since any 2π-phase shift yields an identical wave function of the superconducting electrodes, we can expect even $J_s(\varphi)$ to be a 2π-periodic function, i.e., $J_s(\varphi) = J_s(\varphi + 2\pi\, n)$. Finally, with no current applied $\varphi = 0$, i.e. $\theta_1 = \theta_2$, and thus $J_s(0) = J_s(2\pi\, n) = 0$. Summing up, we can say that

$$J_s(\varphi) = J_c \sin\varphi + \sum_{m=2}^{\infty} J_m \sin(m\varphi), \tag{1.3}$$

with J_c being the Josephson critical current density, which depends on the coupling between the superconductors. In general, time-invariance of the Josephson current requires that $J_s(\varphi) = -J_s(-\varphi)$, and this is why all cosine terms in the Fourier series forming $J_s(\varphi)$ are ignored. Equation (1.3) is usually refereed to as the *1st Josephson relation*, or even as the *current-phase relation* (CPR). Often, only the first term of the expansion is relevant, i.e.,

$$J_s(\varphi) = J_c \sin\varphi, \tag{1.4}$$

as it was derived by Josephson in his original article for an insulating barrier [5, 6].

So far we performed a 1D analysis by considering a uniform supercurrent density: however, the reasoning presented remains still valid if applied to each point (y, z) of the junction area. In particular, we can assume $J_c = J_c(y, z)$ and J_s flowing perpendicularity to the junction area for any given y and z, since the current flows in the x-direction. Thus, the current density may vary with y and z and the CPR has to be generalized to

$$J_s(y, z, t) = J_c(y, z) \sin\varphi(y, z, t). \tag{1.5}$$

A tunnel JJ represents a sort of bottleneck for the current density in the superconducting channel, being the Josephson critical current, i.e., the maximum Josephson current that can flow without triggering a voltage state, far smaller than the usual depairing current in superconducting electrodes [7]; from this aspect comes the term “weak link”, which serves to capture the idea that the JJ is a region where deviations from ideal superconducting behavior are most significant, impacting both the tunneling of Cooper pairs (Josephson effect) and the stability of superconductivity (depairing).

1.1.2 Second Josephson Equation

To derive the link between the voltage drop and the Josephson phase, we start from the time derivative of Eq. (1.2),

$$\frac{\partial \varphi}{\partial t} = \frac{\partial \theta_2}{\partial t} - \frac{\partial \theta_1}{\partial t} - \frac{2\pi}{\Phi_0}\frac{\partial}{\partial t}\int_1^2 \mathbf{A}\, d\mathbf{l}\,. \tag{1.6}$$

Then, we recall that in a bulk superconductor, with $n_s^\star = constant$, the following relation holds:

$$-\hbar\frac{\partial \theta}{\partial t} = \frac{1}{2n_s^\star}\Lambda J_s^2 + q^\star \varphi. \tag{1.7}$$

Inserting this expression into Eq. (1.6) and for a continuous supercurrent density, i.e., $J_s(2) = J_s(1)$, one obtains

$$\frac{\partial \varphi}{\partial t} = \frac{2\pi}{\Phi_0}\int_1^2 \left(-\nabla\phi - \frac{\partial \mathbf{A}}{\partial t}\right) d\mathbf{l}. \tag{1.8}$$

Since the contribution in parentheses is nothing more than the electric field, this equation can be recast as

$$\frac{\partial \varphi}{\partial t} = \frac{2\pi}{\Phi_0}\int_1^2 \mathbf{E}(\mathbf{r}, t) d\mathbf{l}, \tag{1.9}$$

obtaining the so-called *2nd Josephson relation*, or, alternatively, the *voltage-phase relation*. In fact, the integral in this equation is the potential difference V, which turns out in the chemical potential difference, $\Delta\mu = \mu_2 - \mu_1 = eV$, between the two superconductors. In the case of a constant voltage drop applied to the junction, one simply obtain

$$\frac{\partial \varphi}{\partial t} = \frac{2\pi}{\Phi_0}V, \tag{1.10}$$

that is the Josephson phase grows linearly in time according to

$$\varphi(t) = \varphi_0 + \frac{2\pi}{\Phi_0}V\,t. \tag{1.11}$$

In this case, the Josephson current oscillates in time at a specific frequency, $\nu/V = \Phi_0^{-1} \simeq 483.6\,\mathrm{MHz}/\mu\mathrm{V}$. In other words, a JJ can work as a voltage-controlled oscillator capable of producing very high frequencies ($\sim$ 500 GHz at 1 mV). According to these equations, exciting a junction with a signal at a frequency ν would result in the emergence of constant voltage regions on its current-voltage (IV) curve at values $n\Phi_0\nu$. Shapiro experimentally validated this prediction in 1963 [8], and it is now recognized as the *ac Josephson effect*. The practical implications of this phenomenon were promptly applied to metrology, as it establishes a connection between the volt and the second via a proportionality reliant solely on fundamental constants. Initially, this aspect contributed to a refined determination of the h/e ratio. Presently, it serves as the foundation for voltage standards globally [9].

1.1.3 Estimation of the Maximum Josephson Current Density

This section discusses the determination, with general arguments, of the maximum value of the Josephson current density, i.e., the *critical current density* J_c, in the case of a superconductor-insulator-superconductor (SIS) junction, see Fig. 1.1a. In particular, we address this situation with the so-called *wave-matching method*, i.e., solving the Schrödinger problem in both the superconducting electrodes and in the insulating region and looking at the boundary conditions to determine the coefficients for matching the solutions in these regions.

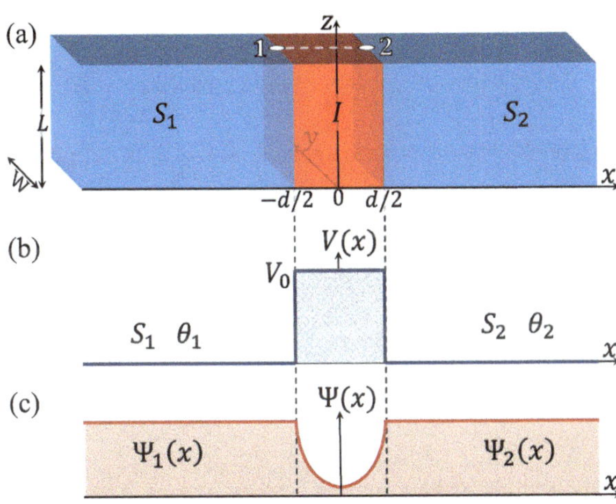

Fig. 1.1 (**a**) Sketch of an SIS junction, formed by sandwiching two superconductors, S_1 and S_2, on an insulating layer, I, of thickness d. The junction area $A = LW$ lies in the (y, z) plane, while the current flow through A occurs along x. (**b**) Potential barrier $V(x) = V_0$ for $|x| \leq d/2$ and 0 otherwise. (**c**) Sketching of the profile of a wave function $\Psi(x)$, see Eq. (1.14)

Looking at the superconducting electrodes, the supercurrent density is given by Eq. (1.1) at the positions $x = \pm d/2$. In the absence of electric/magnetic fields, from Eq. (1.7) we get

$$\frac{\partial \theta}{\partial t} = -\frac{\Lambda J_s^2}{2 n_s^\star \hbar} = -\frac{E_0}{\hbar}, \tag{1.12}$$

with $E_0 = m^\star v_s^2/2$ being the kinetic energy of superelectrons. The time-dependent, macroscopic wave function can be expressed as $\Psi(\mathbf{r}, t) = \Psi(\mathbf{r}) \exp[-i(E_0/\hbar)t]$, where $\Psi(\mathbf{r})$ indicates its time-independent amplitude.

Let's consider now the insulating region. As depicted in Fig. 1.1b, the potential is considered to be zero outside this area and equal to $V_0 > E_0$ inside it. We assume only elastic tunneling processes, that is in which superelectrons conserve energy, so that the time evolution of $\Psi(\mathbf{r}, t)$ is the same outside and inside the barrier and it is therefore sufficient to regard the time-independent part of the problem. Thus, we can write down a time-independent Schrödinger-like equation

$$-\frac{\hbar^2}{2m^\star} \nabla^2 \Psi(\mathbf{r}) = (E_0 - V_0)\Psi(\mathbf{r}). \tag{1.13}$$

At this point, we need to additional assumptions: (1) a uniform tunneling barrier and (2) a small junction area $A = LW$, so that the Josephson current density can be considered uniform within A and Eq. (1.13) reduces just to a one-dimensional problem in the x direction, whose solution is

$$\Psi(x) = A \cosh(\kappa x) + B \sinh(\kappa x), \tag{1.14}$$

where the *characteristic decay constant* κ is defined as

$$\kappa = \sqrt{\frac{2m^\star}{\hbar^2}(V_0 - E_0)}. \tag{1.15}$$

Taking into account the boundary conditions,

$$\Psi(-d/2) = \sqrt{n_1^\star} e^{i\theta_1} \quad \text{and} \quad \Psi(d/2) = \sqrt{n_2^\star} e^{i\theta_2}, \tag{1.16}$$

the coefficients become

$$A = \frac{\sqrt{n_1^\star} e^{i\theta_1} + \sqrt{n_2^\star} e^{i\theta_2}}{\cosh(\kappa d/2)} \quad \text{and} \quad B = \frac{\sqrt{n_1^\star} e^{i\theta_1} - \sqrt{n_2^\star} e^{i\theta_2}}{\sinh(\kappa d/2)}. \tag{1.17}$$

Finally, inserting the wave function expression given in Eq. (1.14) into the supercurrent density equation yields

$$J_s = \frac{q^\star}{m^\star} \mathrm{Re}\left\{\Psi^\star \left(\frac{\hbar}{i}\nabla\right)\Psi\right\} = \frac{q^\star}{m^\star} \mathrm{Im}\left\{A^\star B\right\}. \tag{1.18}$$

Now, by considering the specific coefficients in Eq. (1.17), one obtains the supercurrent density $J_s = J_c \sin(\theta_2 - \theta_1)$, where

$$J_c = -\frac{q^\star \kappa\hbar}{m^\star}\frac{\sqrt{n_1^\star n_2^\star}}{\sinh(2\kappa d)}. \tag{1.19}$$

A V_0 of the order of a few eV gives a decay length, $1/\kappa$, less than a nanometer, so that $\kappa d \gg 1$, if the barrier thickness d is just a few nanometers. Thus, one can approximate $2\sinh(2\kappa d) \simeq \exp(2\kappa d)$ in Eq. (1.19), obtaining

$$J_c \simeq \frac{e\kappa\hbar}{m}\sqrt{n_1^\star n_2^\star e^{-2\kappa d}}, \tag{1.20}$$

where we replaced $q^\star = -2e$ and $m^\star = 2m$. In other words, with simple assumptions we achieved Eq. (1.20) describing a supercurrent density that decays exponentially as the thickness of the insulating layer is increased.

1.1.4 Anomalous Josephson Effect

The CPR represents the way to calculate most of junction properties. Anyway, only in a few cases the CPR reduces to the familiar sinusoidal form $I_s(\phi) = I_c \sin(\phi)$, which is however ordinarily used to study the dynamics and performance of devices based on conventional JJs.

There are several properties of the CPR that are rather general and depend neither on the junction's materials and geometry nor on the theoretical model used to describe the processes in the junction [10]:

1. A change of phase of the order parameter of 2π in any of the electrodes is not accompanied by a change in their physical state. Consequently, this change must not influence the supercurrent across a junction, and $I_s(\varphi)$ should be a 2π periodic function, $I_s(\varphi) = I_s(\varphi + 2\pi)$;
2. Changing the direction of supercurrent entails a change of the sign of φ; therefore $I_s(\varphi) = -I_s(-\varphi)$. Note that this condition is violated in superconductors with broken time-reversal symmetry (TRS), and this can lead to spontaneous currents;

3. A dc supercurrent can flow only if there is a gradient of the order-parameter phase. Hence, in the absence of phase difference, there should be zero supercurrent, $I_s(2\pi n) = 0$, with $n = 0, \pm 1, \pm 2, \ldots$;
4. It follows from 1. and 2. that the supercurrent should also be zero at $\varphi = \pi n$, that is $I_s(\pi n) = 0$, for $n = 0, \pm 1, \pm 2, \ldots$; therefore, it is sufficient to consider $I_s(\varphi)$ only in the interval $0 < \varphi < \pi$.

Thus, $I_s(\varphi)$ can be in general decomposed into a Fourier series, i.e.,

$$I_s(\varphi) = \sum_{n \geq 1} [I_n \sin(n\varphi) + J_n \cos(n\varphi)]. \tag{1.21}$$

The I_n term depends on the barrier transparency D as a D^n power-law and corresponds to the n-multiple Andreev reflection process, thus these terms become more relevant when increasing the barrier transparency. For satisfying the condition 2., the term $J_n \cos(n\varphi)$ has to be zero, that is, in other words, J_n is present only if TRS is broken.

The first Josephson equation, $I_s(\varphi) = I_c \sin(\varphi)$, represents a particular case of the general Eq. (1.21). A quite simple example of deviation from the simple sinusoidal CPR, in particular with the occurrence of the second harmonic sinusoidal contribution, is given by HTS d-wave JJs and by junctions with ferromagnetic barriers [11–14]. In this case:

$$I_s(\varphi) = I_1 \sin(\varphi) + I_2 \sin(2\varphi). \tag{1.22}$$

From this CPR, the Josephson free energy can be written as

$$F_J(\varphi) = \frac{\Phi_0}{2\pi} \int_0^{\varphi} I_s(\varphi')d\varphi' = -\frac{\Phi_0}{2\pi} I_1 \left[\cos(\varphi) + \frac{I_2}{2I_1}\cos(2\varphi)\right]. \tag{1.23}$$

The ground state, $\tilde{\varphi}$, has to satisfy the condition

$$\left|\frac{dF_J(\varphi)}{d\varphi}\right|_{\varphi=\tilde{\varphi}} \propto I_s(\tilde{\varphi}) = 0. \tag{1.24}$$

Since $I_s(\varphi) = \sin(\varphi)[I_1 + 2I_2\cos(\varphi)]$, the previous equation is satisfied if $\sin(\tilde{\varphi}) = 0$ or $\cos(\tilde{\varphi}) = -\frac{I_1}{2I_2}$. This gives rise to the following cases:

1. If $\left|\frac{I_1}{2I_2}\right| > 1$ and $I_2 > 0$:

- $\tilde{\varphi} = 0$ for $I_1 > 0$. If $I_2 \in [0, 1]$ implies that $I_1 \in [2I_2, 2]$;
- $\tilde{\varphi} = \pi$ for $I_1 < 0$. If $I_2 \in [0, 1]$ implies that $I_1 \in [-2, -2I_2]$;

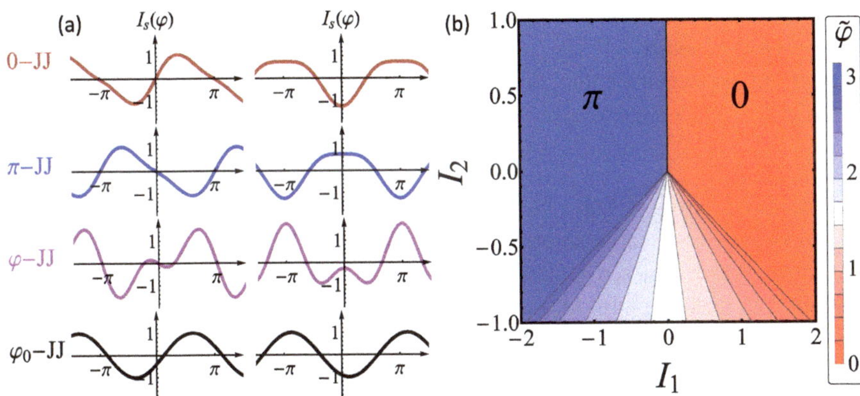

Fig. 1.2 (**a**) CPRs, $I_s(\varphi)$, and Josephson energies, $F_J(\varphi)$, in the case of a 0—(i.e., $I_1 = 1$ and $I_2 = 1/2$), π—(i.e., $I_1 = -1$ and $I_2 = 1/2$), φ—(i.e., $I_1 = 1$ and $I_2 = -3/2$), and φ_0—(i.e., $\varphi_0 = 0.4$) JJ (see the red, blue, purple, and black curves, respectively). For the first three cases refer to Eq. (1.22), while for the last one to Eq. (1.25). (**b**) (I_1, I_2)-parameter space of the ground-state phase, $\tilde{\varphi}$, of the CPR in Eq. (1.22)

2. If $\left|\frac{I_1}{2I_2}\right| < 1$ and $I_2 < 0$:

- $\tilde{\varphi} = \pm \arccos\left(\frac{I_1}{2I_2}\right)$. Since $I_2 < 0$, we can assume $I_2 \in [-1, 0]$, which implies $|I_1| \in [0, 2I_2]$.

These three cases are depicted in Fig. 1.2a, see the red, blue, and purple curves, respectively. Moreover, Fig. 1.2b show the full (I_1, I_2)-parameter space of the ground-state phase, $\tilde{\varphi}$. The regions of (I_1, I_2) values giving the 0-, π-, and φ-junctions are evident. If $\left|\frac{I_1}{2I_2}\right| < 1$ but $I_2 > 0$, the potential profile shows, in addition to the ground state at $\tilde{\varphi} = 0$ (if $I_1 > 0$) or $\tilde{\varphi} = \pi$ (if $I_1 < 0$), an extra metastable minimum at a higher energy at π or 0, respectively [15].

The phase shift $\tilde{\varphi} = \pi$ corresponds to an interesting case, i.e., a junction with $I_c < 0$, that is a negative supercurrent values for $\varphi \in [0 - \pi]$, the so-called *π-junction* [13, 16, 17]. Such a junction has an energy minimum at $\varphi = \pi$, i.e., it provides a phase shift of π in the ground state, and may find a variety of applications in electronic circuits. The crossover from 0- to π-state is observed in superconductor-ferromagnet-superconductor (SFS) junction.

The intrinsic phase shift $\tilde{\varphi} = \pm \arccos\left(\frac{I_1}{2I_2}\right)$ corresponds to a twofold degenerate state and $I_s(\varphi)$ crosses the horizontal φ axis at a position between $\varphi = 0$ and π. This case corresponds to the so-called *φ-junctions* [13, 18, 19].

A different kind of unconventional junction, i.e., the so-called φ_0*-junctions* [20–29], shows a non-trivial ($\varphi_0 \neq 0, \pi$) ground phase and is characterized by a significant even cosine component in the CPR, which transforms into

$$I_s(\varphi) = I_c \sin(\varphi - \varphi_0), \tag{1.25}$$

see the black curve in Fig. 1.2a for $\varphi_0 = 0.4$. Although, as we have seen, a non-trivial ground state can also emerge for the CPR in Eq. (1.22), the CPR of a φ_0-junction gives a non-zero current at $\varphi = 0$. In other words, this kind a junction can give a constant phase bias $\varphi = -\varphi_0$ in an open circuit configuration, or even a current $I_s = I_c \sin(\varphi_0)$, i.e., the *anomalous Josephson current*, if inserted into a closed superconducting loop.

Anomalous CPRs usually emerge in junctions under external magnetic or electric fields. It can also be obtained in the systems undisturbed by external interactions. For instance, SFS junctions with spin-orbit Rashba-type interaction demonstrate a significant shift of CPR, proportional to the product of the exchange field and the strength of spin-orbit coupling, but φ_0-junctions are also predicted in the systems with unconventional superconductors. Ferromagnetic anomalous JJs, resulting from joining superconductivity and ferromagnetic materials, have attracted considerable attention for their intriguing features, such as anomalous phase shifts and magnetization reversal induced by supercurrent flow, positioning them as promising candidates for potential applications in spintronics, quantum technologies, and information processing [30–35]. The appearance of this anomalous phase has been recently predicted even in a multiband scenario [36].

Josephson systems with a non-trivial ground states, i.e., $\tilde{\varphi}$ or φ_0, allows the realization of phase shifter [37] and batteries [25, 38]. In fact, in a conventional JJ, the implementation of a phase battery is prevented by symmetry constraints, either time-reversal or inversion, which impose rigidity on the superconducting phase. Conversely, the break of TRS ($t \to -t$) can lead to a phase shift 0 or π, while the break of both TRS ($t \to -t$) and inversion symmetry ($r \to -r$) can lead to a phase shift $\varphi_0 \in (0 - \pi)$ and to an anomalous Josephson current, $I_s(\varphi) = I_c \sin(\varphi - \varphi_0)$.

1.1.5 Short JJ: The Resistively and Capacitively Shunted Junction (RCSJ) Model

The CPR presented so far exclusively addresses the Cooper pair supercurrent. However, when $V > 0$ and $T > 0$, a quasiparticle current component also traverses the system. The complete time-dependent response of a JJ is appropriately captured by the equivalent circuit illustrated in Fig. 1.3a, representing the so-called *resistively and capacitively shunted junction* (RCSJ) model [39, 40]. This circuit includes a capacitance $C = \epsilon_r \epsilon_0 A / t_{ox}$, where A denotes the junction area, t_{ox} and ϵ_r represent the thickness and relative permittivity of the oxide layer, and ϵ_0 is the vacuum permittivity. The circuit also incorporates the resistance R of the weak link. Owing to capacitive effects, tunnel junctions with typical high R values exhibit hysteretic

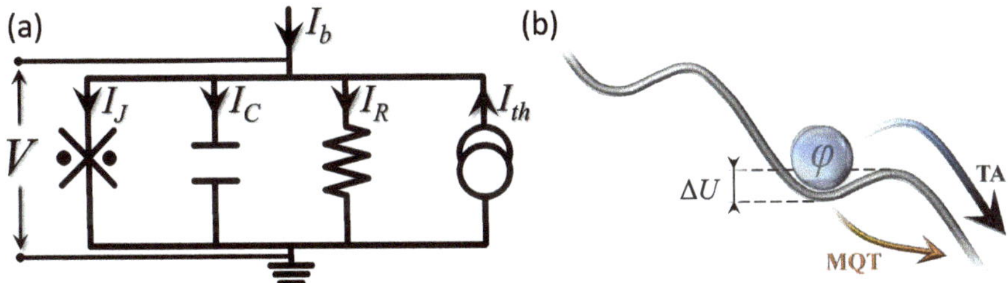

Fig. 1.3 (**a**) Electrical model of a short JJ. The bias current, I_b, and the thermal noise current, I_{th}, are included in the diagram. (**b**) Sketch of the phase particle within a potential minimum of the tilted washboard potential U. The barrier height, ΔU, is also shown

static IV characteristics. If the current surpasses the critical value, the JJ transitions to a state in which a measurable voltage drop appears, i.e., the so-called *voltage state*. To mitigate the arising hysteresis, an additional shunt resistance R_{sh} may be employed, resulting in an effective resistance equal to $RR_{sh}/(R + R_{sh})$.

The resistance R exhibits nonlinear dependencies on both voltage and temperature [41], in particular, $R(V, T) = R_N$ if $V > V_g$ and $R(V, T) = R_{sg}(T)$ if $V \leq V_g$., with R_N and R_{sg} being the normal-state and the subgap resistances [42], respectively, and $V_g = 2\Delta/e$ being the gap voltage (where Δ is the superconducting gap). However, being related to the dissipation in the system, its nonlinearity is often neglected in cases of moderate or weak damping. Alongside the resistor R, Fig. 1.3a shows a thermal noise current source, denoted as I_{th}, and the electric current, I_b, biasing the junction. Notably, the Josephson element can be alternatively viewed as a nonlinear inductance $L_J = L_c/\cos\varphi$, where $L_c = \Phi_0/(2\pi I_c)$.

Therefore, considering all current contributions and Josephson equations, we can easily derive the following Langevin equation using Kirchhoff's laws [2,43],

$$C\frac{\hbar}{2e}\frac{d^2\varphi}{dt^2} + \frac{1}{R}\frac{\hbar}{2e}\frac{d\varphi}{dt} + I_c \sin\varphi = I_b + I_{th}(t)\,. \tag{1.26}$$

By normalizing time with respect to the characteristic frequency $\omega_c = \frac{R}{L_c} = \frac{2e}{\hbar}\frac{I_c}{R}$, i.e., $\tilde{t} = \omega_c t$, and the current contributions on the rhs of this equation to the critical current I_c, previous equation becomes

$$\beta_c\frac{d^2\varphi}{d\tilde{t}^2} + \frac{d\varphi}{d\tilde{t}} + \sin\varphi = i_b + i_{th}(\tilde{t}), \tag{1.27}$$

including the Stewart-McCumber damping parameter [2] $\beta_c = \frac{\omega_c^2}{\omega_p^2}$, with $\omega_p = \frac{1}{\sqrt{L_c C}} = \sqrt{\frac{2e}{\hbar}\frac{I_c}{C}}$ being the Josephson plasma frequency.

Interestingly, Eq. (1.27) can even describe the damped motion of a particle in a tilted washboard-like potential [2], see Fig. 1.3b, expressed by

$$U(\varphi, i_b) = E_{J_0}[1 - \cos(\varphi) - i_b\, \varphi], \tag{1.28}$$

where $E_{J_0} = \Phi_0/(2\pi)I_c$. The current i_b biasing the system tilts $U(\varphi, i_b)$, so that for $i_b < 1$ the potential has metastable wells, with a barrier height equal to

$$\Delta\mathcal{U}(i_b) = \frac{\Delta U(i_b)}{E_{J_0}} = 2\left[\sqrt{1 - i_b^2} - i_b \cos^{-1}(i_b)\right]. \tag{1.29}$$

Instead, for $i_b \geq 1$ the potential profile has no maxima and minima.

The equation of the Josephson phase is also equivalent to the equation that governs the motion of a driven pendulum. Looking at a mechanical analog helps to better visualize the behavior of a JJ: for example, increasing the applied torque will increase the pendulum angle until it reaches $\pi/2$, at which point a minute additional applied twist, or noise, will cause the pendulum to flip over the top and then spin. Delving into this analogy, the phase difference φ corresponds to the angle from vertical, the voltage V to the angular velocity, the critical current I_c to the restoring constant, the conductance R^{-1} to the damping coefficient, the capacitance C to the moment of inertia, and the bias current I to the external torque.

Viewing instead the problem in terms of a phase particle rolling on a rippled potential profile, in the presence of friction, helps to provide insight into the IV characteristic of a JJ. The latter is schematically illustrated in Fig. 1.4. We start with the system in the zero-voltage state, with the phase particle quite in a potential minimum. Then, increasing the bias current, the system remains in the zero-voltage state as long as $I < I_c$, see (a) In this condition, the tilting imposed by the bias current to the washboard potential is not enough to vanish the potential barrier ΔU, so that the phase particle remains confined in a potential well. When I reaches I_c, $\Delta U = 0$ and the particle starts to roll down along the potential profile, i.e., a non-zero voltage drop appears, see (b) This holds even by increasing the current further, see (c) Now, if the bias current is reduced, the system remains in the voltage state, at least as long as the current is greater than I_c. When $I_b < I_c$, the potential barriers appear again, but the following cases can occur:

- Overdamped junction (where the friction term $\dot{\varphi}$ dominates): the combination of low kinetic energy and significant damping, makes the particle to immediately stop, remaining trapped in a potential minimum, so that φ ceases to evolve further. In this case, a non-hysteretic IV characteristic emerges;
- Underdamped junction (where the inertial term $\ddot{\varphi}$ dominates): the low friction and the high kinetic energy, makes the particle to continue moving along the local minima. To stop the particle, the potential slope must be brought close to zero, requiring a very small I_b to restore the zero voltage state. This phenomenon depends on the β_c value. In this case, a hysteretic IV characteristic emerges.

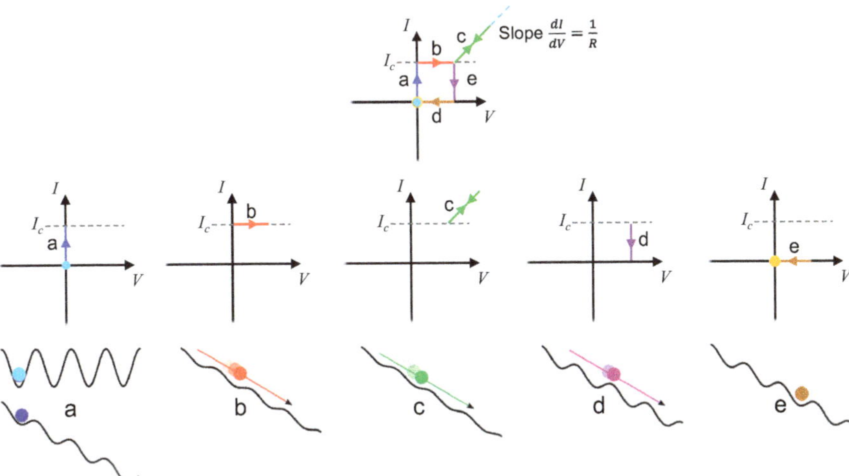

Fig. 1.4 Scheme of the IV characteristic of a JJ. Top panel schematically sketches the whole non-hysteretic path followed by the system when the bias current is first increased above I_c, and then reduced to zero. It can be divided in five distinct branches, represented in the bottom **(a)**–**(e)** panels. Each contains the corresponding characteristic IV branch and a sketch of the potential with the phase particle under stationary or moving (see the arrows) conditions

The example case sketched in Fig. 1.4 is clearly non-hysteretic. Finally, reducing further the bias current, the phase particle remains trapped again in a potential minimum, so the zero-voltage state is restored, and the system returns to the initial condition, see d and e.

The current value at which the system re-enters the zero-voltage state is called "retrapping current". Thus, in an overdamped system, switching and retrapping current coincide, $I_0 = I_r$, whereas in an underdamped system they are different, and in particular $I_r/I_0 \sim 4/(\pi\sqrt{\beta_c})$ [41].

We have intentionally referred to "switching" and not critical current. In fact, if we assume to raise the bias current from zero and record the value at which the system switches to the voltage state, it turns out that this is below the expected critical current value, and that it is temperature-dependent. In fact, this mechanism is governed by thermal fluctuations, which are able to "push" the phase particle out of the metastable state even if the potential barrier is not completely zero, i.e., even if $I_b < I_c$. This point is the core of the next section.

1.1.5.1 Temperature Effects

Thermal effects are accounted by the stochastic noise term, I_{th}, with the statistical properties

$$\langle I_{th}(t)\rangle = 0 \qquad \text{and} \qquad \langle I_{th}(t), I_{th}(t')\rangle = 2\frac{k_B T}{R}\delta(t - t'), \tag{1.30}$$

where k_B is the Boltzmann constant, $\delta()$ is the Dirac delta function, and T is the temperature. In normalized units, i.e., $\tilde{t} = \omega_c t$, previous equations become

$$\langle i_{th}(\tilde{t})\rangle = 0 \qquad \text{and} \qquad \langle i_{th}(\tilde{t}), i_{th}(\tilde{t}')\rangle = 2\Gamma\delta(\tilde{t} - \tilde{t}'), \tag{1.31}$$

where the adimensional noise intensity is

$$\Gamma = \frac{k_B T}{E_{J_0}}, \tag{1.32}$$

i.e., the ratio between the thermal energy and the Josephson coupling energy E_{J_0}.

The presence of the noise term makes escape from the metastable state an inherently stochastic process; therefore, when studying this type of dynamics, one typically reiterates the same "experiment" several times in order to construct distributions, from which to extract, e.g., mean values and variances, as, for instance, in the examples reported in Refs. [44–50]. In particular, Fig. 1.5a shows the average switching time, obtained by averaging over $N = 10^3$ independent repetitions, as a function of the inverse noise intensity, Γ^{-1}, setting $I_b = 0.5$ and $\beta_c = 10^4$.

As mentioned above, thermal fluctuations can drive the system out of a washboard potential minimum even in the case of a non-zero barrier height, i.e., for $I_b < I_c$. In fact, noise makes the phase solution metastable; thus the escape rate via thermal activation (TA), Γ_{TA}, is given by the Kramers approximation [51], which, for moderate damping, can be written as

$$\Gamma_{\mathrm{TA}}(\alpha, i_b, \Gamma) = \frac{1}{2\pi}\left(\sqrt{\frac{\alpha^2}{4} + \sqrt{1 - i_b^2}} - \frac{\alpha}{2}\right) e^{-\frac{\Delta\mathcal{U}(i_b)}{\Gamma}}. \tag{1.33}$$

Interestingly, the prediction of an exponentially scaling average switching time for low noise is verified looking at the numerical results at small noise amplitudes in Fig. 1.5a.

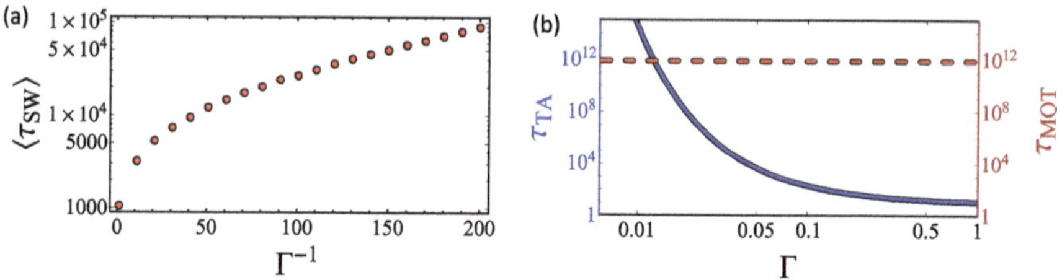

Fig. 1.5 (**a**) MST, τ_{SW} as a function of the inverse noise intensity, Γ^{-1}, obtained by numerical integration of Eq. (1.27), setting $i_b = 0.5$, $\beta = 10^4$, and $N = 10^4$. (**b**) TA (τ_{TA}, left vertical axis, blue solid line) and MQT (τ_{MQT}, right vertical axis, red dashed line) switching times, calculated as the inverse of the rates in Eqs. (1.33) and (1.34), as a function of the noise intensity Γ, at $i_b = 0.7$ and $\beta = 10^2$

The phase particle can leave the metastable state also through a macroscopic quantum tunneling (MQT) mechanism, which rate is given by [52]

$$\Gamma_{\mathrm{MQT}}(\alpha, i_b) = \frac{a_q}{2\pi}\sqrt[4]{1-i_b^2}\,e^{\left[-7.2\frac{\Delta U(i_b)}{\hbar\omega_p(i_b)}(1+0.87\alpha)\right]}, \tag{1.34}$$

with $a_q = \sqrt{120\pi\left[\frac{7.2\Delta U(i_b)}{\hbar\omega_p(i_b)}\right]}$, $\omega_p(i_b) = \omega_p\sqrt[4]{1-i_b^2}$, and $\alpha = 1/\sqrt{\beta_c}$ is the damping parameter. The MQT rate depends only on ΔU, and not on T; this means that, as the temperature decreases, the TA rate can reduce so much that MQT processes dominate the escape dynamics. In this regard, one can define the so-called *crossover temperature* [53], T_{cr}, as the temperature at which these two rates coincides: for $\alpha \ll 1$ and $a_q \approx 1$, it reduces to $T_{\mathrm{cr}} \approx \frac{\hbar\omega_p}{7.2k_B}$. This is evident in Fig. 1.5b, which shows the TA and MQT times, i.e., the inverse of the switching rates in Eqs. (1.33) and (1.34), τ_{TA} and τ_{MQT}, respectively, as a function of the noise amplitude; in fact, the curves meet at $\Gamma \simeq 0.012$, that corresponds to $T \simeq 0.012\frac{\Phi_0}{2\pi k_B}I_c \simeq 30\,\mathrm{mK}$, assuming $I_c = 0.1\,\mu\mathrm{A}$.

1.1.5.2 Magnetic Field Effects

In this section, we focus on the effects produced by an applied magnetic field. So far, we looked at a so-called *short* JJ (SJJ), specifically a junction where the magnetic field induced by the Josephson current is negligible in comparison to the externally applied magnetic field. Under these circumstances, the device dimensions are constrained to be smaller than the characteristic length-scale for such systems, defined as the *Josephson penetration depth* [2], which is derived in the following.

Let's assume a magnetic field applied in the junction plane and a rectangular barrier, like in Fig. 1.6a. The magnetic flux density penetrates also into electrodes within a distance $t_H \approx d + 2\lambda_L$, i.e, the *effective magnetic thickness* (see the yellow and red regions in the figure), assuming electrodes made by the same superconductors with a thickness larger than the London penetration depth λ_L and d being the insulating layer thickness. What is the relation between the magnetic field and the Josephson phase? We now that

$$\mathbf{B} = \nabla \times \mathbf{A} \qquad \text{and} \qquad \varphi = (\theta_2 - \theta_1) - \frac{2\pi}{\Phi_0}\int_1^2 \mathbf{A}\,dl. \tag{1.35}$$

From Eq. (1.1), one can obtain $\nabla\theta = \frac{2\pi}{\Phi_0}(\mu_0\lambda_L^2\mathbf{J}_s + \mathbf{A})$, where $\lambda_L = \sqrt{\frac{m^\star}{\mu_0(q^\star)^2 n^\star}}$ and μ_0 is the vacuum permeability. Then, we integrate $\nabla\theta$ along the white dashed lines in the two electrodes in Fig. 1.6a, i.e.,

- Integration Path $2 \to 1$: $\quad \theta(1) - \theta(2) = \frac{2\pi}{\Phi_0}\mu_0\lambda_L^2\int_2^1 \mathbf{J}_s\,dl + \frac{2\pi}{\Phi_0}\int_2^1 \mathbf{A}\,dl;$
- Integration Path $1' \to 2''$: $\quad \theta(2') - \theta(1') = \frac{2\pi}{\Phi_0}\mu_0\lambda_L^2\int_{1'}^{2'} \mathbf{J}_s\,dl + \frac{2\pi}{\Phi_0}\int_{1'}^{2'} \mathbf{A}\,dl.$

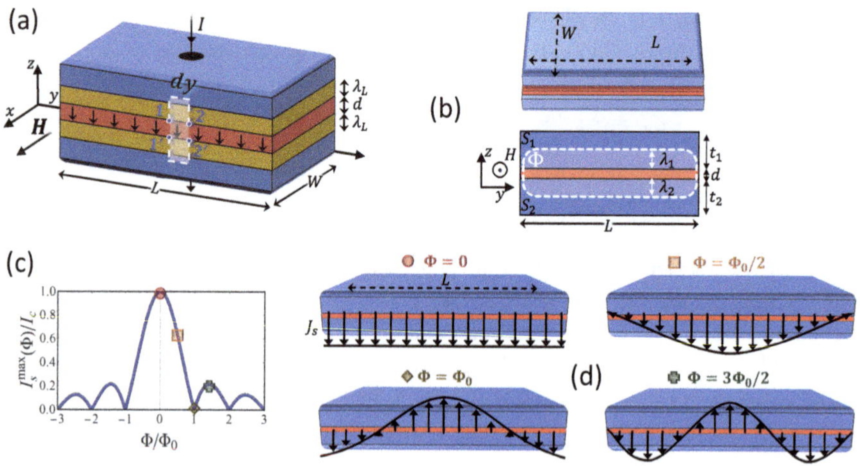

Fig. 1.6 (**a**) Sketch of the SJJ, with indicated the path followed to perform the integrals in Eq. (1.36). (**b**) Rectangular SJJ; the magnetic flux through the junction area is also reported. (**c**) Fraunhofer diffraction pattern for the critical current, with highlighted four situations of interest, corresponding to four different magnetic flux values. (**d**) Supercurrent density distributions along the junction length in the cases highlighted in (**c**)

Summing up these equations and including the integrals of **A** across the barrier, one obtains:

$$\left[\theta(2') - \theta(2) - \frac{2\pi}{\Phi_0}\int_2^{2'} \mathbf{A}dl\right] - \left[\theta(1') - \theta(1) - \frac{2\pi}{\Phi_0}\int_{1'}^{1} \mathbf{A}dl\right]$$
$$= \frac{2\pi}{\Phi_0}\left[\oint \mathbf{A}dl + \mu_0\lambda_L^2\left(\int_2^1 \mathbf{J}_s dl + \int_{1'}^{2'} \mathbf{J}_s dl\right)\right]. \tag{1.36}$$

Since last term in the lhs vanishes, this equation reduces to:

$$\frac{\partial\varphi}{\partial y} = \frac{2\pi}{\Phi_0} t_H B_x. \tag{1.37}$$

According to this equation, an external magnetic field therefore affects the Josephson phase difference, producing measurable effects also in the critical current profile. Indeed, let's assume the geometry sketched in Fig. 1.6b. If $\varphi(x, y)$ indicates the phase difference that is influenced by the external magnetic field, we have [2]:

$$\frac{\partial\varphi}{\partial y} = \frac{2\pi}{\Phi_0} t_H \mu_0 H_x(y) \qquad \text{and} \qquad \frac{\partial\varphi}{\partial x} = 0. \tag{1.38}$$

These equations come from the condition $W \ll \lambda_J$, ensuring that $\varphi(x, y) \equiv \varphi(y)$.

In a rectangular SJJ, since the external magnetic field is spatially homogeneous along the junction, i.e., $H_x(y) \equiv H_{ext}$, the phase exhibits a linear increase, which is given by

$$\varphi(y) = \left(2\pi \frac{\mu_0 t_d H_{ext}}{\Phi_0}\right) y + \varphi_0 = ky + \varphi_0. \tag{1.39}$$

In contrast, in a *long* junction, in which $L \gg \lambda_J$, both the penetrating external field and the self-field generated by the Josephson current contribute, resulting in a nonlinear variation of $\varphi(y)$ along the junction, see Eq. (1.46); this type of system will be faced in the next section.

From Eq. (1.39), one obtains $J_s(x, y, t) = J_c(x, y)\sin(ky + \varphi_0)$, where the oscillation period of J_s is $\Delta y = 2\pi/k = \Phi_0/(\mu_0 H_{ext} t_H)$. Thus, the magnetic flux through the junction within a single oscillation period corresponds to a single flux quantum: $\Phi \equiv \Delta y t_H \mu_0 H_{ext} = \Phi_0$. The Josephson current can be calculated as:

$$\begin{aligned} I_s(H) &= \int_{-\frac{L}{2}}^{\frac{L}{2}} \int_{-\frac{W}{2}}^{\frac{W}{2}} J_c(x, y)\sin(ky + \varphi_0)\, dx\, dy \\ &= \int_{-\frac{L}{2}}^{\frac{L}{2}} i_c(y)\sin(ky + \varphi_0)\, dy = \mathrm{Im}\left\{e^{i\varphi_0}\int_{-\infty}^{\infty} i_c(y)e^{iky}\, dy\right\}, \end{aligned} \tag{1.40}$$

where $i_c(y) = \int_{-W/2}^{W/2} J_c(x, y)\, dx$. In the last step we assumed $i_c(|y| > L/2) = 0$, so as to replace the integral limits with $\pm\infty$. Moreover, if $i_c(|y| \le L/2) = constant = i_c$, the maximum Josephson current becomes

$$\begin{aligned} I_s^{\max}(H) &= \left|\int_{-\infty}^{\infty} i_c(y)e^{iky}\, dy\right| = \left|i_c \int_{-\frac{L}{2}}^{\frac{L}{2}} \cos(ky)\, dy\right| \\ &= \left|i_c \frac{1}{kL}\Big[\sin(ky)\Big]_{-\frac{L}{2}}^{\frac{L}{2}}\right| = i_c L \left|\frac{\sin(kL/2)}{kL/2}\right|. \end{aligned} \tag{1.41}$$

Finally, since $kL/2 = \left(\frac{2\pi}{\Phi_0}\mu_0 H_x t_H\right) L/2 = \pi\Phi/\Phi_0$ and $I_c = i_c L$, from Eq. (1.41) we obtain a *Fraunhofer diffraction pattern for the critical current*

$$\frac{I_s^{\max}(\Phi)}{I_c} = \left|\frac{\sin(\pi\Phi/\Phi_0)}{\pi\Phi/\Phi_0}\right|, \tag{1.42}$$

which is shown in Fig. 1.6c. We stress that $I_c^{\max}$ is not a current itself but only indicates the maximum value of the Josephson current. A different CPR and/or a different geometry can lead to a different magnetic-field dependence of the critical current, e.g., see Refs. [29, 54–58] and even a brief overview on some of the most significant deviations from the standard patterns given in [59].

It is possible to make an intriguing parallel between the critical current distribution within a JJ and the renowned Fraunhofer diffraction pattern observed in optics. Let's start highlighting the Φ-dependence of $\varphi(y)$, i.e.,

$$\varphi(y) = 2\pi \frac{\mu_0 t_d H_{ext} L}{\Phi_0} \frac{y}{L} + \varphi_0 = 2\pi \frac{\Phi}{\Phi_0} \frac{y}{L} + \varphi_0. \tag{1.43}$$

Then, we focus on four situations of interest, corresponding to four different magnetic flux values, giving the critical currents highlighted in Fig. 1.6c with dedicated symbols. The resulting configurations of current density along the junction length are shown in Fig. 1.6d. In particular, we look at the supercurrent density $i_c(y)$ and $J_s(x, y, t) = J_c(x, y) \sin(ky + \varphi_0)$:

- In the zero-field case, $\Phi = 0$ so that $\varphi(y) = \varphi_0$ according to Eq. (1.43), we have $i_c(y) = constant$. The maximum Josephson current (in the negative z direction) is obtained for $\varphi_0 = -\pi/2$, that is $J_s(x, y, t) = -J_c(x, y)$, see top-left panel of Fig. 1.6d.
- If $\Phi = \Phi_0/2$, $\varphi(y) = \pi y/L + \varphi_0$ and $\varphi(L/2) - \varphi(-L/2) = \pi$: since the supercurrent density oscillates sinusoidally, half of a full oscillation period lies within the junction. The choice of φ_0 says which "half period" is considered. In particular, in the top-right panel of Fig. 1.6d is shown the $\varphi_0 = -\pi/2$ situation, which gives $\varphi(-L/2) = -\pi$, $\varphi(L/2) = 0$ and the maximum Josephson current in the negative z direction.
- If $\Phi = \Phi_0$, $\varphi(y) = 2\pi y/L + \varphi_0$ and $\varphi(L/2) - \varphi(-L/2) = 2\pi$, so that a full oscillation period lies within the junction, see bottom-left panel of Fig. 1.6d. In this case, the total Josephson current is zero for each φ_0.
- If $\Phi = 3\Phi_0/2$, $\varphi(y) = 3\pi y/L + \varphi_0$ and $\varphi(L/2) - \varphi(-L/2) = 3\pi$, so that one and a half oscillation period lies within the junction, see bottom-right panel of Fig. 1.6d. Current contributions from the full period are cancelled, so that the overall current is just given by the remaining half period. Naturally, the resulting current is less than that for $\Phi = \Phi_0/2$, where an entire semi-period is comprised into the system. This observation underscores the trend that the Josephson current typically diminishes as the applied magnetic field increases.

1.1.6 Long JJ: The Sine-Gordon Model

In this section we consider a long JJ (LJJ) in which the barrier is in the yz-plane, the magnetic field is applied in x-direction resulting in phase variations along the y-direction, and the bias current is flowing in the negative z-direction, see Fig. 1.7.

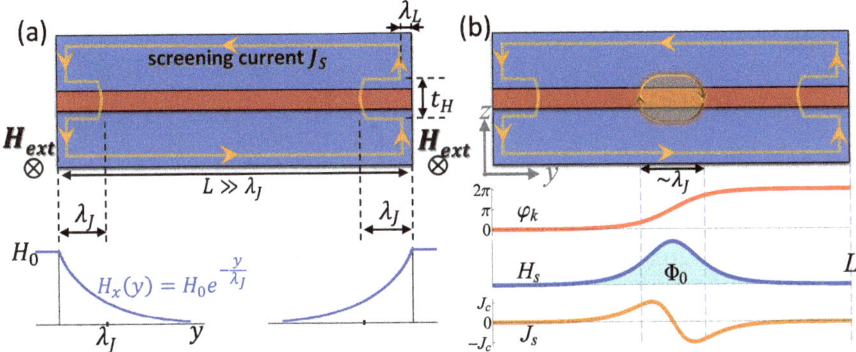

Fig. 1.7 (**a**) Sketch of a LJJ in the case of a small applied field. Bottom panel shows the decaying magnetic field profile. (**b**) Sketch of a LJJ hosting a kink. Bottom panels shows a 2π-phase profile and the corresponding magnetic field and supercurrent

The magnetic flux density results both from the externally applied field and the Josephson current density and must satisfy Ampére's law,

$$\nabla \times \mathbf{H} = \mathbf{J} + \epsilon\epsilon_0 \frac{\partial \mathbf{E}}{\partial t} \qquad \text{that for our geometry becomes}$$

$$\frac{\partial H_x(y,t)}{\partial y} = -J_z(y,t) - \epsilon\epsilon_0 \frac{\partial E_x(y,t)}{\partial t}. \tag{1.44}$$

If a spatial dimension, e.g., L, is larger than the Josephson penetration depth, we have to consider the self-field effect due the supercurrent. In the presence of an external magnetic field, we know that

$$\frac{\partial \varphi}{\partial y} = \frac{2\pi t_H \mu_0}{\Phi_0} H_x. \tag{1.45}$$

In the zero-voltage state, $\frac{\partial E}{\partial t} = 0$, Ampére's law reduces to $\frac{\partial H_x}{\partial y} = -J_z(y)$. Since the current is flowing in the negative z-direction, $J_z(y) = -J_c \sin\varphi(y)$, from Eq. (1.45), we obtain

$$\frac{\partial^2 \varphi}{\partial y^2} = \frac{2\pi t_H \mu_0}{\Phi_0} \frac{\partial H_x}{\partial y} = -\frac{2\pi t_H \mu_0}{\Phi_0} J_c \sin\varphi(y) = \frac{1}{\lambda_J^2} \sin\varphi(y). \tag{1.46}$$

This is often called the *Ferrel-Prange equation* [60], or more generally the *stationary sine-Gordon equation* (SSGE). Here, we defined the Josephson penetration depth $\lambda_J = \sqrt{\frac{\Phi_0}{2\pi \mu_0 t_H J_c}}$: this is the characteristic length scale over which the magnetic field is screened (similar to λ_L in a bulk superconductor).

For small applied fields, Eq. (1.46) becomes $\partial^2\varphi/\partial y^2 = \varphi(y)/\lambda_J^2$, giving a solution

$$\varphi(y) = \varphi(0)e^{-y/\lambda_J} \qquad \text{and} \qquad H_x(y) = -\frac{\Phi_0}{2\pi\lambda_J t_H \mu_0}\varphi(0)e^{-y/\lambda_J} = H_0 e^{-y/\lambda_J}, \tag{1.47}$$

where λ_J clearly represents a decay length for the magnetic field, this justifying the expression "penetration depth", see Fig. 1.7a.

In the case of $\frac{\partial E}{\partial t} \neq 0$, and considering that $E_x = -V/d$, $\frac{\partial \varphi}{\partial t} = \frac{2\pi V}{\Phi_0}$, and $J_x = -J_c \sin(\varphi)$, Eq. (1.44) becomes

$$\frac{\partial^2\varphi(y,t)}{\partial y^2} = \frac{2\pi}{\Phi_0} t_H \mu_0 \left[J_c \sin(\varphi(y,t)) + \epsilon\epsilon_0 \frac{\Phi_0}{2\pi d} \frac{\partial^2\varphi(y,t)}{\partial t^2} \right]. \tag{1.48}$$

Using the definition of the Josephson penetration depth λ_J, one finally obtains the *time-dependent Sine-Gordon equation* (SGE) [61–66]

$$\frac{\partial^2\varphi(y,t)}{\partial y^2} - \frac{1}{\bar{c}^2}\frac{\partial^2\varphi(y,t)}{\partial t^2} - \frac{1}{\lambda_J^2}\sin(\varphi(y,t)) = 0, \tag{1.49}$$

where $\bar{c} = \sqrt{\frac{d}{\epsilon\epsilon_0 t_H \mu_0}} = c\sqrt{\frac{1}{\epsilon(1+2\lambda_L/d)}}$ is the transverse electromagnetic velocity, called *Swihart velocity*. Recalling that $\omega_p = \sqrt{\frac{2\pi\Phi_0 I_c}{C_J}} = \sqrt{\frac{2\pi\Phi_0 I_c}{\epsilon\epsilon_0 A/d}} = \sqrt{\frac{2\pi\Phi_0 J_c d}{\epsilon\epsilon_0}}$, one obtains $\omega_p\lambda_J = \bar{c}$. Therefore, the SGE can be recast as

$$\lambda_J^2 \frac{\partial^2\varphi(y,t)}{\partial y^2} - \frac{1}{\omega_p^2}\frac{\partial^2\varphi(y,t)}{\partial t^2} - \sin(\varphi(y,t)) = 0. \tag{1.50}$$

The SGE admits a kink solution, moving at a speed v_y and centered in y_0, with expression:

$$\varphi_k(y,t) = 4\arctan\left\{\exp\left[\pm\gamma_{\tilde{v}}\left(\frac{y-y_0}{\lambda_J} - \frac{v_y}{\bar{c}}t\right)\right]\right\} \tag{1.51}$$

where $\gamma_{\tilde{v}} = 1/\sqrt{1-\tilde{v}^2}$, with $\tilde{v} = v_y/\bar{c}$. The sign ± 1 distinguishes a kink from an anti-kink. This solitonic solution corresponds to a 2π-step in the phase profile, is stable, travels maintaining its shape (also after collision with other waves), gives rise to step structures in the IV characteristic, produces microwave radiation emission, and transports a clear magnetic flux equal to Φ_0, see Fig. 1.7b.

Equation (1.51) represents a fluxon moving with velocity v_y along the junction. Under the action of the Lorentz force due to an externally applied current, the fluxon shifts along the junction, suffering Lorentz contraction on approaching the Swihart velocity $\bar{c}$. The moving fluxon locally causes a temporal change of φ, which corresponds, according to the 2nd Josephson relation, to a voltage pulse that becomes sharper with increasing velocity due to Lorentz contraction, in order to satisfy the condition $\int V dt = \Phi_0$.

The SGE admits other kinds of solutions, e.g., (in normalized units and imposing $y_0 = 0$):

- kink-antikink collision

$$\varphi_{ka}(\tilde{y}, \tilde{t}) = 4 \arctan \left\{ \frac{1}{\tilde{v}} \frac{\sinh\left(\gamma_{\tilde{v}} \tilde{v} \tilde{t}\right)}{\cosh\left(\tilde{y} \gamma_{\tilde{v}}\right)} \right\}; \tag{1.52}$$

- kink-kink collision

$$\varphi_{kk}(\tilde{y}, \tilde{t}) = 4 \arctan \left\{ \tilde{v} \frac{\sinh\left(\gamma_{\tilde{v}} \tilde{y}\right)}{\cosh\left(\tilde{v} \tilde{t} \gamma_{\tilde{v}}\right)} \right\}; \tag{1.53}$$

- small amplitude oscillations (i.e., *plasma waves*), in which case $\sin[\varphi(y,t)] \approx \varphi(y,t)$, so that

$$\lambda_J^2 \frac{\partial^2 \varphi(y,t)}{\partial y^2} - \frac{1}{\omega_p^2} \frac{\partial^2 \varphi(y,t)}{\partial t^2} = \varphi(y,t). \tag{1.54}$$

  This gives a solution $\varphi(y,t) = \varphi_0 \exp[-i(ky - \omega t)]$, satisfying the dispersion relation $\frac{\omega}{\omega_p} = \sqrt{1 + \left(\frac{k}{\lambda_J}\right)^2}$;
- breather, with ω_b internal oscillation frequency,

$$\varphi_b(\tilde{y}, \tilde{t}) = 4 \arctan \left\{ \frac{\sqrt{1-\omega_b^2}}{\omega_b} \frac{\sin\left[\gamma_{\tilde{v}} \omega_b (\tilde{t} - \tilde{v} \tilde{y})\right]}{\cosh\left[\gamma_{\tilde{v}} \sqrt{1-\omega_b^2} (\tilde{y} - \tilde{v} \tilde{t})\right]} \right\}. \tag{1.55}$$

Unlike kinks, the breather solution is unstable, oscillates periodically with time and decays exponentially in space, does not produce a potential difference, has no manifestations on the IV characteristic, does not produce a measurable magnetic flux through the JJ, must be generated efficiently and trapped in a confined area to allow measurements [67–72].

Figure 1.8 collects different solutions of SGE (top panels) and the corresponding space derivative (bottom panels): (a)–(e) a kink φ_k, (b)–(f) a kink-antikink collision φ_{k-a}, (c)–(g) a kink-kink collision φ_{k-k}, and (d)–(h) a breather φ_b. It is evident that the spatial derivative of the phase allows kinks and antikinks to be easily visualised, since each 2π-step corresponds to a peak in the $\partial\varphi/\partial x$ profile (in particular, positive for a kink and negative for an antikink). In this regard, Fig. 1.8h clearly demonstrates

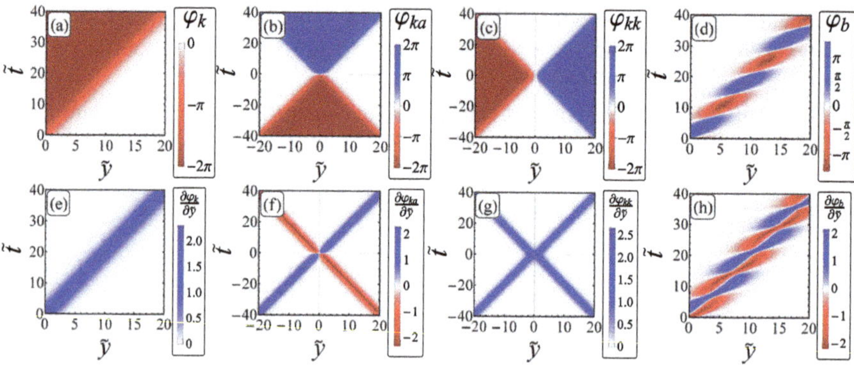

Fig. 1.8 Solutions of SGE (top panels) and corresponding space derivative (bottom panels): (**a**)–(**e**) kink φ_k, (**b**)–(**f**) kink-antikink collision φ_{k-a}, (**c**)–(**g**) kink-kink collision φ_{k-k}, and (**d**)–(**h**) breather φ_b. The other parameters are: $L = 20$, $\tilde{v} = 0.5$, and $\omega_b = 0.5$

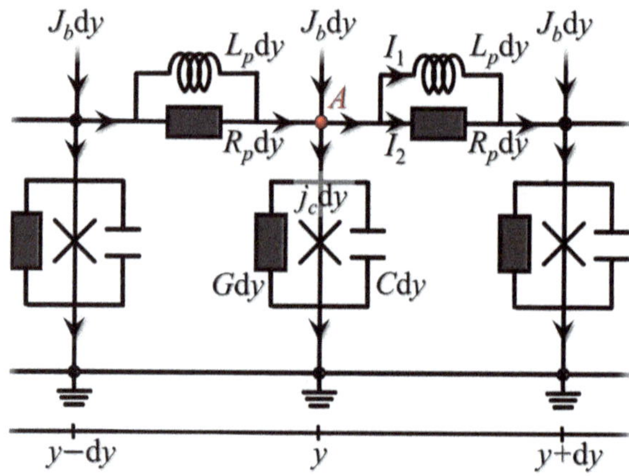

Fig. 1.9 Equivalent circuit elements constituting the LJJ analog [73, 74]

that a breather is composed of a bounded kink-antikink pair oscillating at a proper internal frequency.

In analogy to short junctions, an equivalent circuit can also be designed for LJJs [73, 74], see Fig. 1.9. Focusing on the node A, applying Kirchhoff current law one obtains

$$\frac{\partial (I_1 + I_2)}{\partial y} = J_b - C\frac{\partial V}{\partial t} - \frac{V}{R} - J_c \sin(\varphi). \tag{1.56}$$

Here, L_p is the inductance per unit length representing the magnetic energy stored within one λ_L of the superconducting film, and R_p is a resistance per unit length representing the scattering of quasiparticles in the surface layers of the two superconductors. Using the 2nd Josephson relation and the relation $dV = -R_p I_2 dy = -L_p \frac{\partial I_1}{\partial t} dy$ one obtains

$$\frac{\partial I_1}{\partial t} = -\frac{1}{L_p}\frac{dV}{dy} = -\frac{1}{L_p}\frac{\Phi_0}{2\pi}\frac{\partial^2 \varphi}{\partial y \partial t}, \tag{1.57}$$

where

$$I_1 = -\frac{1}{L_p}\frac{\Phi_0}{2\pi}\frac{\partial \varphi}{\partial y} \quad \text{and} \quad I_2 = -\frac{1}{R_p}\frac{dV}{dy} = -\frac{1}{R_p}\frac{\Phi_0}{2\pi}\frac{\partial^2 \varphi}{\partial y \partial t}, \tag{1.58}$$

so that

$$-\frac{1}{L_p}\frac{\Phi_0}{2\pi}\frac{\partial^2 \varphi}{\partial y^2} - \frac{1}{R_p}\frac{\Phi_0}{2\pi}\frac{\partial^3 \varphi}{\partial y^2 \partial t} = J_b - C\frac{\Phi_0}{2\pi}\frac{\partial^2 \varphi}{\partial t^2} - \frac{1}{R}\frac{\partial \varphi}{\partial t} - J_c \sin(\varphi). \tag{1.59}$$

Then, by normalizing $\tilde{y} = y/\lambda_J$ and $\tilde{t} = \omega_p t$ (note that another choice for normalising time has been made previously; in fact, it is usual to normalise times to the inverse of the plasma or the characteristic frequency), one obtains the *perturbed SGE* in normalized units

$$\frac{\partial^2 \varphi}{\partial \tilde{y}^2} - \frac{\partial^2 \varphi}{\partial \tilde{t}^2} = \sin(\varphi) + \alpha\frac{\partial \varphi}{\partial \tilde{t}} - \beta\frac{\partial^3 \varphi}{\partial \tilde{y}^2 \partial \tilde{t}} - \gamma, \tag{1.60}$$

where $\alpha = 1/(RC\omega_p)$ (tunneling of quasiparticles), $\beta = \omega_p L_p/R_p$ (surface currents damping), and $\gamma = J_b/J_c$ (normalized bias current density).

Equation (1.60) is completed by boundary conditions taking into account the external magnetic field, $H_{ext}(t)$:

$$\frac{\partial \varphi(0,t)}{\partial y} = \frac{2\pi t_H \lambda_J}{\Phi_0} H_{ext}(t) \quad \text{and} \quad \frac{\partial \varphi(L,t)}{\partial y} = \frac{2\pi t_H \lambda_J}{\Phi_0} H_{ext}(t). \tag{1.61}$$

At a small H_{ext} value, we can imagine a sort of "tail" of a fluxon, which is ideally centred outside the junction, penetrating the junction itself. Then, increasing the magnetic field, the center y_0 of this fluxon moves closer to the edge of the junction, according to [75]

$$\frac{2}{\cosh(y_0/\lambda_J)} = \frac{2\pi}{\Phi_0} H_{\text{ext}} \lambda_J = \frac{H_{\text{ext}}}{H_0}, \tag{1.62}$$

where $H_0 = \Phi_0/(2\pi t_{\text{H}} \lambda_J)$. At $H_{\text{ext}} = H_{c_1} = 2H_0$, we have that y_0 vanishes, i.e., a fluxon is set at each edge the junction. At $H_{\text{ext}} > H_{c1}$, fluxons penetrate the LJJ from the edges and fill it with some density which depends on the value of H_{ext}.

Looking at the behavior of the screening current, for small applied fields the junction can screen the applied external field by a circulating screening current, which flows in the opposite direction at both junction edges and adds to the external applied bias current. Increasing the applied field, the screening current enlarges until it reaches the critical value at one edge. Then vortices start to penetrate the junction resulting in an oscillating $J_S(z)$ dependence. In particular, the state with no vortex in the junction is called the *Meissner state*, while for magnetic fields larger than the critical field H_{c_1}, one or more vortices enter the junction.

The magnetic field dependence of critical current density in LJJs leads to diffraction patterns reminiscent of the "Fraunhofer-like" phenomena. In the context of SJJ limits, the diffraction lobes are typically well-separated; however, in the case of LJJs we observe a overlapping lobe structure in the $J_c(H)$ pattern. The interpretation of these patterns involves the insertion of kinks into the junction. Each individual lobe corresponds to a state characterized by a specific and unchanging number of kinks. As the magnetic field strength increases, configurations with a higher kinks number become more energetically favorable: in this case, the system undergoes a transition from a metastable state to a more stable one featuring an increased number of kinks. Within the range of magnetic field strengths where diffraction lobes overlap, different solutions with different N may concurrently coexist, but the system stays in the most energetically stable.

Looking at thermal effects, even in the case of LJJ these can be included in Eq. (1.60) by a stochastic noise term, with the statistical properties

$$\langle J_{th}(y,t)\rangle = 0 \qquad \text{and} \qquad \langle J_{th}(y,t), J_{th}(y,t')\rangle = 2\frac{k_B T}{R}\delta(t-t')\delta(y-y'), \tag{1.63}$$

that in normalized units, i.e., $\tilde{y} = y/\lambda_J$ and $\tilde{t} = \omega_p t$, become

$$\langle j_{th}(\tilde{y},\tilde{t})\rangle = 0 \qquad \text{and} \qquad \langle j_{th}(\tilde{y},\tilde{t}), j_{th}(\tilde{y},\tilde{t}')\rangle = 2\Gamma\delta(\tilde{t}-\tilde{t}')\delta(\tilde{y}-\tilde{y}'), \tag{1.64}$$

where the noise intensity now is given by $\Gamma = \frac{k_B T}{R}\frac{\omega_p}{\lambda_J J_c^2}$. In the literature, there are many examples of long Josephson systems in which the effects of noise are effectively taken into account to better explain the device's response, e.g., see Refs. [76–84].

Finally, we conclude mentioning that a mechanical analog can be given even for a LJJ [2,61], that is a chain of coupled pendula with torque. In particular, the φ_{xx} term is represented by the restoring torque of the spring, the φ_{tt} term by the moments of inertia f the pendula, and the $\sin(\varphi)$ by the gravitational torque.

1.1.7 The Superconducting Quantum Interference Device (SQUID)

Two junctions interrupting a superconducting loop constitute a *direct-current superconducting quantum interference device* (SQUID), i.e., a *dc-SQUID*, see Fig. 1.10. The name suggests the use of an external current biasing the SQUID.

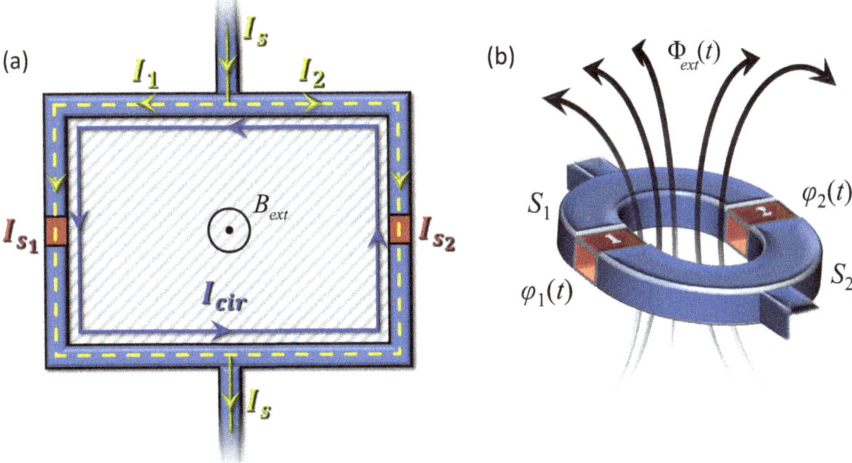

Fig. 1.10 Sketches of a two-junction dc-SQUID: (**a**) including the circulating and the flowing currents through the SQUID loop, and the Josephson currents, and (**b**) highlighting the Josephson phases and the magnetic flux lines through the SQUID loop

For a description of the main properties and usual applications of SQUIDs, see Refs. [85, 86].

Considering an integration path within the superconducting loop, see the yellow dashed line in Fig. 1.10a, the total flux, Φ, through the ring, taking into account also the Josephson phase drop across each JJ, can be written as:

$$2\pi \frac{\Phi}{\Phi_0} = \oint_C \Delta\theta \, dl = 2\pi n + \varphi_2 - \varphi_1. \tag{1.65}$$

The working principles of a SQUID combines two physical phenomena: the flux quantization in the superconducting loop and the Josephson effect. SQUIDs are the most sensitive detectors for magnetic flux Φ, being essentially a Flux-to-Voltage converter giving a flux-dependent output voltage with a period of one flux quantum, Φ_0.

The total current flowing through the SQUID is

$$I_s = I_{s1} + I_{s2} = I_{c_1} \sin(\varphi_1) + I_{c_2} \sin(\varphi_2). \tag{1.66}$$

If, for simplicity, one assumes $I_{c_1} = I_{c_2} = I_c$, the total current can be written as

$$I_s = 2I_c \cos\left(\frac{\varphi_1 - \varphi_2}{2}\right) \sin\left(\frac{\varphi_1 + \varphi_2}{2}\right) = 2I_c \cos\left(\pi \frac{\Phi}{\Phi_0}\right) \sin(\phi), \tag{1.67}$$

by defining $\phi = (\varphi_1 + \varphi_2)/2$. If the inductance of the superconducting loop is negligible, the total flux coincides with the external flux, $\Phi = \Phi_{\text{ext}}$: in this case, the maximum supercurrent can be written as

$$I_s^{\max} = 2I_c \left| \cos\left(\pi \frac{\Phi_{\text{ext}}}{\Phi_0}\right) \right|. \tag{1.68}$$

However, if the inductance L of the superconducting loop is non-negligible, we obtain

$$\frac{\Phi}{\Phi_0} = \frac{\Phi_{\text{ext}}}{\Phi_0} + \beta_L \cos(\phi) \sin\left(\pi \frac{\Phi}{\Phi_0}\right), \tag{1.69}$$

where $\beta_L = 2LI_c/\Phi_0$ is the *screening parameter*, which is one of the most important parameters, since the SQUID characteristic strongly depends on the β_L value [85]. In what follows, we look at $I_s^{\max}(\Phi)$ when I_{c_1} and I_{c_2} are the same or different and β_L is zero or not:

- If $I_{c_1} \neq I_{c_2}$ and $\beta_L = 0$, defining $\alpha = I_{c_2}/I_{c_1}$, the maximum supercurrent can be written as

$$I_s^{\max}(\Phi_{\text{ext}}) = I_{c_1}\sqrt{(1-\alpha)^2 + 4\alpha \cos^2\left(\pi \frac{\Phi_{\text{ext}}}{\Phi_0}\right)}. \tag{1.70}$$

- If $I_{c_1} = I_{c_2}$ and $\beta_L = 0$

$$I_s^{\max}(\Phi_{\text{ext}}) = 2I_c \left| \cos\left(\pi \frac{\Phi_{\text{ext}}}{\Phi_0}\right) \right|. \tag{1.71}$$

- If $I_{c_1} = I_{c_2} = I_c$ and $\beta_L \neq 0$, one has to solve numerically the equations

$$I_s = 2I_c \cos\left(\pi \frac{\Phi}{\Phi_0}\right) \sin(\phi) \qquad \text{and} \qquad \frac{\Phi}{\Phi_0} = \frac{\Phi_{\text{ext}}}{\Phi_0} + \beta_L \cos(\phi) \sin\left(\pi \frac{\Phi}{\Phi_0}\right). \tag{1.72}$$

In the case of identical currents (and very low inductance) the normalized critical current modulates between 2 and 0, see black curve in Fig. 1.11a. Increasing the asymmetry between the JJs, that is reducing α, causes the modulation of the oscillation of the critical current pattern to be progressively reduced, until it is cancelled if the critical current of a junction is zero; in other words, two junctions are required to see an interference pattern, see Fig. 1.11a. A reduction of the modulation depth occurs even for nonzero values of β_L. For $\beta_L = 1$ the critical current modulates by 50%, and for $\beta_L \gg 1$, the modulation $\Delta I_c/I_{c,\max}$ decreases as

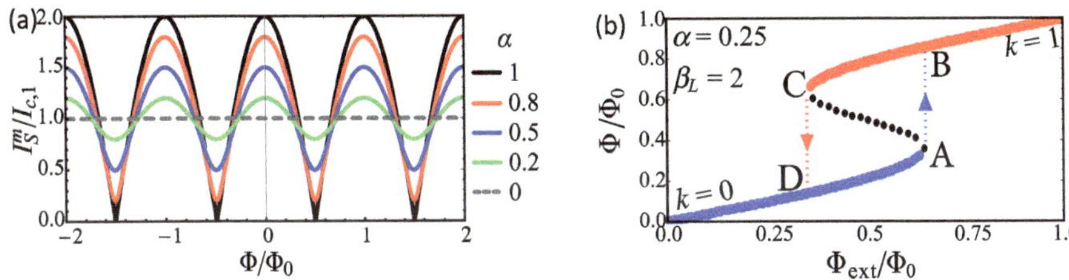

Fig. 1.11 (**a**) Critical current interference pattern of a two-junction SQUID as a function of the critical current ratio $\alpha = I_{c,2}/I_{c,1}$. (**b**) Normalized total magnetic flux through the SQUID loop, Φ/Φ_0, as a function of the normalized external flux, Φ_{ext}/Φ_0. Arrows indicate the transition between quantum states with different numbers k of flux quanta. The dotted part of the curve represents unstable states. Labels A, B, C, and D serve to mark a hysteresis loop followed by the total magnetic flux when sweeping Φ_{ext} back and forth

$1/\beta_L$ [85]. In principle, one could use the strong $I_c(\Phi_{\text{ext}})$ modulation to operate the SQUID as a sensitive flux detector. However, this process takes some time, so it is simpler to operate the SQUID in the voltage state, in which one sends a current I_b and measures the voltage drop V [86].

The SQUID can operate in the so-called *hysteretic mode* [85, 87, 88]. The circulating current I_{cir}, see the blue line in Fig. 1.10a, tends to compensate the applied flux Φ_{ext}; in particular, for a sufficiently high β_L value, the total flux can become multivalued, and abrupt flux transitions occur when changing Φ_{ext}. This is schematically shown in Fig. 1.11b. With increasing Φ_{ext}, the total flux follows a branch with a positive slope; we observe that only the parts with a positive slope are associated to stable states of the system, i.e., the dotted branch in Fig. 1.11b is unstable. If we increase Φ_{ext} further, at a certain moment the supercurrent flowing in the SQUID reaches the critical current of the weak link. Then, the junction momentarily switches to the voltage state, and the SQUID undergoes a $k \to k+1$ transition (the magnetic flux through the SQUID changes by one flux quantum). Now, if the applied flux is decreased, at a certain value a $k+1 \to k$ transition occurs. In other words, by sweeping Φ_{ext} forth and back, the system undergoes transitions from one state to another and back, so that a hysteresis loop (like ABCDA in the figure) is traced on the phase diagram, see Fig. 1.11b.

According to the RCSJ model, the phase evolution of a current-biased dc-SQUID formed by two identical JJs (i.e., the symmetric case) is determined by the following normalized equations of motion, with $k = 1, 2$ and φ_i being the Josephson phase of the k-th junction:

$$\beta_c \frac{\partial^2 \varphi_k}{\partial \tilde{t}^2} + \frac{\partial \varphi_k}{\partial \tilde{t}} = -\frac{\partial \mathcal{U}_{\text{SQUID}}}{\partial \varphi_k}, \tag{1.73}$$

where

$$\mathcal{U}_{\mathrm{SQUID}} = \frac{U_{\mathrm{SQUID}}}{E_{J_0}} = \frac{2}{\pi\beta_L}\underbrace{\left(\frac{\varphi_2-\varphi_1}{2}-\pi\varphi_a\right)^2}_{\text{Magnetic Energy}}$$
$$+\underbrace{\Big(-\cos(\varphi_1)-\cos(\varphi_2)\Big)}_{\text{Josephson Energy}}-\frac{i(\varphi_2+\varphi_1)}{2}. \tag{1.74}$$

Here, i is the normalized bias current, j is the normalized circulating current, and we have defined

$$\beta_L = \frac{E_{J_0}}{E_M}, \qquad \text{where} \qquad E_{J_0} = \frac{\Phi_0 I_c}{2\pi}, \quad \text{and} \quad E_M = \frac{1}{2\pi}\left(\frac{\Phi_0^2}{2L}\right). \tag{1.75}$$

We can generalize the RCSJ equations to the case of different JJs (asymmetric case), where C_k, R_k, and I_{c_k} are the capacitance, the normal resistance, and the critical current of the k-th junction [85]. Moreover, $\Phi = \Phi_{\mathrm{ext}} + L I_{\mathrm{cir}}$, where $L = L_1 + L_2$ (with L_1 and L_2 being the inductances of the two SQUID arms) is the inductance of the superconducting ring. By defining $\bar{I}_c = \frac{1}{2}(I_{c_1} + I_{c_2})$, $R = \frac{2R_1R_2}{R_1+R_2}$, and $\bar{C} = \frac{1}{2}(C_1 + C_2)$, in the case of

$$I_{c_1} \neq I_{c_2}, \quad R_1 \neq R_2, \quad C_1 \neq C_2, \quad \text{and} \quad L_1 \neq L_2, \tag{1.76}$$

one can define asymmetry parameters α_I, α_R, α_C, and α_L, so that

$$I_{c_1} = \bar{I}_c(1-\alpha_I), \quad R_1 = \frac{R}{1-\alpha_R}, \quad C_1 = \bar{C}(1-\alpha_C), \quad L_1 = \frac{L(1-\alpha_L)}{2} \tag{1.77}$$

$$I_{c_2} = \bar{I}_c(1+\alpha_I), \quad R_2 = \frac{R}{1+\alpha_R}, \quad C_2 = \bar{C}(1+\alpha_C), \quad L_2 = \frac{L(1+\alpha_L)}{2}. \tag{1.78}$$

By including also two independent Johnson-Nyquist noise currents, I_{n_1} and I_{n_2}, with the usual white noise features,

$$\langle I_{n_i}(t)\rangle = 0 \qquad \text{and} \qquad \langle I_{n_i}(t) I_{n_i}(t')\rangle = 2\frac{k_B T}{R_i}\delta(t-t'), \tag{1.79}$$

the RCSJ model in normalized units can be therefore written as:

$$\begin{aligned}\frac{i}{2}+j&=(1-\alpha_I)\sin(\varphi_1)+(1-\alpha_R)\frac{\partial\varphi_1}{\partial\tilde{t}}+\beta_c(1-\alpha_C)\frac{\partial^2\varphi_1}{\partial\tilde{t}^2}+i_{n_1}\\ \frac{i}{2}-j&=(1+\alpha_I)\sin(\varphi_2)+(1+\alpha_R)\frac{\partial\varphi_2}{\partial\tilde{t}}+\beta_c(1+\alpha_C)\frac{\partial^2\varphi_2}{\partial\tilde{t}^2}+i_{n_2}.\end{aligned} \tag{1.80}$$

The total flux is given by

$$\Phi_T=\Phi_{\text{ext}}+L_1I_1-L_2I_2=\Phi_{\text{ext}}+L_J-\alpha_L LI/2 \tag{1.81}$$

and

$$\varphi_2-\varphi_1=2\pi\Phi_{\text{ext}}+\pi\beta_L(j+\alpha_L i/2). \tag{1.82}$$

Regarding the thermal fluctuations [85], they can become relevant when the thermal energy, k_BT, approaches:

- the Josephson coupling energy, E_{J_0}, i.e.,

$$\Gamma=\frac{k_BT}{E_{J_0}}=\frac{2\pi/\Phi_0 k_BT}{I_c}=\frac{I_{\text{th}}}{I_c}\to 1. \tag{1.83}$$

- the characteristic magnetic energy, $E_M=E_{J_0}/\beta_L=\Phi_0^2/(2L)/(2\pi)$, i.e.,

$$\Gamma\beta_L=\frac{k_BT}{E_M}=\frac{L}{\Phi_0^2/(4\pi k_BT)}=\frac{L}{L_{\text{th}}}\to 1. \tag{1.84}$$

On the contrary, a small thermal fluctuations regime is established when:

- $\Gamma\ll 1$, that means $I_c\gg I_{\text{th}}=\frac{2\pi}{\Phi_0}k_BT\propto T$. To give a realistic number, $I_{\text{th}}\sim 0.18\ \mu$A at $T=4.2$ K.
- $\Gamma\beta_L\ll 1$, that means $L\ll L_{\text{th}}=\Phi_0^2/(4\pi k_BT)\propto 1/T$. To give a realistic number, $L_{\text{th}}\sim 5.9$ pH at $T=4.2$ K.

Let's look now at the so-called *radio frequency SQUID*, i.e., an *rf-SQUID*, which is a superconducting loop interrupted by one or more JJs with no current bias, but in a non-galvanic coupling, i.e., not in electrical contact, with an LC resonant circuit (i.e., a tank circuit) [85]. In the case of a single-junction rf-SQUID (for a description of a double-junction rf-SQUID, see Ref. [89]), we have $2\pi\Phi/\Phi_0+2\pi n=-\varphi$. The total flowing current and the total magnetic flux through the SQUID are

$$I_s = -I_c \sin\left(2\pi \frac{\Phi}{\Phi_0}\right) \qquad \text{and} \qquad \frac{\Phi}{\Phi_0} = \frac{\Phi_{\text{ext}}}{\Phi_0} + \frac{\beta_{L,\text{rf}}}{2\pi} \sin\left(2\pi \frac{\Phi}{\Phi_0}\right), \tag{1.85}$$

where $\beta_{L,\text{rf}} = 2\pi L I_c/\Phi_0$ is the screening parameter.

The SQUID loop is inductively coupled to the coil, L_T, of the tank circuit, the latter having a quality factor $Q = \frac{R_T}{\omega_{\text{rf}} L_T}$, where $\omega_{\text{rf}} = 1/\sqrt{L_T C_T}$ is its resonance frequency. The tank circuit is excited by an rf-current $I_{\text{rf}} \sin(\omega_{\text{rf}} t)$, which results in an rf-current $I_T = Q I_{\text{rf}}$ flowing in the tank circuit. In the following we look at the frequency response of the tank circuit in both $\beta_{L,\text{rf}} < 1$ and $\beta_{L,\text{rf}} > 1$ cases.

For $\beta_{L,\text{rf}} < 1$, the SQUID response is non-hysteretic and the device behaves as a non-linear inductor. The total flux through L_T is given by

$$\Phi_T = L_T I_T - M I_c \sin\phi_{\text{rf}}, \tag{1.86}$$

where $M = \alpha\sqrt{L_T L}$ is the mutual inductance. However, the flux through the SQUID loop is given by $\Phi = \alpha\Phi_T$, but also by $\Phi = L I_c \sin\phi_{\text{rf}}$, so that

$$I_c \sin\phi_{\text{rf}} = \alpha \frac{\Phi_T}{L} = \alpha \frac{L_T I_T}{L}. \tag{1.87}$$

The total flux through L_T can be written as

$$\Phi_T = \left(L_T - \alpha \frac{M}{L}\right) I_T = L_{T,\text{eff}} I_T, \tag{1.88}$$

where we have defined the effective inductance $L_{T,\text{eff}} = L_T \left(1 - \alpha^2 \sqrt{\frac{L_T}{L}}\right)$, which deviates from L_T due to the coupling to the SQUID loop. It affects the resonance frequency $\omega_{\text{rf,eff}} = \frac{1}{\sqrt{L_{T,\text{eff}} C_T}}$ of the tank circuit periodically with the applied magnetic flux. If the resonant circuit is operated close to its resonance frequency, the change of the resonance frequency causes a strong change of the rf-current and, hence, of the rf-voltage response of the tank circuit.

For $\beta_{L,\text{rf}} > 1$, the situation is different, since one has to deal with hysteretic $\Phi(\Phi_{\text{ext}})$ curves. Anyway, also in this case, the tank voltage is a periodic function of the applied magnetic flux.

1.2 Josephson Devices

In this part of the chapter we will make an overview of the design and technologies associated to circuits involving JJs, and similia. We start with showing how JJs are made and how material properties and technical requirements determine the technologic solution implemented.

1.2.1 JJ Fabrication Technology

As already stated, JJs are made by properly joining two pieces of superconducting materials. This can be made in a variety of different ways. However, the needs to realize useful circuits with the junctions implies that a planar technology, based on thin films of suitable materials, has to be implemented. This is because most of other circuit elements, such as resistors, inductors, and capacitors, have been implemented in planar technology for the realization of integrated circuits. The choice of the superconductor to be used depends mainly on the application foreseen and on the available technology. Historically, superconducting elements with low melting point (the so-called soft materials, such as Aluminium, Lead, Indium and Thin) were used, due to the relatively simple thin-film deposition techniques and to the relatively easy fabrication of an insulating layer separating two films. These materials were easy to fabricate, but have a relatively low critical temperature. A breakthrough in the technology was achieved when full Nb JJs were realized using a in situ trilayer technique, opening the possibility to realize reliable and durable superconducting circuits operating at the liquid Helium temperature of 4.2K. This has been the real starting for the development of superconducting electronics. The discovery of compound superconducting materials having higher critical temperatures, such as NbN, Nb_3Sn, $BKBO$, MgB_2, and the cuprates, above all the $YBCO$, triggered the interest in developing new technologies for the realization of reliable and easier to use, from the point of view of the involved cryogeny, superconducting circuits. However, with the notable exception of NbN and $YBCO$ for different applications, all these new materials showed specific difficulties related to the realization of JJs that impeded the realization of useful circuits. In Table 1.1 is reported a list of different superconducting materials used to realize superconducting circuits, organized with increasing critical temperature and divided in zones corresponding to very low, low, medium, and high critical temperature superconductors (VLTS, LTS, MTS, HTS). Beside the critical temperature other material parameters, relevant for applications, such as the London penetration length and the coherence length, are reported.

In order to have a clean thin film of material, there are different deposition techniques that can be used. Those employed for superconductors are mainly:

Table 1.1 Superconducting materials relevant for electronic applications organized in increasing critical temperature

Class	Material	λ_L [nm]	ξ [nm]	Tc [K]
VLTS	Al	16	1500	1.18
VLTS	In	25	400	3.3
VLTS	Sn	28	300	3.7
LTS	Pb	28	110	7.2
LTS	Nb	32	39	9.2
MTS	NbN	50 (200)	6	17
MTS	Nb_3Sn	50	6	18
MTS	MgB_2	140	3–5	39
HTS	$YBCO$	140	1.5	92

thermal evaporation, e-beam evaporation, and sputtering. A detailed description of these techniques is beyond the scope of this chapter. The interested reader is referred to the book from Milton Ohring [90]. The type of deposition technique, and the deposition parameters, strongly affects the properties of the deposited thin films, their crystalline structure, the eventual gas inclusions, the resulting stress, which determine the lattice constant and ultimately their transport and superconducting properties. Niobium, the workhorse of superconducting electronics, is particularly sensitive to the deposition conditions. After the deposition of a thin film, in order to build the wanted circuits, it is necessary to pattern it into a defined geometrical form. The related technologies, generally knows as photolithography, are very similar to those employed in semiconductor industry. They are based on sequences of photo- (or electron-) sensitive layer deposition (called photo-resist), its exposure (to UV or e-beam), its development and excess film material etching.

In Fig. 1.12 two different commonly used structuring methods are shown: lift-off and direct etching. The etching of the film can be made in different ways: lift-off, wet chemical etching, physical ion etching and reactive ion etching. The lift-off technique essentially consists in predefining, with the resist, of the negative image of the wanted geometry, followed by the deposition of the thin film. By subsequently dissolving the resist in an appropriate solvent, also the thin film on its top is removed, leaving the wanted geometrical pattern on the substrate. The direct etching techniques are somehow opposite. First the thin film is deposited on all the substrate area, it is subsequently covered with a thin layer of resist and then the desired geometry is defined using photolithographic techniques. The remaining

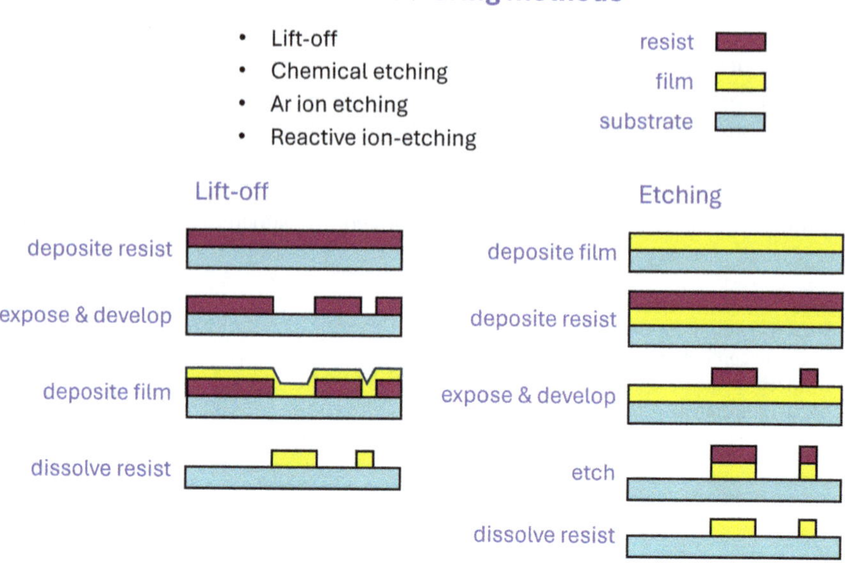

Fig. 1.12 Typical thin film structuring methods

exposed part of the film can be removed in different ways: using and appropriate chemical solution that dissolves the material through a chemical reaction (chemical etching), using a beam of energetic particles, typically Ar ions, that impinging on the film with enough energy, physically remove it (physical etching), or using a reactive plasma, typically CF_4 or SF_6, that chemically reacts with the material producing volatile reaction products, thus eroding the exposed material. Each technique has its pros and cons, in terms of complexity, effects on the material to be patterned, effects on the quality of the patterning, and in particular on the achievable limit resolution. For example, the lift-off technique has the advantage of being gentle with respect to the thin film manipulation and allows the use of shadow evaporation technique [91], which is a key technology for the realization of superconducting qubits.

The realization of a JJ-based device using a planar thin film technology requires a number of fabrication steps in order to realize the wanted coupling between two superconductors in a well defined position and with well defined geometrical shapes. During the several decades of development of JJ technology, different processes have been developed and optimized. Practically all the materials listed in Table 1.1, and some more, have been used to realize JJs for specific applications. However, nowadays most of the electronic applications of JJs are bases on Nb as electrode material and on the Nb trilayer technology for the junction fabrication [92]. A notable exception is the realization of superconducting qubits which, with their specific needs in terms of operating temperature, stray capacitance and low dissipation, are currently mostly realized using the Al as main material. These latter two will be discussed here in some detail.

In Fig. 1.13 it is shown a typical sequence used for the fabrication of Nb-based JJs using the trilayer technique. The grey area represents the substrate, typically a Si

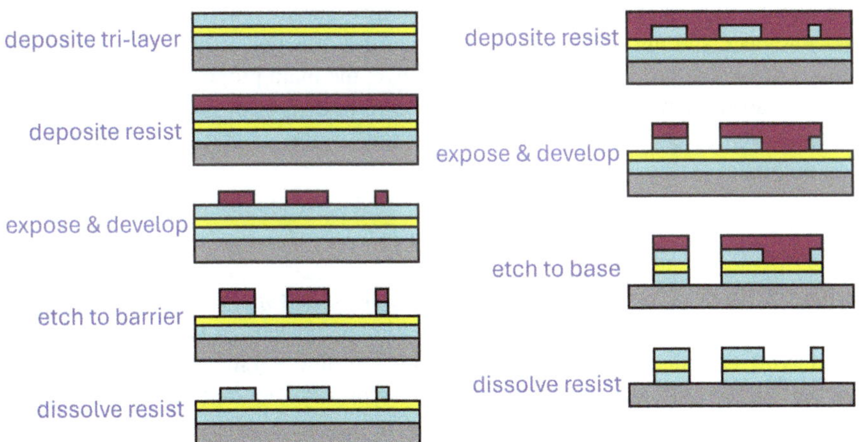

Fig. 1.13 Some steps from the sequence for the fabrication of a Nb-based JJ using the trilayer technique. The sequence reads from top to down and from left to right

wafer covered with a thin (200–300 nm) insulating layer of SiO_2. The aqua colored area represents the Nb layers, while the yellow area is the Al layer. The process starts with the in situ deposition of a Nb layer, typically 100–200 nm in thickness, followed by a thin Al layer of one or two nanometers. The Al layer, which has excellent coverage properties of the underlying Nb film is then oxidized by exposing the wafer to a well-defined O_2 atmosphere for a time ranging from few second to few minutes. The O_2 pressure and the exposure time determine the thickness of the native Al oxide, normally Al_2O_5, formed. This will in turn determine the transparency of the tunnel barrier forming the JJ, so great care is taken to have a controllable and reproducible oxidation process. Subsequently, a second Nb layer, again with a 100 to 200 nm thickness, is deposited on top of the previous ones, thus forming a giant and uniform JJ covering the whole. The rest of the process is devoted to the precise definition of the areas where the junctions are needed and of the relative superconducting wiring. At first a photoresist layer is deposited, using a spinner, on the trilayer and is patterned with the areas where the junctions have to be placed. A selective Nb etching process is then used to remove the uncovered top Nb layer. The etching, being selective, stops at the Al oxide layer. A second photoresist layer is spinned and patterned with the base electrode geometry. At this point, normally, the exposed Al layer is removed using a physical etching process and immediately after the underlaying Nb layer is also removed with a selective etching. After removing the resist, the base electrode pattern is defined as well as the areas where the JJ resides. Successive fabrication steps, not shown in Fig. 1.13, consist in the deposition and patterning of an insulating layer, typically of SiO_2, in order to leave uncovered the top part of the JJs, and a successive Nb deposition and patterning with the top electrode geometry. In this way a complete circuit with JJs can be realized. There are variants of the above process, in the way the insulation layer is formed and in the deposition of additional materials, when resistive components are needed. However, the main advantage of this way of fabricating circuits employing JJs is still the fact that the junction insulating barrier is done once and uniformly for all the junctions, thus guaranteeing a reproducible and controllable process. As mentioned above, the main property of the Josephson tunnel barrier, namely, the maximum pair current density J_c, is determined by the combination of O_2 pressure and oxidation time in the Al layer oxidation process. An empirical expression is:

$$J_c \propto (Pt)^n, \tag{1.89}$$

where J_c is in A/cm^2, P in mbar, t in minutes, and $n \approx 0.4 - 0.5$ [93, 94]. Therefore, by properly choosing the fabrication parameters, it is possible to realize JJs having critical current densities ranging from few A/cm^2 to tens of thousands A/cm^2. Low critical current densities are typically required for the realization of the voltage standard and of ultra sensitive magnetic sensors (SQUIDs), while high critical current densities are required by (classical) digital applications. As mentioned above, the development of quantum circuits has pushed towards the realization of JJs specifically designed for the realization of high-quality qubits. In this context, the fabrication processes must minimize sources of noise and error that can degrade qubit performance. This includes controlling material defects, reducing electromagnetic interference, and optimizing device geometries to minimize unwanted couplings. The preferred fabrication methodology adopted is based on the so called *Dolan* technique [91]. This technique involves using shadow evaporation to create the insulating barrier in the JJ. In this process, a thin layer of insulating material is deposited onto a superconducting film through a shadow mask, creating a localized region where the insulator is present between the superconducting electrodes. Common insulating materials used include aluminum oxide (AlO_x) and silicon dioxide (SiO_2), chosen for their compatibility with superconducting materials and their ability to form high-quality tunnel barriers suitable for JJs. A crucial aspect of the *Dolan* technique is the design of the shadow mask used during deposition. The mask defines the pattern of the insulating barrier, determining the shape and dimensions of the resulting JJs. Further discussion of the technology associated to the development of superconducting quantum circuits is beyond the scope of this chapter. The interested reader is referred to [95] and references therein.

1.2.2 Voltage Standard

Here, we will discuss the application of JJs to the realization of the international voltage standard. Since the discovery of the Josephson effect, it was clear that JJs could be used to realize a quantum device that relates, through universal constants, the frequency of an impinging radiation and the DC voltage across the junction [2], the so-called inverse AC Josephson effect.

This phenomenon has been briefly discussed in the first part of this chapter. Here, it will be recalled with some more details. A simple way to understand this situation is to consider a junction voltage biased with a DC and an RF source:

$$V(t) = V_0 + V_1 \sin(2\pi f_1 t). \tag{1.90}$$

From the second Josephson equation we have then:

$$\varphi = \int \frac{2\pi}{\Phi_0} V\, dt = \varphi_0 + \frac{2\pi}{\Phi_0} V_0 t + \frac{V_1}{\Phi_0 f_1} \cos(2\pi f_1 t), \tag{1.91}$$

where $\Phi_0 = 2.067833831\ 10^{-15}$ Wb $= 2.067833831\,\mu\text{V/GHz}$ is the flux quantum, an universal constant. From the first Josephson equation, the overall flowing current is:

$$\begin{aligned} I = I_c \sin\varphi &= I_c \sin\left(\varphi_0 + \frac{2\pi}{\Phi_0} V_0 t + \frac{V_1}{\Phi_0 f_1} \cos(2\pi f_1 t)\right) \\ &= I_c \sin\left(\varphi_0 + \frac{2\pi}{\Phi_0} V_0 t\right) \cos\left(\frac{V_1}{\Phi_0 f_1} \cos(2\pi f_1 t)\right) \end{aligned}$$

$$+I_c \cos\left(\varphi_0 + \frac{2\pi}{\Phi_0} V_0 t\right) \sin\left(\frac{V_1}{\Phi_0 f_1} \cos\left(2\pi f_1 t\right)\right). \tag{1.92}$$

By using well known series expansion of $\sin(z \cos x)$ and $\cos(z \cos x)$ in terms of the Bessel functions, it is straightforward to derive:

$$I = I_c \sum_{n=0}^{\infty} (-1)^n J_n\left(\frac{V_1}{\Phi_0 f_1}\right) sin\left(\varphi_0 + \frac{2\pi}{\Phi_0} V_0 t - 2\pi f_1 t\right). \tag{1.93}$$

This expression represents the current as an infinite sum of oscillations with zero average value. However, if the DC voltage assumes one of the following values:

$$V_n = n \Phi_0 f_1, \qquad \text{with} \qquad n = 0, \pm 1, \pm 2, \ldots \tag{1.94}$$

there is a net DC current flowing through the junction, appearing as a constant-voltage current step, called *Shapiro step* [8], with amplitudes given by $I_c J_n\left(\frac{V_1}{\Phi_0 f_1}\right)$, in the current voltage characteristic curve.

In Fig. 1.14b such situation is shown. The relation (1.94) allows to connect the frequency of the impinging electromagnetic radiation, which can be known with a precision of one part in 10^{11}, to the measured DC voltage across the junction and, through an universal constant Φ_0, to transfer such precision to it. The international voltage standard used at the time of the discovery of the Josephson effect was based on the Weston cells, and had a precision of the was one part in 10^6. Therefore, it is clear the advantage of using the Josephson effect to define a new voltage standard.

The technological route to reach this goal was, however, difficult for two main reasons. Firstly, to obtain clean steps the voltage-bias conditions have to be fulfilled. This requires that most of the RF current flows in the junction through its capacitance and therefore, that the applied RF frequency f_1 must be greater than the

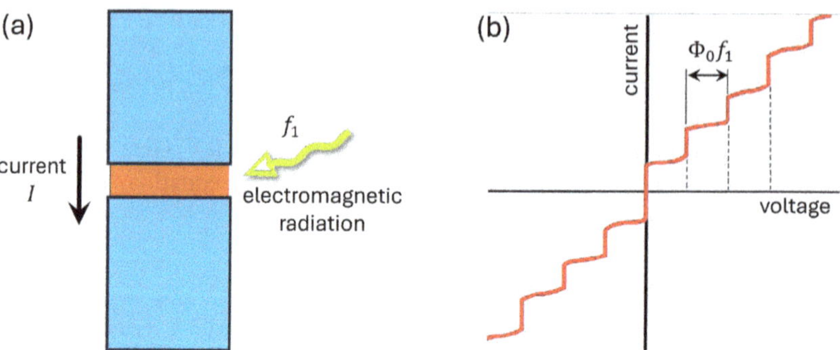

Fig. 1.14 (**a**) Sketch of a JJ irradiated with an electromagnetic radiation at frequency f_1 producing a flowing current I. (**b**) current voltage characteristic of a JJ irradiated by an electromagnetic wave at frequency f_1. Constant voltage steps appear at voltages given by Eq. (1.94)

junction plasma frequency $f_p = \sqrt{J_c/(2\pi C)}$, where C is the junction capacitance per unit area. As typical values of f_p are in the GHz range and, in order to use available RF equipment, this implies that the junction critical current density has to be kept as small as possible. Secondly, the voltage of each step is rather small, in the few μV range. To have a practical voltage standard, voltages of the order of 1 V are necessary.

Consequently, in the last 40 years arrays employing a large number of JJs, in series from the DC point of view and in parallel for the RF signal, have been developed. Current technology, based on trilayer Nb fabrication with low Josephson critical current densities, has developed devices that can provide DC voltages up to 10 V and also low frequency AC voltages with fundamental accuracy [96, 97].

1.2.3 SQUIDs

The Superconducting QUantum Interference Device (SQUID) is perhaps the most successful electronic device employing JJs. Its basic theory has been already discussed in the first part of the chapter. Here we will start from there and discuss the consequences relevant for the applications. More specifically we will concentrate on one version of the SQUID, the dc-SQUID, which is formed by a superconducting loop with to *equal* junctions. In Fig. 1.15a such situation is shown. A bias current I is applied to the device and the voltage drop across it is recorded. In Fig. 1.15b the dc current-voltage characteristic of a dc-SQUID is shown. The curve depends also on the amount of external magnetic field threading the SQUID loop, so that it oscillates from a larger current state when there is an even number of half flux quanta in the loop to a lower current state for an odd number. Such oscillation is periodic on the scale of one flux quantum and results in a periodic oscillation of the SQUID voltage, when it is biased at a dc current value slightly above the maximum zero voltage current, shown in Fig. 1.15c. This is the key factor for the use of a

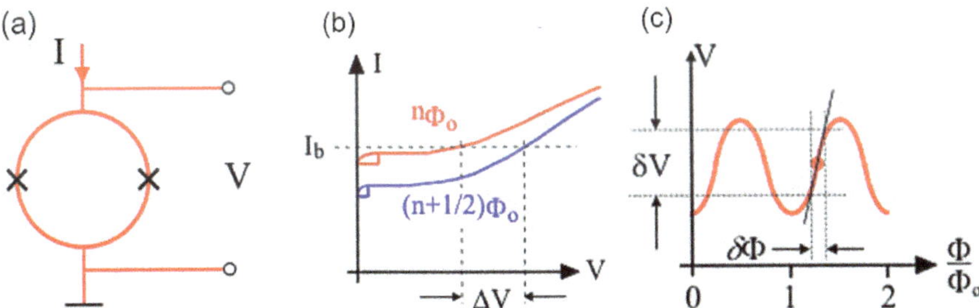

Fig. 1.15 (**a**) Circuit representation of a dc-SQUID. (**b**) Current voltage characteristic of a dc-SQUID when an even (upper red curve), or odd (lower blue curve) number of half flux quanta are threading the loop. I_b is the dc-bias current and ΔV is the voltage swing between the two flux states. (**c**) Dependence of the dc voltage across the SQUID on the applied magnetic flux. The red dot indicates the point of maximum responsivity

SQUID as magnetic sensor. The red point marked in Fig. 1.15c indeed indicates the point of maximum responsivity of the SQUID, i.e., the bias point where there is the larges variation of dc voltage in response of a small change in the applied magnetic flux.

Through an analysis of the SQUID equations, it is possible to derive an expression of its responsivity, defined as:

$$V_\Phi \equiv \left.\frac{\delta V}{\delta \Phi}\right|_I \approx \frac{R}{L}, \tag{1.95}$$

where R is the junction shunt resistance and L the SQUID loop inductance. Therefore:

$$\delta V \approx \frac{R}{L}\,\delta\Phi. \tag{1.96}$$

This result indicates that large R values and small L values can amplify the SQUID sensitivity. As with all sensors, the ultimate sensitivity is given by the intrinsic noise level. In SQUIDs, noise is generated by thermal fluctuations of the shunt resistors. The Nyquist voltage noise spectral density of a resistor R is given by $S_V(f) = 4\,k_B T\,R$, where k_B is the Boltzmann constant and T the absolute temperature. This can be translated into magnetic flux spectral noise density as:

$$S_\Phi(f) = \frac{4\,k_B T\,R}{V_\Phi^2} \approx \frac{16\,k_B T\,L^2}{R}, \tag{1.97}$$

where the additional factor 4 on the rhs stems from more accurate numerical studies of the SQUID equations in the optimal conditions. Using realistic values for the parameters ($L = 200\ pH,\ R = 6\ \Omega,\ T = 4.2\ K$) a value for the magnetic flux noise of about $1.2\ \mu\Phi_0/\sqrt{Hz}$ is obtained, which translates in a noise energy $\varepsilon(f) = \frac{S_\Phi(f)}{2L} \approx 10^{-32}\frac{J}{Hz} \approx 100\,\hbar$, very close to the quantum limit.

In order to translate this extreme sensitivity to magnetic flux into sensitivity to magnetic field, it is important to maximize the area of the loop that is threaded by the magnetic field to be measured. This would imply a strong increase of the loop inductance, while Eq. (1.97) calls for small values of it. A solution is the use of a superconducting flux transformer, as shown in Fig. 1.16, which, by means of its multiple windings and large pick-up area, can reach a field sensitivity of about $1\ fT/\sqrt{Hz}$.

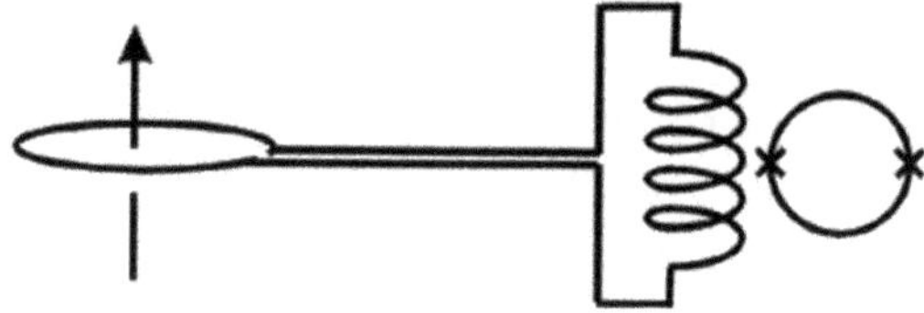

Fig. 1.16 example of a flux transformer coupled to a dc-SQUID

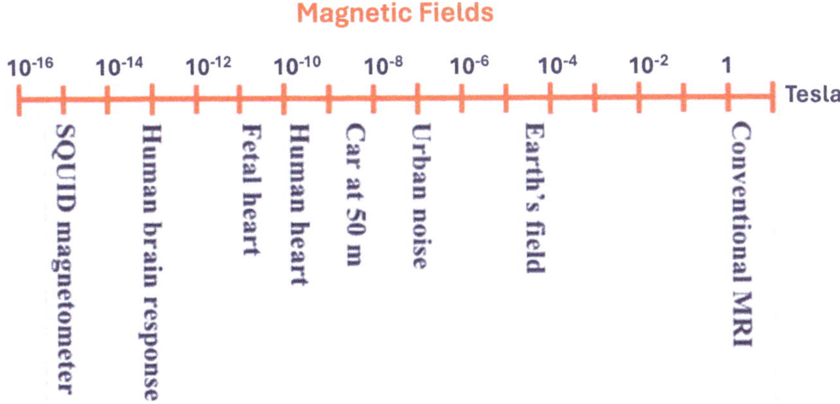

Fig. 1.17 Typical amplitudes of magnetic fields and sensitivity level of SQUID magnetometers

Such high sensitivity paves the way to numerous applications, but also requires proper shielding from unwanted signals, that can easily saturate the SQUID response. This is achieved using a number of techniques, from the employment of passive magnetically shielded chambers to the use of active shielding, to the use of higher order (gradiometric) pickup coil configurations. Moreover, the exploitation of the nonlinear responsivity of the SQUID into a linear response device requires the use of closed loop electronics with active feedback. All these aspects are rather technical and will not be treated here. Figure 1.17 shows a logarithmic scale of typical magnetic fields encountered. At the bottom there is the SQUID sensitivity. With their extreme sensitivity, down to quantum level, SQUIDs have found in the years numerous applications in several fields. In medical imaging and diagnostics, they are at the base of Magnetoencephalography (MEG) where SQUIDs detect the weak magnetic fields generated by neural currents in the brain and help in mapping brain activity with high spatial and temporal resolution. SQUIS are also used in magnetocardiography (MCG), where they measure the magnetic fields produced by electrical currents in the heart, aiding in diagnosing cardiac conditions. In material research SQUIDS are used to investigate the magnetic properties of materials, such as superconductors, magnetic nanoparticles, and other materials at low temperatures. They also help to explore vortex dynamics in superconductors. In geophysics and environmental studies, SQUIDs play a role in geophysical prospecting. They map variations in Earth's magnetic field, assisting in mineral exploration, detecting underground water reservoirs and studying subsurface structures. In astronomy and cosmology SQUIDS are widely used as amplifiers for ultrasensitive superconducting detectors in earth and space telescopes. Finally, SQUID are integral to quantum bits (qubits) in quantum computers where their sensitivity allows precise manipulation of qubits. The interested reader can refer to the review by Kleiner et al. and references therein [98]. More recent applications to quantum computing can be found in [99].

1.2.4 Classical Digital Circuits

Since the prediction and subsequent experimental confirmation of the Josephson effect, it was clear that JJs could be used as switching elements. The presence of two clearly distinguishable states, i.e., zero and finite, although small, voltage, and a sub ns response time for the passage between these states, made them ideal candidates as binary logic elements. Already in 1967 the IBM Corporation started a Josephson computer project that, in few years and based on the technologies then available, was able to develop a prototype processor running at the astonishing clock of 1 GHz [100]. The history of Josephson computing has had several ups and down, due to internal reasons: new technologies and new designs, and to external ones: the everlasting competition with semiconductors [101]. The Moore's law is universally known and has led the performances of semiconducting digital circuits through an exponential growth in the last almost 60 years. What was considered an enormous advantage in terms of speed and packaging density back in 1980 has now been largely overtaken by modern semiconducting processors. However, JJs have demonstrated a surprising vitality, thanks to new ways to exploit its intrinsic quantum nature.

Figure 1.18 shows the current-voltage characteristics of two different types of junctions used to realize digital devices: a hysteretic junction and a non-hysteretic one. In the first case the binary state is associated to two stable bias points, shown as orange dots in the figure. The binary "0" is associated the zero-voltage state and the binary "1" to the finite voltage state (about 2.5 mV for a Nb junction). The transition between the states is achieved by a positive current pulse, for the $0 \longrightarrow 1$ case, and a negative current pulse for the $1 \longleftarrow 0$ case. It is worth noting that the curves in Fig. 1.18 represent the average voltage measured across the junction, as the Josephson oscillations, being of the order of few GHz, cannot be recorded by ordinary electronic equipment. The time it takes for the junction to stabilize in the new bias point after a transition determines the maximum speed of operation of the logic device. For typical Nb-based junctions, its dynamics limits such speed

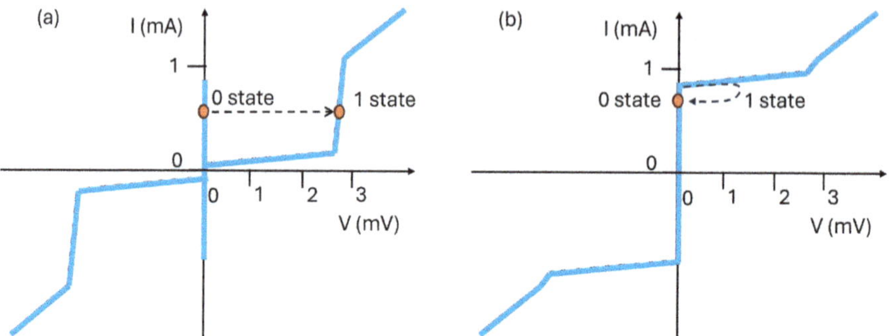

Fig. 1.18 Current-voltage characteristics of a hysteretic (**a**) and a non-hysteretic one (**b**). The dots represent the stable bias states

to a rate of about 1 GHz. This sets the maximum operating clock frequency of the device. Such value, which was at least one order of magnitude higher than of that of the large IBM mainframes in the 1980s, is nowadays easily overcome by any portable computing device. This, together with other material related reliability problems lead IBM to abandon the Josephson computer project in 1983. Meanwhile other institutions, in US and in Japan kept working on Josephson digital circuits development, also thanks to the advent of the trilayer Nb technology, discussed above. In 1991 the group of K. Likharev proposed a new mechanism for logic operations using JJs [102], which was not affected by the speed limitation of the voltage state mechanism. The new logic operation was called Rapid Single Flux Quantum (RSFQ) and has been demonstrated to operate up to frequencies of few hundreds of GHz, an unreachable value for semiconducting electronics.

The operation of RSFQ logic is quite different from the voltage state one. It is a clocked type of logic, that is it requires a reference train of pulses (the clock). The Josephson gate, in certain conditions, can generate a Single Flux Quantum (SFQ) pulse and the logical 1 and 0 states are encoded as the presence or absence of an SFQ pulse between two clock pulses. The junction is resistively shunted, in order to show a non-hysteretic current voltage characteristic (see Sect. 1.1 of this chapter), inserted in a superconductive loop and dc biased near it critical current value (the orange dot in Fig. 1.18b). Upon arrival of a current pulse, the junction switches momentarily to the voltage state, but, being the current-voltage curve single valued, it has to self-reset to its initial state (see dashed curve in Fig. 1.18b). This occurs in few ps and corresponds to a 2π phase jump across the junction. From the second Josephson equation a phase jump corresponds to a voltage pulse with a quantized area corresponding to exactly one flux quantum, hence the name SFQ pulse.

$$\int_{V(\varphi=\varphi_0)}^{V(\varphi=\varphi_0+2\,\pi)} V(t)\,dt = \int_{\varphi_0}^{\varphi_0+2\pi} \frac{\Phi_0}{2\,\pi}\frac{d\varphi}{dt}\,dt = \Phi_0 = 2.07\,mV\;ps \tag{1.98}$$

Such flux quantum can be trapped in the superconducting loop or moved, through JJs, from one loop to another as millivolt picosecond SFQ pulses. In this way this basic bit of information can be manipulated and perform logic operations in a few picoseconds. Such short pulses require impedance matched microstrip lines to propagate over adjacent gates, which represents a design challenge, only recently addressed by the semiconductor industry.

A detailed description of the operation of Josephson logic gates is beyond the scope of this chapter. Many details regarding both types of logics can be found in standard textbooks [2, 41, 103], while a more recent survey on superconducting computing technology can be found in [104].

Overall, the superconducting digital technology has demonstrated a good capability to withstand the overwhelming increase of performances exhibited by semiconductive circuits. From one side, the level of integration reached in modern processors is unreachable by superconductors for many reasons, also economic. However, superconductive circuits outperforms in terms of speed and low dissipated power and, in perspective, can effectively team with semiconductors in applications where ultra-fast signal processing is needed while leaving slower tasks, like storage, to the ultra-dense semiconductive circuits.

1.3 Towards Quantum Computing with Superconducting Qubits

In this Section the basics of quantum computing will be described. Starting from the Bloch sphere representation for the qubit, we will introduce the quantum gates and their operations, to conclude with an overview on the quantum supremacy and the state of the art on Quantum computers.

1.3.1 Bloch Sphere Representation

As we have seen in the previous Chapter, the superconducting qubit is a type of qubit that utilizes superconducting circuits to exploit the unique properties of superconductors for quantum computation and its basic building block is a JJ. Generically the qubit is a two-level system that can be represented as a superposition of two states (*quantum state*). The *Bloch sphere* is a unit sphere used to represent the quantum state of the qubit. In Fig. 1.19 the Bloch sphere is shown with a *Bloch vector* representing the state $|\psi\rangle = \alpha|0\rangle + \beta|1\rangle$. By convention, the north pole represents state $|0\rangle$ and the south pole state $|1\rangle$. For pure quantum states such as $|\psi\rangle$, the Bloch vector is of unit length, $|\alpha|^2 + |\beta|^2 = 1$, connecting the center of the sphere to any point on the surface.

The z-axis connects the north and south poles. It represents the *qubit quantization axis* for the states $|0\rangle$ and $|1\rangle$ in the qubit eigenbasis. Following our convention, state $|0\rangle$ at the north pole is associated with $+1$, and state $|1\rangle$ (the south pole) with -1. In

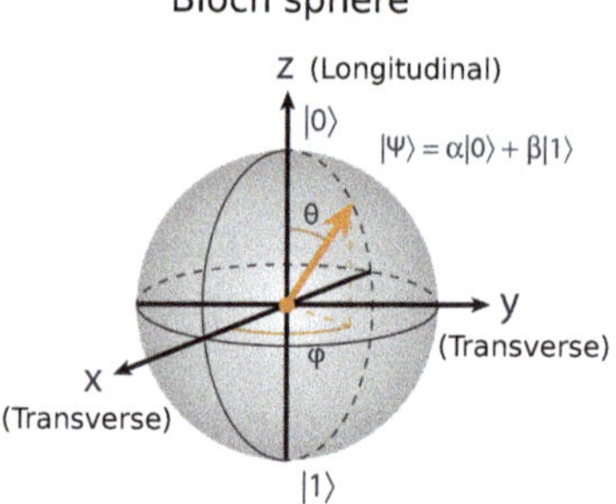

Fig. 1.19 Bloch sphere representation of the quantum state $|\psi\rangle = \alpha|0\rangle + \beta|1\rangle$. The qubit quantization axis—the z axis—is *longitudinal* in the qubit frame, corresponding to σ_z terms in the qubit Hamiltonian. The x-y plane is *transverse* in the qubit frame, corresponding to σ_x and σ_y terms in the qubit Hamiltonian

turn, the x-y plane is the *transverse plane* and x and y are called *transverse axes*. In a parametric representation, the unit Bloch vector $\boldsymbol{a} = (\sin\theta\cos\phi, \sin\theta\sin\phi, \cos\theta)$ is represented using the polar angle $0 \le \theta \le \pi$ and the azimuthal angle $0 \le \phi < 2\pi$, as illustrated in Fig. 1.19. We can similarly represent the quantum state using the angles θ and ϕ,

$$|\psi\rangle = \alpha|0\rangle + \beta|1\rangle = \cos\frac{\theta}{2}|0\rangle + e^{i\phi}\sin\frac{\theta}{2}|1\rangle. \tag{1.99}$$

The Bloch vector is stationary on the Bloch sphere in the *rotating frame picture*. If state $|1\rangle$ has a higher energy than state $|0\rangle$ (as it generally does in superconducting qubits), then in a stationary frame, the Bloch vector would precess around the z-axis at the qubit frequency $(E_1 - E_0)/\hbar$. Without loss of generality (and much easier to visualize), we instead *choose* to view the Bloch sphere in a reference frame where the x and y-axes also rotate around the z-axis at the qubit frequency. In this *rotating frame*, the Bloch vector appears stationary as written in Eq. (1.99).

We also note that the density matrix $\rho = |\psi\rangle\langle\psi|$ for a pure state $|\psi\rangle$ is equivalently written

$$\rho = \frac{1}{2}(I + \boldsymbol{a}\cdot\boldsymbol{\sigma}) = \frac{1}{2}\begin{pmatrix} 1+\cos\theta & e^{-i\phi}\sin\theta \\ e^{i\phi}\sin\theta & 1+\sin\theta \end{pmatrix} \tag{1.100}$$

$$= \begin{pmatrix} \cos^2\frac{\theta}{2} & e^{-i\phi}\cos\frac{\theta}{2}\sin\frac{\theta}{2} \\ e^{i\phi}\cos\frac{\theta}{2}\sin\frac{\theta}{2} & \sin^2\frac{\theta}{2} \end{pmatrix} \tag{1.101}$$

$$= \begin{pmatrix} |\alpha|^2 & \alpha\beta^* \\ \alpha^*\beta & |\beta|^2 \end{pmatrix}, \tag{1.102}$$

where I is the identity matrix, and $\boldsymbol{\sigma} = [\sigma_x, \sigma_y, \sigma_z]$ is a vector of Pauli matrices. If the Bloch vector $\boldsymbol{a}$ is a unit vector, then ρ represents a pure state, ψ, and $\mathrm{Tr}(\rho^2) = 1$. More generally, the Bloch sphere can be used to represent *mixed states*, for which $|\boldsymbol{a}| < 1$; in this case, the Bloch vector terminates at points *inside* the unit sphere, and $0 \le \mathrm{Tr}(\rho^2) < 1$. To summarize, the surface of the unit sphere represents pure states, and its interior represents mixed states [105].

1.3.2 Qubit Control

In this section, we will introduce how superconducting qubits are manipulated to implement quantum algorithms. The transmon-like variant of superconducting qubits has emerged as the most extensively utilized mean for implementing quantum programs. Consequently, the discussion in this section will concentrate on methodologies tailored for transmons. However, that the techniques presented herein can be extended to all categories of superconducting qubits.

Thus, we make a brief review on the gates used in classical computing as well as quantum computing, and the concept of universality. In the latter part of this section, we review some of the most common implementations of two-qubit gates in both tunable and fixed-frequenzy transmon qubits. The single-qubit and two-qubit operations together form the basis of many of the medium-scale superconducting quantum processors that exist today.

Throughout this section, we write everything in the computational basis $\{|0\rangle, |1\rangle\}$ where $|0\rangle$ is the $+1$ eigenstate of σ_z and $|1\rangle$ is the -1 eigenstate. To indicate the rotation operator of a qubit state, e.g. rotations around the x-axis by an angle θ we use the notation

$$\mathsf{X}_\theta = R_X(\theta) = e^{-i\frac{\theta}{2}\sigma_x} = \cos(\theta/2)1 - i\sin(\theta/2)\sigma_x \tag{1.103}$$

and we use the shorthand notation 'X' for a full π rotation about the x axis (and similarly for $\mathsf{Y} := \mathsf{Y}_\pi$ and $\mathsf{Z} := \mathsf{Z}_\pi$).

1.3.3 Boolean Logic Gates in Classical Computers

Boolean logic can be implemented on classical computers using a small set of single-bit and two-bit gates. In Fig. 1.20 several classical logic gates are shown along with their truth tables. In classical boolean logic, bits can take on one of two values: state 0 or state 1. The state 0 represents logical FALSE, and state 1 represents logical TRUE.

Beyond the trivial *identity operation*, which simply gives a boolean bit unchanged, the only other possible single-bit boolean logic gate is the NOT gate. As shown in Fig. 1.20, the NOT gate flips the bit: $0 \rightarrow 1$ and $1 \rightarrow 0$. This gate is *reversible*, because it is trivial to determine the input bit value given the output bit values. As we will see, for two-bit gates, this is not the case.

There are several two-bit gates shown in Fig. 1.20. A two-bit gate takes two bits as inputs, and it gives as an output the result of a boolean operation. One common example is the AND gate, for which the output is 1 if and only if both inputs are 1; otherwise, the output is 0. The AND gate, and the other two-bit gates shown in Fig. 1.20, are all examples of *irreversible* gates; that is, the input bit values cannot be inferred from the output values. For example, for the AND gate, an output of logical 1 uniquely identifies the input 11, but an output of 0 could be associated with 00, 01, or 10. Once the operation is performed, in general, it cannot be undone and the input information is lost. There are several variants of two-bit gates, including,

- AND and OR;
- NAND (a combination of NOT and AND) and NOR (a combination of NOT and OR);
- XOR (exclusive OR) and NXOR (NOT XOR).

The XOR gate is also known *parity* gate. That is, it returns a logical 0 if the two inputs are the same values (i.e., they have the same parity), and it returns a logical

GATE NAME	SYMBOL	TRUTH TABLE
NOT		Input – Output 0 – 1 1 – 0
AND		Input – Output 0 0 – 0 0 1 – 0 1 0 – 0 1 1 – 1
OR		Input – Output 0 0 – 0 0 1 – 1 1 0 – 1 1 1 – 1
NAND		Input – Output 0 0 – 1 0 1 – 1 1 0 – 1 1 1 – 0
NOR		Input – Output 0 0 – 1 0 1 – 0 1 0 – 0 1 1 – 0
XOR		Input – Output 0 0 – 0 0 1 – 1 1 0 – 1 1 1 – 0
XNOR	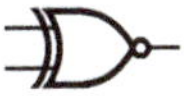	Input – Output 0 0 – 1 0 1 – 0 1 0 – 0 1 1 – 1

Fig. 1.20 Classical single-bit and two-bit boolean logic gates. For each gate, the name, a short description, circuit representation, and input/output truth tables are presented. The numerical values in the truth table correspond to the classical bit values 0 and 1. Modified from [106], © The Authors, MIT xPro Quantum Curriculum, https://learn-xpro.mit.edu/quantum-computing, 2018. All rights reserved

1 if the two inputs have different values (i.e., different parity). Still, the XOR and NXOR gates are not reversible, because knowledge of the output does not allow one to uniquely identify the input bit values.

Universality, in the context of quantum computing, pertains to the capability of executing any boolean logic algorithm using a small set of single-bit and two-bit gates. A universal gate set can in principle transform any state to any other state in the state space represented by the classical bits. The set of gates that enable universal computation is not unique, and may be represented by a small set of gates. For example, the NOT gate and the AND gate together form a universal gate set. Similarly, the NAND gate itself is universal, as is the NOR gate. The efficiency of implementation depends on the choice of the gate set.

1.3.4 Quantum Logic Gates Used in Quantum Computers

Quantum logic can similarly be performed by a small set of single-qubit and two-qubit gates. Qubits can of course assume the classical states $|0\rangle$ and $|1\rangle$, at the north pole and south pole of the Bloch sphere, but they can also assume arbitrary superpositions $\alpha|0\rangle + \beta|1\rangle$, corresponding to any other position on the sphere.

Single-qubit operations translate an arbitrary quantum state from one point on the Bloch sphere to another point by rotating the Bloch vector (spin) a certain angle about a particular axis. As shown in Fig. 1.21, there are several single-qubit operations, each represented by a matrix that describes the quantum operation in the computational basis represented by the eigenvectors of the σ_z operator, i.e. $|0\rangle \equiv [1\ 0]^T$ and $|1\rangle \equiv [0\ 1]^T$.

For example, the *identity gate* performs no rotation on the state of the qubit. This is represented by a two-by-two identity matrix. The X-gate performs a π rotation about the x axis. Similarly, the Y-gate and Z-gate perform a π rotation about the y axis and z axis, respectively. The S-gate performs a $\pi/2$ rotation about the z axis, and the T-gate performs a rotation of $\pi/4$ about the z axis. The Hadamard gate H is also a common single-qubit gate that performs a π rotation around an axis diagonal in the x-z plane, see Fig. 1.21.

Two-qubit quantum-logic gates are generally *conditional* gates and take two qubits as inputs. Typically, the first qubit is the *control* qubit, and the second is the *target* qubit. A unitary operator is applied to the target qubit, dependent on the state of the control qubit. The two common examples are the controlled NOT (CNOT-gate) and controlled phase (CZ or CPHASE gate). The CNOT-gate flips the state of the target qubit conditioned on the control qubit being in state $|1\rangle$. The CPHASE-gate applies a Z gate to the target qubit, conditioned on the control qubit being in state $|1\rangle$. As we will shown later, the iSWAP gate—another two-qubit gate—can be built from the CNOT-gate and single-qubit gates.

GATE NAME	CIRCUIT REPRESENTATION	MATRIX REPRESENTATION	TRUTH TABLE
I Identity-gate: no rotation	I	$I = \begin{pmatrix} 1 & 0 \\ 0 & 1 \end{pmatrix}$	Input / Output: $\vert 0\rangle$ → $\vert 0\rangle$; $\vert 1\rangle$ → $\vert 1\rangle$
X gate: rotation by π radiants about the x-axis.	X	$X = \begin{pmatrix} 0 & 1 \\ 1 & 0 \end{pmatrix}$	Input / Output: $\vert 0\rangle$ → $\vert 1\rangle$; $\vert 1\rangle$ → $\vert 0\rangle$
Y gate: rotation by π radiants about the y-axis.	Y	$Y = \begin{pmatrix} 0 & -i \\ i & 0 \end{pmatrix}$	Input / Output: $\vert 0\rangle$ → $i\vert 1\rangle$; $\vert 1\rangle$ → $-i\vert 0\rangle$
Z gate: rotation by π radiants about the z-axis.	Z	$Z = \begin{pmatrix} 1 & 0 \\ 0 & -1 \end{pmatrix}$	Input / Output: $\vert 0\rangle$ → $\vert 0\rangle$; $\vert 1\rangle$ → $-\vert 1\rangle$
S gate: rotation by π/2 radiants about the z-axis.	S	$S = \begin{pmatrix} 1 & 0 \\ 0 & e^{i\frac{\pi}{2}} \end{pmatrix}$	Input / Output: $\vert 0\rangle$ → $\vert 0\rangle$; $\vert 1\rangle$ → $e^{i\frac{\pi}{2}}\vert 1\rangle$
T gate: rotation by π/4 radiants about the z-axis.	T	$T = \begin{pmatrix} 1 & 0 \\ 0 & e^{i\frac{\pi}{4}} \end{pmatrix}$	Input / Output: $\vert 0\rangle$ → $\vert 0\rangle$; $\vert 1\rangle$ → $e^{i\frac{\pi}{4}}\vert 1\rangle$
H gate: rotation by π radiants about an axis diagonal in the x-z plane.	H	$H = \frac{1}{\sqrt{2}}\begin{pmatrix} 1 & 1 \\ 1 & -1 \end{pmatrix}$	Input / Output: $\vert 0\rangle$ → $\frac{\vert 0\rangle + \vert 1\rangle}{\sqrt{2}}$; $\vert 1\rangle$ → $\frac{\vert 0\rangle - \vert 1\rangle}{\sqrt{2}}$

Fig. 1.21 Quantum single-qubit gates. For each gate, the name, circuit representation, matrix representation, input/output truth tables are presented. Matrices are defined in the basis spanned by the state vectors $|0\rangle \equiv [1\ 0]^T$ and $|1\rangle \equiv [0\ 1]^T$. The numerical values in the truth table correspond to the quantum states $|0\rangle$ and $|1\rangle$. Modified from [106], © The Authors, MIT xPro Quantum Curriculum, https://learn-xpro.mit.edu/quantum-computing, 2018. All rights reserved

The unitary operator of the CNOT gate can be written in a useful way, highlighting that it applies an X depending on the state of the control qubit.

$$U_{\mathsf{CNOT}} = \begin{bmatrix} 1 & 0 & 0 & 0 \\ 0 & 1 & 0 & 0 \\ 0 & 0 & 0 & 1 \\ 0 & 0 & 1 & 0 \end{bmatrix} = |0\rangle\langle 0| \otimes 1 + |1\rangle\langle 1| \otimes \mathsf{X} \tag{1.104}$$

and similarly for the CPHASE gate,

$$U_{\mathsf{CPHASE}} = \begin{bmatrix} 1 & 0 & 0 & 0 \\ 0 & 1 & 0 & 0 \\ 0 & 0 & 1 & 0 \\ 0 & 0 & 0 & -1 \end{bmatrix} = |0\rangle\langle 0| \otimes 1 + |1\rangle\langle 1| \otimes \mathsf{Z}\,. \tag{1.105}$$

Comparing the last equality above with the unitary for the CNOT (1.104), it is clear that the two gates are closely related. Indeed, a CNOT can be generated from a CPHASE by applying two Hadamard gates,

$$U_{\mathsf{CNOT}} = (1 \otimes \mathsf{H})U_{\mathsf{CPHASE}}(1 \otimes \mathsf{H}), \tag{1.106}$$

since $\mathsf{HZH} = \mathsf{X}$. Due to the form of (1.105), the CPHASE gate is also denoted the CZ gate, since it applies a controlled Z operator, by analogy with CNOT (a controlled application of X operator).

Some two-qubit gates such as CNOT and CPHASE are also called *entangling gates*, because they can take product states as inputs and output entangled states. They are thus an indispensable component of a universal gate set for quantum logic. For example, consider two qubits A and B in the following state:

$$|\psi\rangle = \frac{1}{\sqrt{2}}\left(|0\rangle + |1\rangle\right)_A |0\rangle_B. \tag{1.107}$$

If we perform a CNOT gate, U_{CNOT}, on this state, with qubit A the control qubit, and qubit B the target qubit, the resulting state is:

$$U_{\mathsf{CNOT}}|\psi\rangle = \frac{1}{\sqrt{2}}\left(|0\rangle_A|0\rangle_B + |1\rangle_A|1\rangle_B\right) \neq (\ldots)_A(\ldots)_B, \tag{1.108}$$

which is a state that cannot be factored into an isolated qubit-A component and a qubit-B component. This is one of the two-qubit entangled *Bell states*, a manifestly quantum mechanical state.

A universal set of single-qubit and two-qubit gates is sufficient to implement arbitrary quantum logic. This means that this gate set can in principle reach *any* state in the multi-qubit state-space. How efficiently this is done depends on the choice of quantum gates that comprise the gate set. We also note that each of the single-qubit and two-qubit gates is *reversible*, that is, given the output state, one can uniquely determine the input state. As we discuss further, this distinction between classical and quantum gates arises, because quantum gates are based on *unitary* operations U. If a unitary operation U is a particular gate applied to a qubit, then its hermitian conjugate $U^\dagger$ can be applied to recover the original state, since $U^\dagger U = I$ resolves an identity operation.

1.3.5 Classical Versus Quantum Gates

The gate-sequences used to represent quantum algorithms have certain similarities to those used in classical computing, with a few striking differences. As an example, we consider first the classical NOT gate (discussed previously), and the related quantum circuit version, shown in Fig. 1.22.

While the classic bit-flip gate inverts any input state, the quantum bit-flip does not in general produce the antipodal vector, but rather exchanges the prefactors of the wavefunction. The X operator is sometimes referred to as 'the quantum NOT', but we note that X only acts similar to the classical NOT gate in the case of classical data stored in the quantum bit, i.e. $\mathsf{X}|g\rangle = |\bar{g}\rangle$ for $g \in \{0, 1\}$.

As briefly mentioned in Sect. 1.3.4, *all* quantum gates are *reversible*, due to the underlying unitary nature of the operators implementing the logical operations. Certain other processes used in quantum information processing, however, are irreversible. Namely, measurements and energy loss to the environment. These processes are modeled by the *amplitude damping* or *phase damping* operator, but we refer the interested reader, e.g., to Ref. [105]. Let us finally note that quantum circuits are written left-to-right (in order of application), while the calculation of the result of a gate-sequences, e.g the circuit

$$|\psi_{\text{in}}\rangle \; -\boxed{U_0}-\boxed{U_1}- \cdots -\boxed{U_n}- \; |\psi_{\text{out}}\rangle \tag{1.109}$$

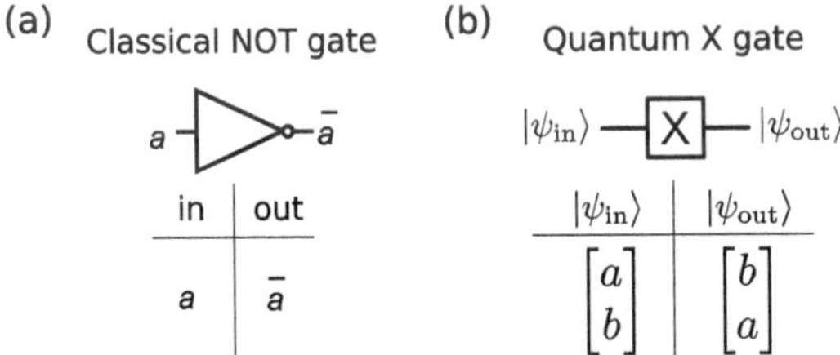

Fig. 1.22 Comparison of the classical (NOT) gate and quantum (X) gate. **(a)** The classical NOT gate that inverts the state of a classical bit. **(b)** The quantum X gate, which flips the coefficients of the two components of the quantum bit

is performed right-to-left, i.e.

$$|\psi_{\text{out}}\rangle = U_n \cdots U_1 U_0 |\psi_{\text{in}}\rangle. \tag{1.110}$$

As discussed in Sect. 1.3.3, the NOR and NAND gates are each individually universal gates for classical computing. Since both of these gates have no direct quantum analogue (because they are not reversible), it is natural to ask which gates *are* needed to build a universal quantum computer. It turns out that the ability to rotate about arbitrary axes on the Bloch-sphere (i.e., a complete single-qubit gate set), supplemented with any entangling 2-qubit operation will suffice for universality [105, 107]. By using what is known as the 'Krauss-Cirac decomposition', any two-qubit gate can be decomposed into a series of CNOT operations [105].

1.3.5.1 Minimum Gate Sets

A universal quantum gate set is

$$\mathcal{G}_0 = \{\mathsf{X}_\theta, \mathsf{Y}_\theta, \mathsf{Z}_\theta, \mathrm{Ph}_\theta, \mathsf{CNOT}\} \tag{1.111}$$

where $\mathrm{Ph}_\theta = e^{i\theta}\mathbb{1}$ applies an overall phase θ to a single qubit. For completeness we mention another universal gate set, which is of particular interest from a theoretical perspective, namely

$$\mathcal{G}_1 = \{\mathsf{H}, \mathsf{S}, \mathsf{T}, \mathsf{CNOT}\}. \tag{1.112}$$

Let us note that the restriction to a discrete gate set still gives rise to universality. This fact relies on using the so-called Solovay-Kitaev [108, 109] theorem, which states that any other single-qubit gate can be approximated to an error ϵ using only $O(\log^g(1/\epsilon))$ (where $g > 0$) single-qubit gates from $\mathcal{G}_1$. The gate-set $\mathcal{G}_1$ is typically referred to as the 'Clifford + T' set, where H, S and CNOT are all Clifford gates.

Each quantum computing architecture will have certain gates that are simpler to implement at the hardware level than others (these are referred to as 'native' gates). These are typically the gates for which the Hamiltonian governing the gate-implementation gives rise to a unitary propagator that corresponds to the gate itself. Regardless of which gates are natively available, as long as one has a complete gate set, one can use the Solovay-Kitaev theorem to synthesize any other set efficiently. In general one wants to keep the overall number of time steps in which gates are applied (known as *depth* of a quantum circuit) as low as possible, and one wants to use as many of the native gates as possible. Moreover, running a quantum algorithm also depends on the qubit connectivity of the device. The process of designing a quantum gate sequence that efficiently implements a specific algorithm is known as *gate synthesis* and *gate compilation*, respectively. A full discussion of this research line can be found in Refs. [110, 111] and references therein as a starting point. As a concrete (and trivial) example of how gate identities can be used, in (1.113) we illustrate how the Hadamard gate from $\mathcal{G}_1$ can be generated by two single-qubit gates (from $\mathcal{G}_0$) and an overall phase gate,

$$\mathsf{H} = \mathsf{Ph}_{\frac{\pi}{2}}\mathsf{Y}_{\frac{\pi}{2}}\mathsf{Z}_{\pi} = i\frac{1}{\sqrt{2}}\begin{bmatrix} 1 & -1 \\ 1 & 1 \end{bmatrix}\begin{bmatrix} -i & 0 \\ 0 & i \end{bmatrix} = \frac{1}{\sqrt{2}}\begin{bmatrix} 1 & 1 \\ 1 & -1 \end{bmatrix} \tag{1.113}$$

The gates X_{θ}, Y_{θ} and Z_{θ} are all natively available in a superconducting quantum processor. The question on how single qubit rotations and two-qubit operations are implemented in transmon-based superconducting quantum processors have already been addressed in the previous Chapter.

1.3.6 Further Developments for Superconducting Qubits

As discussed up to now, the planar superconducting qubits represents a promising platform for realizing medium scale quantum processors. While we have focused here on how superconducting qubits can be used for quantum information processing, there has of course also been tremendous activity in related fields, that briefly summarized below. One of these is **quantum annealing**. Superconducting qubits also form the basis for certain quantum annealing platforms [112]. Quantum annealing operates by finding the ground state of a given Hamiltonian (typically a classical Ising Hamiltonian), and this state will correspond to the solution of an optimization problem. By utilizing a flux-qubit type design, the company D-Wave has demonstrated quantum annealing processors [113] which have now reached beyond 2000 qubits [114].

Another effort to the planar superconducting qubits has been the development of **3D cavity-based superconducting qubits**. In these systems, quantum information is encoded in superpositions of coherent photonic modes of the cavity [115]. The cat states can be highly coherent due to the inherently high quality factors associated with 3D cavities [116]. This platform is characterized by a small hardware overhead for encoding a logical qubit [117], and it's kin to certain implementations of asymmetric error-correcting codes due to the fact that errors due to single-photon loss in the cavity is a tractable observable to decode. Using this architecture, several important advances were recently demonstrated including extending the lifetime of an error-corrected qubit [118].

Despite significant advancements in the improvement of qubit lifetimes and gate fidelity over the past decades, the need for error correction persists, especially as quantum processors move towards larger scales. Incorporating error correction into the quantum data necessitates the adoption of an **error-correcting scheme**. Certain elements of the surface code quantum error correction framework have already been successfully demonstrated in the context of superconducting qubits, e.g. see Refs. [119, 120]). However, a considerable challenge that remains is the demonstration of a logical qubit with a longer lifespan than its underlying physical qubits. Despite the surface code's potential as a quantum error-correcting code due to its relatively robust fault tolerance threshold, it faces limitations in implementing a *universal* gate set in a fault-tolerant manner. Consequently, error-corrected gates need supplementation, such as incorporating a T gate through techniques like *state distillation* [121]. Although gate-teleportation, a precursor to magic state distillation, has been demonstrated using planar superconducting qubits [122], demonstration of distillation and injecting into a surface code logical state remains an unresolved challenge.

The continual evolution of new quantum codes represents a rapidly developing field. For more detailed insights, readers are encouraged to consult recent reviews, such as Ref. [111]. Another pivotal step in the progression towards large-scale quantum processor architecture is the establishment of remote entanglement, facilitating the distribution of quantum information across different nodes within a quantum processing network [123].

In the next year, one of the significant challenges facing superconducting qubits is the attainment of **quantum computational supremacy**, [124]. The fundamental concept involves showcasing a computation utilizing qubits and algorithmic gates that surpasses the capabilities of classical computers, assuming certain plausible computational complexity conjectures (see review [125]). A recent advancement toward achieving quantum supremacy was documented, utilizing 9 tunable transmons [126]. The realization of quantum supremacy with a number of qubits ranging in number of hundreds would represent a remarkable achievement for the field of quantum computing.

References

1. R.P. Feynman, R.B. Leighton, M. Sands, *The Schrödinger Equation in a Classical Context: A Seminar on Superconductivity*. The Feynman Lectures on Physics, vol. III (Addison-Wesley, Boston, 1965) 2
2. A. Barone, G. Paternó, *Physics and Applications of the Josephson Effect* (Wiley, New York, 1982) 2, 11, 12, 15, 16, 24, 35, 41
3. L.D. Landau, E.M. Lifschitz, *Lehrbuch der Theoretischen Physik*, Bd. IX (Akademie-Verlag, Berlin, 1980) 2
4. R. Gross, A. Marx, F. Deppe, *Applied Superconductivity: Josephson Effect and Superconducting Electronics*. De Gruyter Textbook (De Gruyter, Berlin/Boston, 2016) 2
5. B.D. Josephson, Phys. Lett. **1**, 251 (1962). https://doi.org/10.1016/0031-9163(62)91369-0. https://www.sciencedirect.com/science/article/abs/pii/0031916362913690 3
6. B.D. Josephson, Rev. Mod. Phys. **46**, 251 (1974). https://doi.org/10.1103/RevModPhys.46.251. https://journals.aps.org/rmp/abstract/10.1103/RevModPhys.46.251 3
7. K.K. Likharev, Rev. Mod. Phys. **51**, 101 (1979). https://doi.org/10.1103/RevModPhys.51.101. https://journals.aps.org/rmp/abstract/10.1103/RevModPhys.51.101 3
8. S. Shapiro, Phys. Rev. Lett. **11**, 80 (1963). https://doi.org/10.1103/PhysRevLett.11.80. https://journals.aps.org/prl/abstract/10.1103/PhysRevLett.11.80 5, 36
9. C.A. Hamilton, Rev. Sci. Instrum. **71**, 3611 (2000). https://doi.org/10.1063/1.1289507. https://pubs.aip.org/aip/rsi/article-abstract/71/10/3611/437471/Josephson-voltage-standards 5
10. A.A. Golubov, M.Y. Kupriyanov, E. Il'ichev, Rev. Mod. Phys. **76**, 411 (2000). https://doi.org/10.1103/RevModPhys.76.411. https://journals.aps.org/rmp/abstract/10.1103/RevModPhys.76.411 7

11. Y.S. Barash, A.V. Galaktionov, A.D. Zaikin, Phys. Rev. B **52**, 665 (1995). https://doi.org/10.1103/PhysRevB.52.665. https://journals.aps.org/prb/abstract/10.1103/PhysRevB.52.665 8
12. Y. Tanaka, S. Kashiwaya, Phys. Rev. B **53**, R11957 (1996). https://doi.org/10.1103/PhysRevB.53.R11957. https://journals.aps.org/prb/abstract/10.1103/PhysRevB.53.R11957
13. A. Buzdin, A.E. Koshelev, Phys. Rev. B **67**, 220504(R) (2003). https://doi.org/10.1103/PhysRevB.67.220504. https://journals.aps.org/prb/abstract/10.1103/PhysRevB.67.220504 9
14. M.J.A. Stoutimore, A.N. Rossolenko, V.V. Bolginov, V.A. Oboznov, A.Y. Rusanov, D.S. Baranov, N. Pugach, S.M. Frolov, V.V. Ryazanov, D.J. Van Harlingen, Phys. Rev. Lett. **121**, 177702 (2018). https://doi.org/10.1103/PhysRevLett.121.177702. https://journals.aps.org/prl/abstract/10.1103/PhysRevLett.121.177702 8
15. E. Goldobin, D. Koelle, R. Kleiner, A. Buzdin, Phys. Rev. B **76**, 224523 (2007). https://doi.org/10.1103/PhysRevB.76.224523. https://journals.aps.org/prb/abstract/10.1103/PhysRevB.76.224523 9
16. V.V. Ryazanov, V.A. Oboznov, A.Y. Rusanov, A.V. Veretennikov, A.A. Golubov, J. Aarts, Phys. Rev. Lett. **86**, 2427 (2001). https://doi.org/10.1103/PhysRevLett.86.2427. https://journals.aps.org/prl/abstract/10.1103/PhysRevLett.86.2427 9
17. E. Goldobin, D. Koelle, R. Kleiner, R.G. Mints, Phys. Rev. Lett. **107**, 227001 (2011). https://doi.org/10.1103/PhysRevLett.107.227001. https://journals.aps.org/prl/abstract/10.1103/PhysRevLett.107.227001 9
18. E. Goldobin, D. Koelle, R. Kleiner, Phys. Rev. B **91**, 214511 (2015). https://doi.org/10.1103/PhysRevB.91.214511. https://journals.aps.org/prb/abstract/10.1103/PhysRevB.91.214511 9
19. S.V. Bakurskiy, N.V. Klenov, I.I. Soloviev, A. Sidorenko, M.Y. Kupriyanov, A.A. Golubov, in *Functional Nanostructures and Metamaterials for Superconducting Spintronics: From Superconducting Qubits to Self-Organized Nanostructures*, ed. by Sidorenko (Springer International Publishing, Cham, 2018), pp. 49–71. https://doi.org/10.1007/978-3-319-90481-8_3. https://link.springer.com/chapter/10.1007/978-3-319-90481-8_3 9
20. F.S. Bergeret, I.V. Tokatly, Europhys. Lett. **110**, 57005 (2015). https://doi.org/10.1209/0295-5075/110/57005. https://iopscience.iop.org/article/10.1209/0295-5075/110/57005 10
21. D. Szombati, S. Nadj-Perge, D. Car, S. Plissard, E. Bakkers, L. Kouwenhoven, Nat. Phys. **12**, 568 (2016). https://doi.org/10.1038/nphys3742. https://www.nature.com/articles/nphys3742
22. A. Murani, A. Kasumov, S. Sengupta, Y.A. Kasumov, V.T. Volkov, I.I. Khodos, F. Brisset, R. Delagrange, A. Chepelianskii, R. Deblock, H. Bouchiat, S. Guéron, Nat. Commun. **8**, 15941 (2017). https://doi.org/10.1038/ncomms15941
23. A. Assouline, C. Feuillet-Palma, N. Bergeal, T. Zhang, A. Mottaghizadeh, A. Zimmers, E. Lhuillier, M. Eddrie, P. Atkinson, M. Aprili, H. Aubin, Nat. Commun. **10**, 126 (2019). https://doi.org/10.1038/s41467-018-08022-y. https://www.nature.com/articles/s41467-018-08022-y
24. W. Mayer, M.C. Dartiailh, J. Yuan, K.S. Wickramasinghe, E. Rossi, J. Shabani, Nat. Commun. **11**, 212 (2020). https://doi.org/10.1038/s41467-019-14094-1. https://www.nature.com/articles/s41467-019-14094-1
25. E. Strambini, A. Iorio, O. Durante, R. Citro, C. Sanz-Fernández, C. Guarcello, I.V. Tokatly, A. Braggio, M. Rocci, N. Ligato, V. Zannier, L. Sorba, F.S. Bergeret, F. Giazotto, Nat. Nanotechnol. **15**, 656 (2020). https://doi.org/10.1038/s41565-020-0712-7. https://www.nature.com/articles/s41565-020-0712-7 10
26. C. Guarcello, R. Citro, O. Durante, F.S. Bergeret, A. Iorio, C. Sanz-Fernández, E. Strambini, F. Giazotto, A. Braggio, Phys. Rev. Res. **2**, 023165 (2020). https://doi.org/10.1103/PhysRevResearch.2.023165. https://link.aps.org/doi/10.1103/PhysRevResearch.2.023165
27. H. Idzuchi, F. Pientka, K.-F. Huang, K. Harada, Ö. Gül, Y.J. Shin, L.T. Nguyen, N.H. Jo, D. Shindo, R.J. Cava, P.C. Canfield, P. Kim, Nat. Commun. **12**, 5332 (2021). https://doi.org/10.1038/s41467-021-25608-1
28. D. Margineda, J.S. Claydon, F. Qejvanaj, C. Checkley, Phys. Rev. B **107**, L100502 (2023). https://doi.org/10.1103/PhysRevB.107.L100502. https://link.aps.org/doi/10.1103/PhysRevB.107.L100502

29. A. Maiellaro, M. Trama, J. Settino, C. Guarcello, F. Romeo, R. Citro, Engineered Josephson diode effect in kinked Rashba nanochannels. Preprint (2024). arXiv:2405.17269. https://doi.org/10.48550/arXiv.2405.17269. https://arxiv.org/abs/2405.17269 10, 18
30. Y.M. Shukrinov, I.R. Rahmonov, K. Sengupta, A. Buzdin, Appl. Phys. Lett. **110**, 182407 (2017). https://doi.org/10.1063/1.4983090. https://pubs.aip.org/aip/apl/article/110/18/182407/286634/Magnetization-reversal-by-superconducting-current 10
31. C. Guarcello, F. Bergeret, Phys. Rev. Appl. **13**, 034012 (2020). https://doi.org/10.1103/PhysRevApplied.13.034012. https://journals.aps.org/prapplied/abstract/10.1103/PhysRevApplied.13.034012
32. C. Guarcello, F.S. Bergeret, Chaos, Solitons Fractals **142**, 110384 (2021). https://doi.org/10.1016/j.chaos.2020.110384. https://www.sciencedirect.com/science/article/pii/S0960077920307785
33. Y.M. Shukrinov, Phys.-Usp. **65**, 317 (2022). https://doi.org/10.3367/UFNe.2020.11.038894. https://iopscience.iop.org/article/10.3367/UFNe.2020.11.038894
34. I.V. Bobkova, A.M. Bobkov, M.A. Silaev, J. Phys. Condens. Matter **34**, 353001 (2022). https://doi.org/10.1088/1361-648X/ac7994. https://iopscience.iop.org/article/10.1088/1361-648X/ac7994
35. C. Guarcello, F.S. Bergeret, R. Citro, Appl. Phys. Lett. **123**, 152602 (2023). https://doi.org/doi.org/10.1063/5.0167769. https://pubs.aip.org/aip/apl/article/123/15/152602/2916053/Switching-current-distributions-in-ferromagnetic 10
36. C. Guarcello, L. Chirolli, M.T. Mercaldo, F. Giazotto, M. Cuoco, Phys. Rev. B **105**, 144513 (2022). https://doi.org/10.1103/PhysRevB.105.134503. https://journals.aps.org/prb/abstract/10.1103/PhysRevB.105.134503 10
37. T. Golod, R.A. Hovhannisyan, O.M. Kapran, V.V. Dremov, Nano Lett. **21**, 5240 (2021). https://doi.org/10.1021/acs.nanolett.1c01366. https://pubs.acs.org/doi/full/10.1021/acs.nanolett.1c01366 10
38. S. Pal, C. Benjamin, Europhys. Lett. **126**, 57002 (2019). https://doi.org/10.1209/0295-5075/126/57002. https://iopscience.iop.org/article/10.1209/0295-5075/126/57002 10
39. C. Stewart, Appl. Phys. Lett. **12**, 277 (1968). https://doi.org/10.1063/1.16519911. https://pubs.aip.org/aip/apl/article-abstract/12/8/277/41050/CURRENT-VOLTAGE-CHARACTERISTICS-OF-Josephson 10
40. D.E. McCumber, J. Appl. Phys. **39**, 3113 (1968). https://doi.org/10.1063/1.1656743. https://pubs.aip.org/aip/jap/article-abstract/39/7/3113/507231/Effect-of-ac-Impedance-on-dc-Voltage-Current 10
41. K. Likharev, *Dynamics of Josephson Junctions and Circuits* (Gordon and Breach, New York, 1986) 11, 13, 41
42. M. Tinkham, *Introduction to Superconductivity*. Dover Books on Physics Series (Dover Publications, Mineola, 2004) 11
43. E. Ben-Jacob, Phys. Rev. A **29**, 2021 (1984). https://doi.org/10.1103/PhysRevA.29.2021. https://link.aps.org/doi/10.1103/PhysRevA.29.2021 11
44. C. Guarcello, D. Valenti, B. Spagnolo, V. Pierro, G. Filatrella, Phys. Rev. Appl. **11**, 044078 (2019). https://doi.org/10.1103/PhysRevApplied.11.044078. https://journals.aps.org/prapplied/abstract/10.1103/PhysRevApplied.11.044078 14
45. C. Guarcello, G. Filatrella, B. Spagnolo, V. Pierro, D. Valenti, Phys. Rev. Res. **2**, 043332 (2020). https://doi.org/10.1103/PhysRevResearch.2.043332. https://journals.aps.org/prresearch/abstract/10.1103/PhysRevResearch.2.043332
46. A.S. Piedjou Komnang, C. Guarcello, C. Barone, C. Gatti, S. Pagano, V. Pierro, A. Rettaroli, G. Filatrella, Chaos Solitons Fractals **142**, 110496 (2021). https://doi.org/10.1016/j.chaos.2020.110496. https://www.sciencedirect.com/science/article/pii/S0960077920308882
47. C. Guarcello, A.S. Piedjou Komnang, C. Barone, A. Rettaroli, C. Gatti, S. Pagano, G. Filatrella, Phys. Rev. Appl. **16**, 054015 (2021). https://doi.org/10.1103/PhysRevApplied.16.054015. https://journals.aps.org/prapplied/abstract/10.1103/PhysRevApplied.16.054015
48. C. Guarcello, Chaos, Solitons Fractals **153**, 111531 (2021). https://doi.org/10.1016/j.chaos.2021.111531. https://www.sciencedirect.com/science/article/pii/S0960077921008857

49. R. Grimaudo, D. Valenti, B. Spagnolo, G. Filatrella, C. Guarcello, Phys. Rev. D **105**, 033007 (2022). https://doi.org/10.1103/PhysRevD.105.033007. https://journals.aps.org/prd/abstract/10.1103/PhysRevD.105.033007
50. R. Grimaudo, D. Valenti, G. Filatrella, B. Spagnolo, C. Guarcello, Chaos Solitons Fractals **173**, 113745 (2023). https://doi.org/10.1016/j.chaos.2023.113745. https://www.sciencedirect.com/science/article/pii/S096007792300646X 14
51. H. Kramers, Physica **7**, 284 (1940). https://doi.org/10.1016/S0031-8914(40)90098-2. http://www.sciencedirect.com/science/article/pii/S0031891440900982 14
52. M.H. Devoret, M.H. Devoret, J. Clarke, Phys. Rev. Lett., **55**, 1908 (1985). https://doi.org/10.1103/PhysRevLett.55.1908. https://link.aps.org/doi/10.1103/PhysRevLett.55.1908 15
53. H. Grabert, U. Weiss, Phys. Rev. Lett. **53**, 1787 (1984). https://doi.org/10.1103/PhysRevLett.53.1787. http://link.aps.org/doi/10.1103/PhysRevLett.53.1787 15
54. M. Weides, M. Kemmler, H. Kohlstedt, R. Waser, D. Koelle, R. Kleiner, E. Goldobin, Phys. Rev. Lett. **97**, 247001 (2006). https://doi.org/10.1103/PhysRevLett.97.247001. https://journals.aps.org/prl/abstract/10.1103/PhysRevLett.97.247001 18
55. M. Kemmler, M. Weides, M. Weiler, M. Opel, S.T.B. Goennenwein, A.S. Vasenko, A.A. Golubov, H. Kohlstedt, D. Koelle, R. Kleiner, E. Goldobin, Phys. Rev. B **81**, 054522 (2010). https://doi.org/10.1103/PhysRevB.81.054522. https://journals.aps.org/prb/abstract/10.1103/PhysRevB.81.054522
56. H.J. Suominen, J. Danon, M. Kjaergaard, K. Flensberg, J. Shabani, C.J. Palmstrøm, F. Nichele, C.M. Marcus, Phys. Rev. B **95**, 035307 (2017). https://doi.org/10.1103/PhysRevB.95.035307. https://journals.aps.org/prb/abstract/10.1103/PhysRevB.95.035307
57. G. Singh, C. Guarcello, E. Lesne, D. Winkler, T. Claeson, T. Bauch, F. Lombardi, A. Caviglia, R. Citro, M. Cuoco, A. Kalaboukhov, npj Quantum Mater. **7**, 2 (2022). https://doi.org/10.1038/s41535-021-00406-6. https://www.nature.com/articles/s41535-021-00406-6
58. A. Maiellaro, J. Settino, C. Guarcello, F. Romeo, R. Citro, Hallmarks of orbital-flavored Majorana states in Josephson junctions based on oxide nanochannels Phys. Rev. B **107**, L201405 (2023). https://doi.org/10.1103/PhysRevB.107.L201405. https://journals.aps.org/prb/abstract/10.1103/PhysRevB.107.L201405 18
59. F. Tafuri, *Fundamentals and Frontiers of the Josephson Effect* (Springer, Cham, 2019) 18
60. V.V. Schmidt, *The Physics of Superconductors*, ed. by P. Müller, A.V. Ustinov (Springer, Berlin, 1997) 19
61. A. Barone, F. Esposito, C.J. Magee, A.C. Scott, La Riv. Nuovo Cimento **1**, 227–267 (1971). https://doi.org/10.1007/BF02820622. https://link.springer.com/article/10.1007/BF02820622 20, 24
62. R. Parmentier, *Fluxons in Long Josephson Junctions* (Academic, Cambridge, 1978), pp 173–99
63. D.W. McLaughlin, A.C. Scott, Phys. Rev. A **18**, 1652 (1978). https://doi.org/10.1103/PhysRevA.18.1652
64. P.S. Lomdahl, J. Stat. Phys. **39**, 551 (1985). https://doi.org/10.1007/BF01008351. https://link.springer.com/article/10.1007/BF01008351
65. A.V. Ustinov, Physica D **123**, 315–29 (1998). https://doi.org/10.1016/S0167-2789(98)00131-6. https://www.sciencedirect.com/science/article/pii/S0167278998001316
66. J. Cuevas-Maraver, P.G. Kevrekidis, F. Williams, *The sine-Gordon Model and Its Applications* (Springer, Cham, 2014) 20
67. D.R. Gulevich, M.B. Gaifullin, F.V. Kusmartsev, Eur. Phys. J. B. **85**, 24 (2012). https://doi.org/10.1140/epjb/e2011-20689-4. https://link.springer.com/article/10.1140/epjb/e2011-20689-4 21
68. D. De Santis, C. Guarcello, B. Spagnolo, A. Carollo, D. Valenti, Chaos, Solitons Fractals **158**, 112039 (2022). https://doi.org/10.1016/j.chaos.2022.112039. https://www.sciencedirect.com/science/article/pii/S0960077922002491
69. D. De Santis, C. Guarcello, B. Spagnolo, A. Carollo, D. Valenti, Commun. Nonlinear Sci. Numer. Simul. **115**, 106736 (2022). https://doi.org/10.1016/j.cnsns.2022.106736. https://www.sciencedirect.com/science/article/pii/S100757042200274X

70. D. De Santis, C. Guarcello, B. Spagnolo, A. Carollo, D. Valenti, Chaos, Solitons Fractals **168**, 113115 (2023). https://doi.org/10.1016/j.chaos.2023.113115. https://www.sciencedirect.com/science/article/pii/S0960077923000164
71. D. De Santis, C. Guarcello, B. Spagnolo, A. Carollo, D. Valenti, Chaos, Solitons Fractals **170**, 113382 (2023). https://doi.org/10.1016/j.chaos.2023.113382. https://www.sciencedirect.com/science/article/pii/S0960077923002837
72. D. De Santis, B. Spagnolo, A. Carollo, D. Valenti, C. Guarcello, Heat-transfer fingerprint of Josephson breathers. Chaos, Solitons Fractals, **185**, 115088 (2024). https://doi.org/10.1016/j.chaos.2024.115088. https://www.sciencedirect.com/science/article/pii/S0960077924006404 21
73. A.C. Scott, F.Y.F. Chu, S.A. Reible, J. Appl. Phys. **47**, 3272–3286 (1976). https://doi.org/10.1063/1.323126. https://pubs.aip.org/aip/jap/article/47/7/3272/171154/Magnetic-flux-propagation-on-a-Josephson 22
74. P.S. Lomdahl, O.H. Soerensen, P.L. Christiansen, Phys. Rev. B **25**, 5737 (1982). https://doi.org/10.1103/PhysRevB.25.5737. https://journals.aps.org/prb/abstract/10.1103/PhysRevB.25.5737 22
75. E. Goldobin, *Flux-Flow Oscillators and Phenomenon of Cherenkov Radiation from Fast Moving Fluxons*. Microwave Superconductivity (Springer, Dordrecht, 2001), pp. 581–614 23
76. K.G. Fedorov, A.L. Pankratov, Phys. Rev. B **76**, 024504 (2007). https://doi.org/10.1103/PhysRevB.76.024504. https://journals.aps.org/prb/abstract/10.1103/PhysRevB.76.024504 24
77. K.G. Fedorov, A.L. Pankratov, B. Spagnolo, Int. J. Bifurcat. Chaos **18**, 2857–62 (2008). https://doi.org/10.1142/S0218127408022111. https://www.worldscientific.com/doi/abs/10.1142/S0218127408022111
78. K.G. Fedorov, A.L. Pankratov, Phys. Rev. Lett. **103**, 260601 (2009). https://doi.org/10.1103/PhysRevLett.103.260601. https://journals.aps.org/prl/abstract/10.1103/PhysRevLett.103.260601
79. G. Augello, D. Valenti, A.L. Pankratov, B. Spagnolo, Eur. Phys. J. B. **70**, 145–51 (2009). https://doi.org/10.1140/epjb/e2009-00155-x. https://link.springer.com/article/10.1140/epjb/e2009-00155-x
80. A.L. Pankratov, A.V. Gordeeva, L.S. Kuzmin, Phys. Rev. Lett. **109**, 087003 (2012). https://doi.org/10.1103/PhysRevLett.109.087003. https://journals.aps.org/prl/abstract/10.1103/PhysRevLett.109.087003
81. D. Valenti, C. Guarcello, B. Spagnolo, Phys. Rev. B **89**, 214510 (2014). https://doi.org/10.1103/PhysRevB.89.214510. https://journals.aps.org/prb/abstract/10.1103/PhysRevB.89.214510
82. C. Guarcello, D. Valenti, A. Carollo, B. Spagnolo, Entropy **17**, 2862 (2015). https://doi.org/10.3390/e17052862. http://www.mdpi.com/1099-4300/17/5/2862
83. C. Guarcello, D. Valenti, A. Carollo, B. Spagnolo, J. Stat. Mech. Theory Exp. **2016**, 054012 (2016). https://doi.org/10.1088/1742-5468/2016/05/054012. https://iopscience.iop.org/article/10.1088/1742-5468/2016/05/054012
84. A.L. Pankratov, E.V. Pankratova, V.A. Shamporov, S.V. Shitov, Appl. Phys. Lett. **110**, 112601 (2017). https://doi.org/10.1063/1.4978514. https://pubs.aip.org/aip/apl/article/110/11/112601/32784/Oscillations-in-Josephson-transmission-line 24
85. J. Clarke, A. Braginski, *The SQUID Handbook: Fundamentals and Technology of SQUIDs and SQUID Systems*, vol. 1 (Wiley, New York, 2004) 25, 26, 27, 28, 29
86. C. Granata, A. Vettoliere, Phys. Rep. **614**, 1 (2016). https://doi.org/10.1016/j.physrep.2015.12.001. https://www.sciencedirect.com/science/article/pii/S0370157315004998 25, 27
87. C. Guarcello, P. Solinas, M. Di Ventra, F. Giazotto, Phys. Rev. Appl. **7**, 044021 (2017). https://doi.org/10.1103/PhysRevApplied.7.044021. https://journals.aps.org/prapplied/abstract/10.1103/PhysRevApplied.7.044021 27
88. C. Guarcello, P. Solinas, A. Braggio, M. Di Ventra, F. Giazotto, Phys. Rev. Appl. **9**, 014021 (2018). https://doi.org/10.1103/PhysRevApplied.9.014021. https://journals.aps.org/prapplied/abstract/10.1103/PhysRevApplied.9.014021 27

89. M. Bo, T. Zhong-Kui, M. Shu-Chao, D. Yuan-Dong, W. Fu-Ren, Chin. Phys. **13**, 1226 (2004). https://doi.org/10.1088/1009-1963/13/8/008. https://iopscience.iop.org/article/10.1088/1009-1963/13/8/008 29
90. M. Ohring, *Materials Science of Thin Films: Depositon and Structure* (Elsevier, Amsterdam, 2001). https://www.sciencedirect.com/book/9780125249751/materials-science-of-thin-films 32
91. G.J. Dolan, Offset masks for lift–off photoprocessing. Appl. Phys. Lett. **31**(5), 337–339 (1977). https://doi.org/10.1063/1.89690 33, 35
92. M. Gurvitch, M.A. Washington, H.A. Huggins, High quality refractory Josephson tunnel junctions utilizing thin aluminum layers. Appl. Phys. Lett. **42**(5), 472–474 (1983). https://doi.org/10.1063/1.93974. https://pubs.aip.org/aip/apl/article/42/5/472/48168/High-quality-refractory-Josephson-tunnel-junctions 33
93. L. Fritzsch, H.-J. Köhler, F. Thrum, G. Wende, H.-G. Meyer, Preparation of Nb/Al–AlOx/Nb Josephson junctions with critical current densities down to 1 A/cm2. Phys. C: Supercond. **296**(3–4), 319–324 (1998). https://doi.org/10.1016/S0921-4534(97)01829-7. https://www.sciencedirect.com/science/article/pii/S0921453497018297 34
94. A.W. Kleinsasser, R.E. Miller, W.H. Mallison, Dependence of critical current density on oxygen exposure in Nb-AlO_x-Nb tunnel junctions. IEEE Trans. Appl. Supercond. **5**(1), 26–30 (1995). https://doi.org/10.1109/77.384565. https://ieeexplore.ieee.org/document/384565 34
95. J. Verjauw, R. Acharya, J. Van Damme, T. Ivanov, D. Perez Lozano, F.A. Mohiyaddin, D. Wan, J. Jussot, A.M. Vadiraj, M. Mongillo, et al., Path toward manufacturable superconducting qubits with relaxation times exceeding 0.1 ms. npj Quantum Inf. **8**(1), 93 (2022). https://doi.org/10.1038/s41534-022-00600-9. https://www.nature.com/articles/s41534-022-00600-9 35
96. Y.-H.Tang, J. Wachter, A. Rüfenacht, G.J. FitzPatrick, S.P. Benz, Application of a 10 v programmable Josephson voltage standard in direct comparison with conventional Josephson voltage standards. IEEE Trans. Instrum. Meas. **64**(12), 3458–3466 (2015). https://doi.org/10.1109/TIM.2015.2463392. https://ieeexplore.ieee.org/document/7206594 37
97. N.E. Flowers-Jacobs, A.E. Fox, P.D. Dresselhaus, R.E. Schwall, S.P. Benz, Two-volt Josephson arbitrary waveform synthesizer using Wilkinson dividers. IEEE Trans. Appl. Supercond. **26**(6), 1–7 (2016). https://doi.org/10.1109/TASC.2016.2532798. https://ieeexplore.ieee.org/document/7414432 37
98. R. Kleiner, D. Koelle, F. Ludwig, J. Clarke, Superconducting quantum interference devices: State of the art and applications. Proc. IEEE **92**(10), 1534–1548 (2004). https://doi.org/10.1109/JPROC.2004.833655. https://ieeexplore.ieee.org/document/1335547 39
99. M. Kjaergaard, M.E. Schwartz, J. Braumüller, P. Krantz, J.I.-J. Wang, S. Gustavsson, W.D. Oliver, Superconducting qubits: Current state of play. Ann. Rev. Condensed Matter Phys. **11**, 369–395 (2020). https://doi.org/10.1146/annurev-conmatphys-031119-050605. https://www.annualreviews.org/content/journals/10.1146/annurev-conmatphys-031119-050605 39
100. W. Anacker, Josephson computer technology: an IBM research project. IBM J. Res. Dev. **24**(2), 107–112 (1980) https://doi.org/10.1147/rd.242.0107. https://ieeexplore.ieee.org/abstract/document/5390721 40
101. M. Dorojevets, D. Strukov, A. Silver, A. Kleinsasser, F. Bedard, P. Bunyk, Q. Herr, G. Kerber, L. Abelson, On the road towards superconductor computers: twenty years later, in *Future Trends in Microelectronics: The Nano, the Giga, the Ultra, and the Bio*, ed. by S. Luryi, J. Xu, A. Zaslavsky (John Wiley & Sons, Hoboken, 2004), pp. 214–225 40
102. K.K. Likharev, V.K. Semenov, RSFQ logic/memory family: A new Josephson-junction technology for sub-Terahertz-clock-frequency digital systems. IEEE Trans. Appl. Supercond. **1**(1), 3–28 (1991). https://doi.org/10.1109/77.80745. https://ieeexplore.ieee.org/document/80745 41
103. Van Duzer, C.W. Turner, *Principles of Superconductive Devices and Circuits*, 2nd edn. (Prentice Hall, Philadelphia,1981) 41

104. J. Huang, R. Fu, X. Ye, D. Fan, A survey on superconducting computing technology: circuits, architectures and design tools. CCF Trans. High Performance Comput. **4**(1), 1–22 (2022). https://doi.org/10.1007/s42514-022-00089-w. https://link.springer.com/article/10.1007/s42514-022-00089-w 41
105. M.A. Nielsen, I.L. Chuang, *Quantum Computation and Quantum Information: 10th Anniversary Edition* (Cambridge Press, Cambridge, 2011) 43, 49, 50
106. P. Krantz, M. Kjaergaard, F. Yan, T.P. Orlando, S. Gustavsson, W.D. Oliver, A quantum engineer's guide to superconducting qubits. Appl. Phys. Rev. **6**(2), 021318 (2019). https://doi.org/10.1063/1.5089550. https://pubs.aip.org/aip/apr/article-pdf/doi/10.1063/1.5089550/16667201/021318_1_online.pdf 45, 47
107. A. Barenco, D. Deutsch, A. Ekert, R. Jozsa, Phys. Rev. Lett. **74**, 4083 (1995). https://doi.org/10.1103/PhysRevLett.74.4083. https://link.aps.org/doi/10.1103/PhysRevLett.74.4083 50
108. Y.A. Kitaev, Quantum computations: algorithms and error correction. Russ. Math. Surv. **52**, 1191 (1997). https://doi.org/10.1070/RM1997v052n06ABEH002155. https://iopscience.iop.org/article/10.1070/RM1997v052n06ABEH002155 50
109. C.M. Dawson, M.A. Nielsen, The Solovay-Kitaev algorithm. Quantum Info. Comput. **6**, 81–95 (2006). https://dl.acm.org/doi/10.5555/2011679.2011685 50
110. F.T. Chong, D. Franklin, M. Martonosi, Programming languages and compiler design for realistic quantum hardware. Nature **549**, 180–187 (2017). https://doi.org/10.1038/nature23459. https://www.nature.com/articles/nature23459 50
111. E.T. Campbell, M. Howard, Unified framework for magic state distillation and multiqubit gate synthesis with reduced resource cost. Phys. Rev. A **95**, 022316 (2017). https://doi.org/10.1103/PhysRevA.95.022316. https://journals.aps.org/pra/abstract/10.1103/PhysRevA.95.022316 50, 52
112. E. Farhi, J. Goldstone, S. Gutmann, J. Lapan, A. Lundgren, D. Preda, A quantum adiabatic evolution algorithm applied to random instances of an NP-complete problem. Science **292**, 472–475 (2001). https://doi.org/10.1126/science.1057726. https://science.sciencemag.org/content/292/5516/472 51
113. M.W. Johnson, et al., Quantum annealing with manufactured spins. Nature **473**, 194-198 (2011). https://doi.org/10.1038/nature10012. https://www.nature.com/articles/nature10012 51
114. DWaveSite, "The D-Wave 2000Q System." https://www.dwavesys.com/dwave-two-system 51
115. C. Wang, Y.Y. Gao, P. Reinhold, R.W. Heeres, N. Ofek, K. Chou, C. Axline, M. Reagor, J. Blumoff, K.M. Sliwa, L. Frunzio, S.M. Girvin, L. Jiang, M. Mirrahimi, M.H. Devoret, R.J. Schoelkopf, A Schrdinger cat living in two boxes. Science **352**, 1087–1091 (2016). https://doi.org/10.1126/science.aaf2941. https://science.sciencemag.org/content/352/6289/1087 51
116. W. Pfaff, C.J. Axline, L.D. Burkhart, U. Vool, P. Reinhold, L. Frunzio, L. Jiang, M.H. Devoret, R.J. Schoelkopf, Controlled release of multiphoton quantum states from a microwave cavity memory. Nature Phys. **13**, 882 (2017). https://doi.org/10.1038/nphys4143. https://www.nature.com/articles/nphys4143 51
117. R.W. Heeres, P. Reinhold, N. Ofek, L. Frunzio, L. Jiang, M.H. Devoret, R.J. Schoelkopf, Implementing a universal gate set on a logical qubit encoded in an oscillator. Nature Commun. **8**, 94 (2017). https://doi.org/10.1038/s41467-017-00045-1. https://www.nature.com/articles/s41467-017-00045-1 51
118. N. Ofek, A. Petrenko, R. Heeres, P. Reinhold, Z. Leghtas, B. Vlastakis, Y. Liu, L. Frunzio, S.M. Girvin, L. Jiang, M. Mirrahimi, M.H. Devoret, R.J. Schoelkopf, Extending the lifetime of a quantum bit with error correction in superconducting circuits. Nature **536**, 441–445 (2016). https://doi.org/10.1038/nature18949. HTTP://www.nature.com/articles/nature18949 51
119. D. Ristè, S. Poletto, M.-Z. Huang, A. Bruno, V. Vesterinen, O.P. Saira, L. DiCarlo, Detecting bit-flip errors in a logical qubit using stabilizer measurements. Nature Commun. **6**, 6983 (2015). https://doi.org/10.1038/ncomms7983. https://www.nature.com/articles/ncomms7983 51

120. M. Takita, A. Córcoles, E. Magesan, B. Abdo, M. Brink, A. Cross, J.M. Chow, J.M. Gambetta, Demonstration of weight-four parity measurements in the surface code architecture. Phys. Rev. Lett. **117**, 210505 (2016). https://doi.org/10.1103/PhysRevLett.117.210505. https://journals.aps.org/prl/abstract/10.1103/PhysRevLett.117.210505 51
121. S. Bravyi, A. Kitaev, Universal quantum computation with ideal Clifford gates and noisy ancillas. Phys. Rev. A **71**, 022316 (2005). https://doi.org/10.1103/PhysRevA.71.022316. https://journals.aps.org/pra/abstract/10.1103/PhysRevA.71.022316 52
122. C.A. Ryan, B.R. Johnson, D. Riste, B. Donovan, T.A. Ohki, Rev. Sci. Instrum. **88**, 104703 (2017). https://doi.org/10.1063/1.5006525. https://pubs.aip.org/aip/rsi/article/88/10/104703/836456/Hardware-for-dynamic-quantum-computing 52
123. C.J. Axline, L.D. Burkhart, W. Pfaff, M. Zhang, K. Chou, P. Campagne-Ibarcq, P. Reinhold, L. Frunzio, S.M. Girvin, L. Jiang, M.H. Devoret, R.J. Schoelkopf, On-demand quantum state transfer and entanglement between remote microwave cavity memories. Nature Phys. **14**, 705–710 (2018). https://doi.org/10.1038/s41567-018-0115-y. https://www.nature.com/articles/s41567-018-0115-y 52
124. J. Preskill, *Quantum Computing and the Entanglement Frontier*. Preprint (2012). arXiv:1203.5813. https://doi.org/10.48550/arXiv.1203.5813. https://arxiv.org/abs/1203.5813 52
125. A.W. Harrow, A. Montanaro, Quantum computational supremacy. Nature **549**, 203–209 (2017). https://doi.org/10.1038/nature23458. https://www.nature.com/articles/nature23458 52
126. C. Neill, P. Roushan, K. Kechedzhi, S. Boixo, S.V. Isakov, V. Smelyanskiy, A. Megrant, B. Chiaro, A. Dunsworth, K. Arya, R. Barends, B. Burkett, Y. Chen, Z. Chen, A. Fowler, B. Foxen, M. Giustina, R. Graff, E. Jeffrey, T. Huang, J. Kelly, P. Klimov, E. Lucero, J. Mutus, M. Neeley, C. Quintana, D. Sank, A. Vainsencher, J. Wenner, T.C. White, H. Neven, J.M. Martinis, A blueprint for demonstrating quantum supremacy with superconducting qubits. Science **360**, 195–199 (2018). https://doi.org/10.1126/science.aao4309. https://science.sciencemag.org/content/360/6385/195 52

Introduction to Superconducting Quantum Circuits

2

Michael Stern

Abstract

Standard textbooks on quantum mechanics typically illustrate the theory using examples from the microscopic world, such as atoms, electrons or molecules. At this scale, quantum effects are striking and easily noticeable. At the macroscopic level, quantum mechanics seems however often counter-intuitive. Features like state superposition and entanglement lead to well-known logical paradoxes, challenging our understanding of what we call 'reality'. Controlling quantum features in a macroscopic physical object could open the way for building a new generation of quantum machines with tremendous computational power. Superconducting electrical circuits are an example of such a macroscopic quantum system. As of today, the cutting-edge level of control exhibited by these circuits has led them to be considered as one of the foremost technologies for physically implementing quantum computers. Moreover, it is possible to make hybrid systems in which the quantum variables of an electrical circuit are coupled to various microscopic degrees of freedom, thereby demonstrating that these circuits constitute a general interface to the quantum world. The purpose of this chapter is to provide an introduction to superconducting quantum circuits, elucidating how such systems can exhibit quantum behavior and how they can be controlled to serve as a building block of quantum processors.

M. Stern (✉)
Quantum Nanoelectronics Laboratory, Department of Physics, Bar Ilan University, Ramat Gan, Israel
e-mail: michael.stern@biu.ac.il

R. Aguado et al. (eds.), *New Trends and Platforms for Quantum Technologies*,
Lecture Notes in Physics 1025, https://doi.org/10.1007/978-3-031-55657-9_2

2.1 Quantization of an Electrical Circuit

2.1.1 The Lumped-Element Circuit Model

We will first begin with some reminders concerning classical electrical circuits in the radio-frequency or microwave domain ($f \sim 1$ MHz–100 GHz). For simplicity, we consider a circuit formed by a planar network of electrical dipoles. This lumped element description is valid when the physical size of the circuit is much smaller than the wavelength λ of the signal.

2.1.1.1 Constitutive Relations

At any instant t, the classical state of a dipole can be fully determined by knowing a single dynamical variable. One can measure either $V(t)$ which represents the voltage drop across the dipole or $I(t)$ the current flowing through it. These two dynamical variables are connected to each other by a constitutive relation. This constitutive relation may be linear (e.g. Ohm law) or non linear and characterizes the dipole element. It is often more convenient to describe the state of the dipole by the charge and flux variables, namely $Q(t)$ and $\Phi(t)$ defined as

$$\boxed{\begin{aligned}\dot{Q}(t) &= I(t) \\ \dot{\Phi}(t) &= V(t)\end{aligned}} \tag{2.1}$$

Capacitive elements such as capacitors ($V = Q/C$) have constitutive relations where the voltage drop depends only on charge. Their energy *only* depends on charge:

$$E = \int V(t)I(t)\,dt = \int V(Q)\dot{Q}(t)\,dt = \int V(Q)\,dQ \tag{2.2}$$

Inductive elements have constitutive relations where the current depends only on flux. This function can be linear ($I = \Phi/L$) or non-linear (e.g. $I = I_0 \sin(\Phi/\varphi_0)$). Their energy *only* depends on flux:

$$E = \int I(t)V(t)\,dt = \int I(\Phi)\dot{\Phi}(t)\,dt = \int I(\Phi)\,d\Phi \tag{2.3}$$

2.1.1.2 Defining the Spanning Tree of a Circuit

To solve the circuit, one needs first to define a set of independent variables taking into account these constitutive relations. According to graph theory, a planar network of dipoles with similar constitutive relations (i.e. only resistors or only capacitors) can be reduced to a single equivalent dipole. One can therefore find systematically a set of independent variables by the so-called *node method.* This method consists of:

1. finding first all the nodes of the circuit which connect at least two elements with *distinct* constitutive relations, and associating to each of them an electrical potential.
2. Defining a set of independent fluxes $(\Phi_1, \Phi_2, \ldots, \Phi_n)$ by drawing a *spanning tree* which access every node without forming a loop, preferably passing through inductors only.

To write the equation of motion of the circuit, we use Kirchhoff's node law, which expresses charge conservation, and which states that the sum of currents flowing into a node is equal to the sum of currents flowing out of that node. The hypothesis of charge conservation is well verified in usual metals below their plasma frequency, which is usually in the deep UV range, far above microwave frequencies.

2.1.1.3 A Simple Example

To illustrate our point, let us consider the example shown in Fig. 2.1. The points A and B are nodes of the circuit and are characterized by electrical potentials V_A and V_B . Point C connects three purely resistive elements and thus the elements connected to this point can be reduced to a single equivalent dipole of resistance $R_{eq} = R_1 + \frac{R_2 R_3}{R_2 + R3}$. The spanning tree connecting node A and B defines here a single independent flux variable Φ, the flux threading the inductance L.

We write the equation of motion of the circuit by writing Kirchhoff law at node A and using the constitutive relations of each element:

$$I(t) = i_L(t) + i_C(t) + i_R(t) = \frac{\Phi}{L} + C\ddot{\Phi} + \frac{\dot{\Phi}}{R_{eq}} \tag{2.4}$$

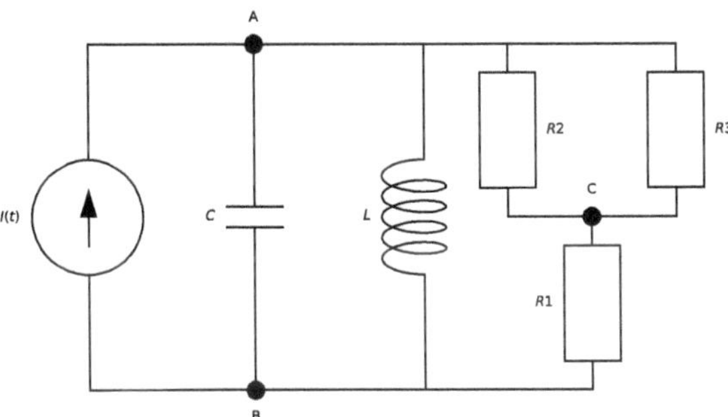

Fig. 2.1 Simple example of a lumped element circuit. A current source $I(t)$ is connected to a network of dipole elements. Points A and B are nodes of the circuit. Point C connects three purely resistive elements and thus can be reduced to a single resistive dipole. The flux threading the inductance $\dot{\Phi}(t) = V_A(t) - V_B(t)$ can be chosen as the dynamical variable of the system

where $\dot{\Phi}(t) = V_A(t) - V_B(t)$, $i_L(t) = \frac{\Phi}{L}$ is the current flowing in the inductor branch, $i_C(t) = C\ddot{\Phi}$ the current flowing in the capacitor branch, and $i_R(t) = \frac{\dot{\Phi}}{R_{eq}}$ the current flowing in the resistive branch.

2.1.2 Quantization of a Lumped Element Circuit

To ensure the quantum mechanical behavior of a circuit, the first requirement is the *absence of dissipation*. Specifically, all metallic components must be constructed from materials that exhibit zero resistance at the circuit's frequency and operating temperature. Zero resistance is obtained by fabricating the circuit out of a metal, which becomes superconducting at low temperatures. Superconductivity arises from the pairing and condensation of electrons with opposite spins into a special ground state [1]. This ground state possesses an excitation gap, 2Δ, which is the energy required to disrupt one of the electron pairs and create an excited state. It is this excitation gap that enables current to flow through a superconductor without dissipation. Additionally, this gap reduces the number of effective degrees of freedom in the circuit, allowing the construction of circuits that behave quantum mechanically, despite being composed of approximately 10^{12} atoms.

In addition, the remaining degrees of freedom of the circuit must be cooled to temperatures where the typical energy of thermal fluctuations is much less that the energy associated with the transition frequency of the circuit. For instance, if the circuit operates at 5 GHz, the required operating temperature should be approximately 20 mK (keeping in mind that 5 GHz corresponds to about 0.25 K). Achieving such temperatures can be accomplished by cooling the circuit using a dilution refrigerator. However, it is equally crucial to cool the wires connected to the circuit's control and readout ports, which can bring a substantive amount of heat to the system. This last point requires meticulous electromagnetic filtering.

2.1.2.1 Definition of the Conjugate Variables

For an arbitrary circuit composed of non dissipative elements, one obtains the equation of motion by first identifying the independent variables as stated in Sect. 1.1 and writing the Lagrangian of the system

$$\mathcal{L} = K(\dot{\Phi}_1, \dot{\Phi}_2 \ldots, \dot{\Phi}_n) - U(\Phi_1, \Phi_2, \ldots, \Phi_n) \tag{2.5}$$

where K is the capacitive energy and U the inductive energy of the circuit. The conjugate momenta of our system are given by

$$Q_i \equiv \frac{\partial \mathcal{L}}{\partial \dot{\Phi}_i} \tag{2.6}$$

Finally, one obtains the Hamiltonian of the system by writing $\mathcal{H} = \sum \dot{\Phi}_i Q_i - \mathcal{L}$. From this point, the equation of motion can be directly obtained.

$$\dot{\Phi}_i = \frac{\partial \mathcal{H}}{\partial Q_i}$$

$$\dot{Q}_i = -\frac{\partial \mathcal{H}}{\partial \Phi_i}$$

The principle of correspondence of Dirac stipulates that one can quantize the system by introducing the operators $\hat{\Phi}_i$ and $\hat{Q}_i$ which obey commutation relations

$$\boxed{\left[\hat{\Phi}_i, \hat{Q}_i\right] = i\hbar} \tag{2.7}$$

2.1.2.2 From the Capacitance Matrix to the Hamiltonian of the Circuit

Let us now consider the example shown in Fig. 2.2. The points A, B and C are nodes of the circuit and are characterized by electrical potentials V_A, V_B and V_C. We define a spanning tree by choosing the flux Φ_1 and Φ_2 shown in the figure and connecting these three nodes.

The inductive energy U of the system is simply given by

$$U = \frac{\Phi_1^2}{2L_1} + \frac{\Phi_2^2}{2L_2} \tag{2.8}$$

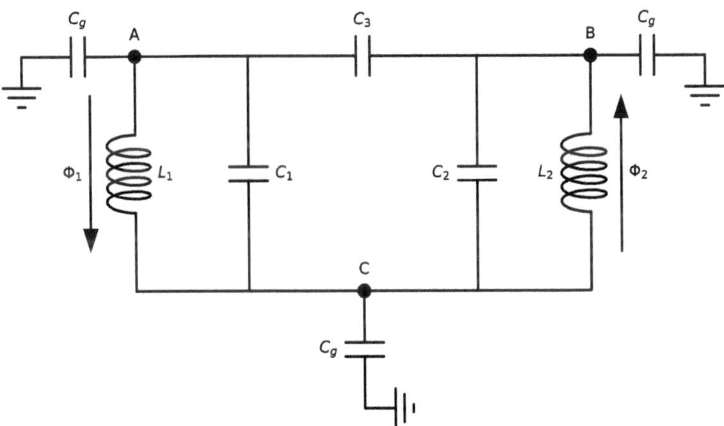

Fig. 2.2 Quantization of two coupled resonators. Each bare resonator is composed of an inductor of inductance $L_{1,2}$ in parallel with a capacitor of capacitance $C_{1,2}$. They are coupled directly by the capacitor C_3 and indirectly via their coupling to the ground. The points A, B and C are nodes of the circuit. The fluxes Φ_1 and Φ_2 define a spanning tree connecting all the three nodes

The capacitive energy K of the system is given by

$$K = \frac{1}{2}C_1(V_A - V_C)^2 + \frac{1}{2}C_2(V_B - V_C)^2 + \frac{1}{2}C_3(V_A - V_B)^2 + \frac{1}{2}C_g V_A^2 + \frac{1}{2}C_g V_B^2 + \frac{1}{2}C_g V_C^2$$

It is possible to write the capacitive energy as $K = \frac{1}{2}\mathbf{V}^T\mathbf{C}\mathbf{V}$ where $\mathbf{V}^\mathbf{T} = (V_A, V_B, V_C)$ and $\mathbf{C}$ is a 3×3 matrix which we will refer in the following as the *capacitance matrix*. In our case,

$$\mathbf{C} = \begin{pmatrix} C_1 + C_3 + C_g & -C_3 & -C_1 \\ -C_3 & C_2 + C_3 + C_g & -C_2 \\ -C_1 & -C_2 & C_1 + C_2 + C_g \end{pmatrix} \tag{2.9}$$

Using the definitions $\dot{\Phi}_1 = V_C - V_A$, $\dot{\Phi}_2 = V_B - V_C$ and Millman theorem [2] for the ground voltage $V_g = V_A + V_B + V_C \equiv 0$, we have

$$\begin{cases} V_A &= -\frac{2}{3}\dot{\Phi}_1 - \frac{1}{3}\dot{\Phi}_2 \\ V_B &= +\frac{1}{3}\dot{\Phi}_1 + \frac{2}{3}\dot{\Phi}_2 \\ V_C &= +\frac{1}{3}\dot{\Phi}_1 - \frac{1}{3}\dot{\Phi}_2 \end{cases}$$

Thus, one can define a passage matrix $\mathbf{P}$ that expresses $\mathbf{V}$ as a function of $\dot{\mathbf{\Phi}} = (\dot{\Phi}_1, \dot{\Phi}_2)$, i.e. $\mathbf{V} = \mathbf{P}\dot{\mathbf{\Phi}}$. It is therefore possible to write the Lagrangian $\mathcal{L}$ as

$$\mathcal{L} = \frac{1}{2}\dot{\mathbf{\Phi}}^T\widetilde{\mathbf{C}}\dot{\mathbf{\Phi}} - \frac{1}{2}\mathbf{\Phi}^T L^{-1}\mathbf{\Phi} \tag{2.10}$$

where $\boldsymbol{L}^{-1} = \begin{pmatrix} 1/L_1 & 0 \\ 0 & 1/L_2 \end{pmatrix}$ and $\widetilde{\boldsymbol{C}} = \mathbf{P}^T\mathbf{C}\mathbf{P}$. The conjugate momenta of our system are given by $\boldsymbol{Q} = (Q_1, Q_2) \equiv \widetilde{\boldsymbol{C}}\dot{\mathbf{\Phi}}$ and the Hamiltonian is thus given by

$$\boxed{\mathcal{H} = \frac{1}{2}\boldsymbol{Q}^T\widetilde{\boldsymbol{C}}^{-1}\boldsymbol{Q} + \frac{1}{2}\mathbf{\Phi}^T\boldsymbol{L}^{-1}\mathbf{\Phi}} \tag{2.11}$$

2.1.2.3 Coupling Between Two Resonators

The Hamiltonian herein above can be greatly simplified if one assumes that $C_3, C_g \ll C_1, C_2$. In this case, one can write easily $\widetilde{\boldsymbol{C}}^{-1}$ as

$$\widetilde{\boldsymbol{C}}^{-1} \simeq \frac{1}{C_1C_2}\begin{pmatrix} C_2 + C_3 + \frac{2}{3}C_g & C_3 + \frac{1}{3}C_g \\ C_3 + \frac{1}{3}C_g & C_1 + C_3 + \frac{2}{3}C_g \end{pmatrix} \tag{2.12}$$

By grouping the quadratic terms of each independent variable with its conjugate, it is straightforward to show that one can write $\mathcal{H}$ as the sum of two harmonic oscillators with a coupling term V

$$\mathcal{H} = \mathcal{H}_1 + \mathcal{H}_2 + V \tag{2.13}$$

where

$$\mathcal{H}_1 = \hbar\omega_1(a_1^+ a_1 + \frac{1}{2}) \tag{2.14}$$

$$\mathcal{H}_2 = \hbar\omega_2(a_2^+ a_2 + \frac{1}{2}) \tag{2.15}$$

$$\omega_i^2 = \frac{1}{L_i}\left[\widetilde{C}^{-1}\right]_{ii} \tag{2.16}$$

and

$$V = \frac{C_3 + \frac{1}{3}C_g}{C_1 C_2} Q_1 Q_2 = \hbar\eta\sqrt{\omega_1\omega_2}(a_1 - a_1^+)(a_2^+ - a_2) \tag{2.17}$$

with $\eta \simeq \frac{C_3+\frac{1}{3}C_g}{\sqrt{C_1 C_2}}$ and where $a_{1,2}$ and $a_{1,2}^+$ are the creation and annihilation operators of each harmonic oscillator. Even without a direct coupling capacitance $C_3 \equiv 0$, an indirect coupling between the resonators is established via their coupling to the ground. This point illustrates a general difficulty in the design of superconducting quantum circuits. Indeed, isolating circuits is difficult due to to their large coupling with the surrounding environment.

2.1.3 Transmission Lines

Contrary to lumped element circuits, the physical dimensions of transmission lines are comparable to the wavelength λ of the signal [3]. Thus, a transmission line is a distributed-parameter network, where voltage and currents can vary in magnitude and phase over its length.

2.1.3.1 Definition of the Propagation Wave Amplitudes

A transmission line can be modeled by a series of discrete lumped elements as shown in Fig. 2.3. The inductance per unit cell u is L_u and the capacitance to the ground per unit cell is C_u.

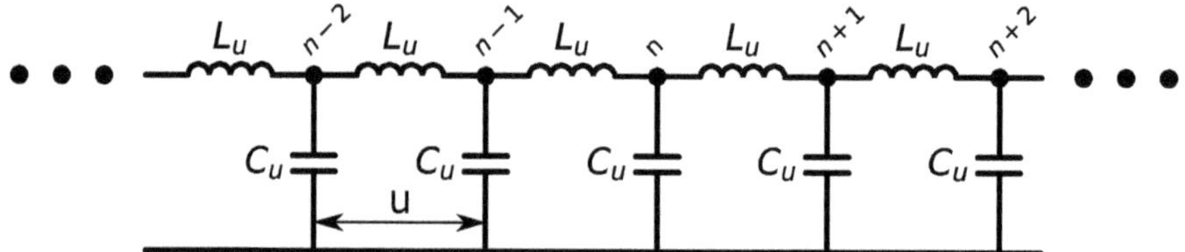

Fig. 2.3 Circuit model of a transmission line. The transmission line is modelled by a series of discrete lumped elements. The unit cell length is u. The inductance per unit cell is L_u and the capacitance per unit cell is C_u

We write the equations for the voltage and currents in the transmission line using the constitutive relations of the dipole elements and Kirchhoff charge conservation

$$V_{n+1} - V_n = -L_u \partial_t I_{n \to n+1}$$
$$-C_u \partial_t V_n = I_{n \to n+1} - I_{n-1 \to n}$$

Going to the continuum limit where $L_u/u \to \mathscr{L}$ and $C_u/u \to \mathscr{C}$, we get

$$\partial_x V(x,t) = -\mathscr{L} \partial_t I(x,t)$$
$$-\mathscr{C} \partial_t V(x,t) = \partial_x I(x,t)$$

To solve these coupled differential equations, we define the propagation wave amplitudes $A^{\rightarrow}$ and $A^{\leftarrow}$ by

$$\boxed{\begin{aligned} A^{\rightarrow} &= \frac{1}{2}\left(V/\sqrt{Z_0} + I\sqrt{Z_0}\right) \\ A^{\leftarrow} &= \frac{1}{2}\left(V/\sqrt{Z_0} - I\sqrt{Z_0}\right) \end{aligned}} \tag{2.18}$$

where $Z_0 = \sqrt{\mathscr{L}/\mathscr{C}}$ and obtain two decoupled first order differential equations

$$\partial_t A^{\rightarrow} + c\partial_x A^{\rightarrow} = 0$$
$$\partial_t A^{\leftarrow} - c\partial_x A^{\leftarrow} = 0$$

where $c = 1/\sqrt{\mathscr{L}\mathscr{C}}$ is the propagation velocity in the transmission line. The solutions are of the form $A^{\rightarrow}(x,t) = \mathscr{A}_{out}(x - ct)$ and $A^{\leftarrow}(x,t) = \mathscr{A}_{in}(x + ct)$ where $\mathscr{A}_{in}$ and $\mathscr{A}_{out}$ are arbitrary functions of their arguments. For an infinite transmission line, $\mathscr{A}_{in}$ and $\mathscr{A}_{out}$ are completely independent. Interestingly, the instantaneous power $\mathcal{P}(x, t)$ is directly related to the square of the propagation wave amplitudes

$$\boxed{\mathcal{P}(x, t) = I(x, t) \times V(x, t) = \mathscr{A}_{out}^2 - \mathscr{A}_{in}^2} \tag{2.19}$$

2.1.3.2 Fourier Components of the Propagation Wave Amplitudes

Since the equations are linear, it is possible to look at individual Fourier components[1] of $A^{\rightarrow/\leftarrow}(x, t)$ at any given point in space x.

$$A^{\rightarrow/\leftarrow}(x, t) = \sum_k A_k^{\rightarrow/\leftarrow}(x)\, e^{-i\omega_k t} + \text{c.c.} \tag{2.20}$$

In the following sections, we will consider monochromatic waves only, thus dropping the sum and index k systematically. We write

$$A^{\rightarrow/\leftarrow}(x, t) = A^{\rightarrow/\leftarrow}(x)\, e^{-i\omega t} + \text{c.c.} \tag{2.21}$$

Assuming $\mathscr{A}_{in} = 0$ and using Eq. (2.19), we get

$$\begin{aligned}\mathcal{P}(x, t) = \mathscr{A}_{out}^2 &= \left(A^{\rightarrow}(x)\, e^{-i\omega t} + \text{c.c.}\right)^2 \\ &= 2\left|A^{\rightarrow}(x)\right|^2 + \left((A^{\rightarrow}(x))^2\, e^{-2i\omega t} + \text{c.c.}\right)\end{aligned}$$

Thus, the modulus of $|A^{\rightarrow}(x)|^2$ is proportional to the average power $\langle P(x)\rangle = \lim_{T\to\infty} \frac{1}{T}\int_0^T \mathcal{P}(x, t)\, dt$

$$\boxed{\langle P(x)\rangle = 2\left|A^{\rightarrow}(x)\right|^2} \tag{2.22}$$

2.1.3.3 Semi-Infinite Transmission Line

When a semi-infinite transmission line is terminated at $x = 0$ by some system S the two solutions $\mathscr{A}_{in}$ and $\mathscr{A}_{out}$ are related by boundary conditions imposed by the system.

$$\begin{aligned}V(x = 0, t) &= \sqrt{Z_0}(\mathscr{A}_{out}(t) + \mathscr{A}_{in}(t)) \\ I(x = 0, t) &= \frac{1}{\sqrt{Z_0}}(\mathscr{A}_{out}(t) - \mathscr{A}_{in}(t))\end{aligned}$$

[1] We adopt the quantum convention for wave propagation (i.e. $e^{i(kx-\omega t)}$), which differs by a sign from the one found typically in the microwave textbooks (i.e. $e^{i(\omega t-kx)}$).

If the system S is an open circuit, $I(x = 0, t) = 0$ and thus $\mathscr{A}_{out}(t) = \mathscr{A}_{in}(t)$. If the system S is short circuit, $V(x = 0, t) = 0$ and thus $\mathscr{A}_{out}(t) = -\mathscr{A}_{in}(t)$. The outgoing waves are simply the result of the incoming wave reflecting from the open/short circuit termination.

In the absence of an incoming wave $\mathscr{A}_{in}(t) = 0$, we have $V(x = 0, t) = Z_0 I(x = 0, t)$ indicating that the transmission line acts as a resistance which instead of dissipating energy by heat carries the energy away from the system as propagating waves.

2.1.4 Quantization of a Transmission Line

Hamiltonian dynamics is inherently reversible and thus dissipationless. Irreversibility however arises when the number of degrees of freedom grows to infinity. In the quantum framework, it was shown that a dissipative impedance can be rigourously taken into account by using the so-called Caldeira Leggett decomposition [4], which consists of modelling any dissipator by an infinite collection of LC resonators. Another way to model dissipation for non-dissipative elements in electrical circuits is to consider an ideal semi-infinite transmission line [5]. As shown in the classical approach herein above, any signal sent down the line will never come back and thus there is a loss of information and entropy creation.

2.1.4.1 Hamiltonian of a Transmission Line

In order to illustrate this point, let us consider a transmission line of length Λ formed by a series of N cells as shown in Fig. 2.4. We assume periodic boundary conditions such that $V_N = V_0$. For each cell of size u, the inductive (potential) energy can be written as $U_n = \frac{\Phi_n^2}{2L_u}$ and the capacitive (kinetic) energy as $K_n = \frac{1}{2} C_u V_n^2$.

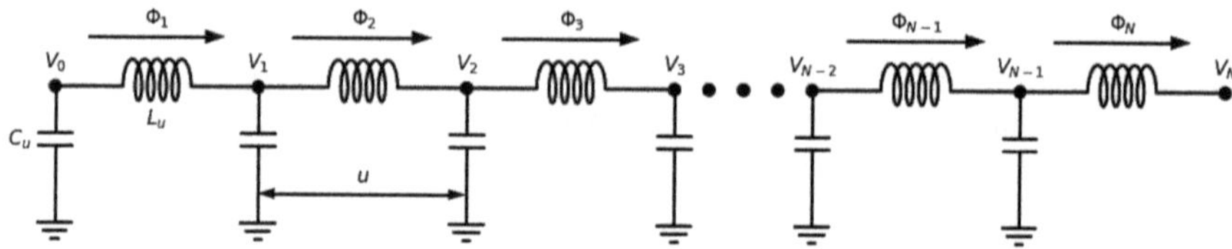

Fig. 2.4 Discrete model for quantization of a transmission line. We denote as V_n the voltage of node n and Φ_n the flux through the inductor between node $n - 1$ and n. The cell unit length is u, the inductance per unit cell is L_u and capacitance per unit cell is C_u. We assume periodic boundary conditions such that $V_N = V_0$

Using $\dot{\Phi}_{n+1} = V_{n+1} - V_n$, one can write

$$V_i = \sum_{j=1}^{i} \dot{\Phi}_j + V_0$$

By summing all the equations together and using Millman theorem [2] $\left(\sum_{n=0}^{N-1} V_n = 0\right)$, we get $V_0 = -\sum_{n=1}^{N-1} V_n = -\big((N-1)\dot{\Phi}_1 + (N-2)\dot{\Phi}_2 + \ldots + \dot{\Phi}_{N-1} + (N-1)V_0\big)$. These equations define a passage matrix $\mathbf{P}$ that expresses the vector $\mathbf{V} = (V_0, \ldots . V_{N-1})$ as a function of $\dot{\mathbf{\Phi}} = (\dot{\Phi}_1, \ldots, \dot{\Phi}_N)$, i.e. $\mathbf{V} = \mathbf{P}\dot{\mathbf{\Phi}}$.

It is thus possible to write the Lagrangian of the system $\mathcal{L}$ as

$$\mathcal{L} = \frac{1}{2} C_u \dot{\Phi}^T \mathbf{P}^T \mathbf{P} \dot{\Phi} - \frac{1}{2L_u} \Phi^T \Phi \tag{2.23}$$

The conjugate momenta of our system are given by $Q = (Q_1, \ldots, Q_N) \equiv \frac{\partial \mathcal{L}}{\partial \dot{\mathbf{\Phi}}}$ and the Hamiltonian is thus given by

$$\mathcal{H} = \frac{1}{2C_u} Q^T \begin{pmatrix} 2 & -1 & 0 & \ldots & -1 \\ -1 & 2 & -1 & \ddots & \vdots \\ 0 & -1 & \ddots & \ddots & 0 \\ \vdots & \ddots & \ddots & 2 & -1 \\ -1 & \cdots & 0 & -1 & 2 \end{pmatrix} Q + \frac{1}{2L_u} \mathbf{\Phi}^T \mathbf{\Phi} \tag{2.24}$$

2.1.4.2 Representation of the Hamiltonian in the Fourier space

This Hamiltonian can be easily diagonalized by introducing a unitary transformation U such that

$$u_{kn} = \frac{1}{\sqrt{N}} \exp\left[\boldsymbol{i}(2k\pi n/N)\right] \tag{2.25}$$

where k, n are integers comprised between 1 and N. Applying the unitary operator U on operators Q_n and Φ_n define a new set of *non-hermitian* operators

$$\widetilde{Q}_k = \sum_{n=1}^{N} u_{kn} Q_n = \frac{1}{\sqrt{N}} \sum_{n=1}^{N} \exp\left[\boldsymbol{i}(2k\pi n/N)\right] Q_n \tag{2.26}$$

$$\widetilde{\Phi}_k = \sum_{n=1}^{N} u_{kn} \Phi_n = \frac{1}{\sqrt{N}} \sum_{n=1}^{N} \exp\left[\boldsymbol{i}(2k\pi n/N)\right] \Phi_n \tag{2.27}$$

These operators follow commutation relations of conjugate variables

$$\left[\widetilde{\Phi}_k, \widetilde{Q}_{k'}^{+}\right] = \frac{1}{N}\sum_{n=1}^{N}\sum_{m=1}^{N} \exp\left[\boldsymbol{i}(2\pi(kn - k'm)/N)\right][\Phi_n, Q_m]$$

$$= \frac{1}{N}\sum_{n=1}^{N} i\hbar \exp\left[\boldsymbol{i}(2\pi n(k - k')/N)\right] = i\hbar\delta_{kk'}$$

Indeed when $k = k'$, $\exp\left[\boldsymbol{i}(2\pi j(k - k')/N)\right] = 1$ and the sum $\sum_{n=1}^{N} \exp\left[\boldsymbol{i}(2\pi n (k - k')/N)\right]$ is equal to N, while if $k \neq k'$,

$$\sum_{n=1}^{N} \exp\left[\boldsymbol{i}(2\pi n(k - k')/N)\right] = \exp\left[\boldsymbol{i}(2\pi(k - k')/N)\right] \left(\frac{\exp\left[\boldsymbol{i}(2\pi(k - k')\right] - 1}{\exp\left[\boldsymbol{i}(2\pi(k - k')/N)\right] - 1}\right) \tag{2.28}$$

$\exp\left[\boldsymbol{i}(2\pi(k - k')\right] = 1$ and thus the sum is equal to zero.

In this new basis, the non-diagonal elements of the Hamiltonian of Eq. (2.24) can be written as

$$\sum_{n=1}^{N} Q_n(Q_{n-1} + Q_{n+1}) = 2\sum_{k=1}^{N} \cos\left[2k\pi/N\right] \widetilde{Q}_k^{+}\widetilde{Q}_k \tag{2.29}$$

Thus, the Hamiltonian of Eq. (2.24) can be written as

$$\mathcal{H} = \sum_{k=1}^{N} \frac{2 - 2\cos\left[2\pi k/N\right]}{2C_u} \widetilde{Q}_k^{+}\widetilde{Q}_k + \frac{1}{2L_u}\widetilde{\Phi}_k^{+}\widetilde{\Phi}_k = \sum_{k=1}^{N} \frac{\hbar\omega_k}{2}\left(q_k^{+}q_k + \varphi_k^{+}\varphi_k\right) \tag{2.30}$$

where

$$\omega_k = \sqrt{\frac{(2 - 2\cos\left[2k\pi/N\right])}{L_u C_u}} \tag{2.31}$$

and

$$[\varphi_k, q_{k'}^{+}] = i\delta_{kk'} \tag{2.32}$$

2.1.4.3 Transmission Line Viewed as an External Bath

For each mode, it is possible to introduce creation and annihilation operators

$$a_k^{\rightarrow} = \frac{1}{\sqrt{2}}(\varphi_k + i q_k)$$
$$a_k^{\leftarrow} = \frac{1}{\sqrt{2}}(\varphi_k - i q_k)$$
$$(a_k^{\rightarrow})^+ = \frac{1}{\sqrt{2}}(\varphi_k^+ - i q_k^+)$$
$$(a_k^{\leftarrow})^+ = \frac{1}{\sqrt{2}}(\varphi_k^+ + i q_k^+)$$

The commutation relations of the $a_k^{\rightarrow}$ operators are such that

$$[a_k^{\rightarrow}, (a_{k'}^{\rightarrow})^+] = \delta_{kk'}$$
$$[a_k^{\leftarrow}, (a_{k'}^{\leftarrow})^+] = \delta_{kk'}$$

If N is even, the system has exactly $N/2$ different eigenenergies. Each mode is doubly degenerate and thus

$$\boxed{\mathcal{H} = \sum_{k=1}^{N/2} \hbar\omega_k \left((a_k^{\rightarrow})^+ a_k^{\rightarrow} + (a_k^{\leftarrow})^+ a_k^{\leftarrow}\right)} \tag{2.33}$$

As we increase the size Λ of the transmission line, the density of modes increases. As we decrease the size of the unit cell, the bandwidth $\sqrt{1/L_u C_u}$ increases. One can therefore safely consider that $k \ll N$ in a realistic situation. This allows to make the approximation that $\cos[x] \simeq 1 - x^2/2$ and thus

$$\omega_k \simeq \sqrt{\frac{1}{L_u C_u}} \frac{2k\pi}{N} \tag{2.34}$$

Using $\sqrt{1/L_u C_u} = \sqrt{1/\mathcal{L}\mathcal{C}u^2} = c/u$ and $\Lambda = Nu$ we get

$$\boxed{\omega_k = k.\frac{2\pi c}{\Lambda}} \tag{2.35}$$

2.1.4.4 Link Between Propagation Amplitudes and Photon Operators

The connection between the photon operators and the propagation amplitudes introduced in the previous section is directly obtained by comparing the incoming power carried by the influx of photons with a well-defined wavevector k to the modulus of the Fourier transform of the propagation amplitude using Eq. (2.22)

$$\langle P \rangle = 2\left|A_k^{\rightarrow}\right|^2 = \left(\frac{c}{\Lambda}\right)\hbar\omega_k \left\langle a_k^{\rightarrow +} a_k^{\rightarrow}\right\rangle$$

The expressions of $A_k^{\rightarrow}$ are thus given by

$$\boxed{A_k^{\rightarrow} = \sqrt{\frac{c}{2\Lambda}\hbar\omega_k a_k^{\rightarrow}}} \tag{2.36}$$

2.1.5 Transmission Line Resonators

In this section, we will study the use of transmission line sections with various lengths and terminations to form resonators.

2.1.5.1 Scattering Matrix

Let us consider an interface of two transmission lines with different characteristic impedance $Z_1|Z_2$. The transmission line is separated into two separate regions, namely the left side and the right side. When an incoming wave impinges on the interface, the propagation wave amplitude can be transmitted and/or reflected partially. We thus write the scattering matrix S.

$$\begin{pmatrix} A_L^{\leftarrow} \\ A_R^{\rightarrow} \end{pmatrix} = \overbrace{\begin{pmatrix} r_{\hookleftarrow} & t_{\leftarrow} \\ t_{\rightarrow} & r_{\hookrightarrow} \end{pmatrix}}^{S} \begin{pmatrix} A_L^{\rightarrow} \\ A_R^{\leftarrow} \end{pmatrix} \tag{2.37}$$

We calculate the scattering coefficients by writing the Kirchhoff equations of voltage and current at the interface assuming $A_R^{\leftarrow} = 0$.

$$V(x^-, t) = V(x^+, t) = \sqrt{Z_1}\left(A_L^{\rightarrow} + A_L^{\leftarrow}\right) = \sqrt{Z_2}A_R^{\rightarrow}$$
$$I(x^-, t) = I(x^+, t) = \left(A_L^{\rightarrow} - A_L^{\leftarrow}\right)/\sqrt{Z_1} = \left(A_R^{\rightarrow}\right)/\sqrt{Z_2}$$

which we solve to get

$$t_{\rightarrow} = \frac{2\sqrt{Z_1 Z_2}}{Z_1 + Z_2}$$
$$r_{\hookleftarrow} = \frac{Z_2 - Z_1}{Z_1 + Z_2}$$

Similarly, two other coefficients can be established by a swap operation $Z_1 \leftrightarrow Z_2$.

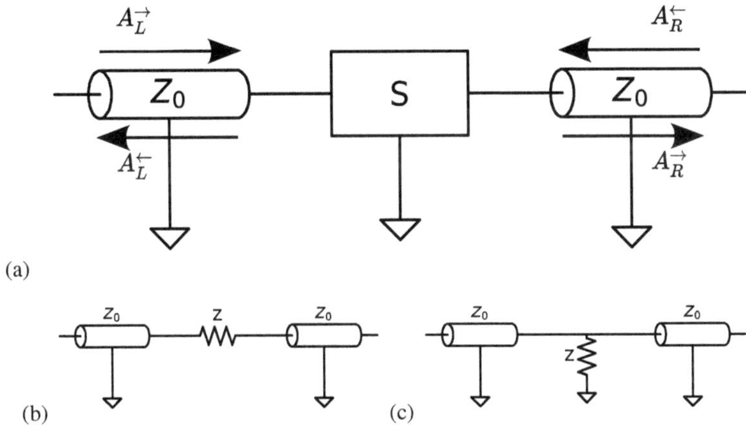

Fig. 2.5 Transmission and reflection coefficients for simple lumped elements. (**a**) Generic lumped element intersecting a transmission line. (**b**) Circuit element of impedance Z in series with the transmission line. (**c**) Circuit element of impedance Z in parallel with the transmission line

2.1.5.2 Calculating Transmission and Reflection Coefficients for Simple Lumped Elements

Let us consider the circuit described in Fig. 2.5a. The transmission line is now intersected by a lumped element system S.

For instance, we consider in Fig. 2.5b a transmission line intersected by an impedance Z in series. We get

$$
\begin{aligned}
I(x^-,t) = I(x^+,t) &= \frac{1}{\sqrt{Z_0}}\left(A_L^{\rightarrow} - A_L^{\leftarrow}\right) = \frac{1}{\sqrt{Z_0}} A_R^{\rightarrow} \\
ZI(x,t) &= V(x^-,t) - V(x^+,t) \\
&= \sqrt{Z_0}\left[\left(A_L^{\rightarrow} + A_L^{\leftarrow}\right) - A_R^{\rightarrow}\right]
\end{aligned}
$$

using the definitions of the scattering matrix, we have $A_R^{\rightarrow} = tA_L^{\rightarrow}$ and $A_L^{\leftarrow} = rA_L^{\rightarrow}$ and thus we get

$$
\boxed{\begin{aligned} r &= z/(2+z) \\ t &= 2/(2+z) \end{aligned}} \tag{2.38}
$$

with $z = Z/Z_0$. For instance, if the scatterer is a capacitor[2] $Z = 1/(-i\omega C)$, we get

$$t = \frac{2}{2 + 1/(-i\omega C Z_0)} \tag{2.39}$$
$$r = \frac{1/(-i\omega C Z_0)}{2 + 1/(-i\omega C Z_0)}$$

Another interesting case to consider is a shorting circuit element as shown in Fig. 2.5c. In that case, the Kirchhoff equations gives

$$V(x^-, t) = V(x^+, t) = \sqrt{Z_0}\left(A_L^{\rightarrow} + A_L^{\leftarrow}\right) = \sqrt{Z_0} A_R^{\rightarrow}$$
$$V(x^{\pm}, t) = Z(I(x^-, t) - I(x^+, t))$$
$$= \frac{Z}{\sqrt{Z_0}}\left(A_L^{\rightarrow} - A_L^{\leftarrow}\right) - \frac{Z}{\sqrt{Z_0}} A_R^{\rightarrow}$$

Thus we get

$$\boxed{\begin{array}{l} r = -1/(2z+1) \\ t = 2z/(2z+1) \end{array}} \tag{2.40}$$

2.1.5.3 λ/2 Resonators with Symmetrical Terminations

The use of symmetrical terminations on both ends of a segment of length $L_{res} = \lambda/2$ ensures that exactly at the resonant frequency $\omega_r = \pi c/L_{res}$, a continuous wave signal is fully transmitted, and no reflection is observed. This results from coherent interference of transmission amplitudes which converges to a unitary transmission coefficient

$$\tau = \sum_{j=0}^{\infty} t e^{ikL_{res}} \left(r^2 e^{2ikL_{res}}\right)^j t = e^{\pi i\omega/\omega_r} \sum_{j=0}^{\infty} t^2 \left(r^2 e^{2\pi i\omega/\omega_r}\right)^j \tag{2.41}$$

$$\tau = \frac{t^2 e^{\pi i\omega/\omega_r(1+i/Q_{int})}}{1 - r^2 e^{2\pi i\omega/\omega_r(1+i/Q_{int})}} \tag{2.42}$$

where $kL_{res} = \frac{\omega}{c}(1 + i/Q_{int}) \times L_{res} = \pi\frac{\omega}{\omega_r}(1 + i/Q_{int})$, Q_{int} representing the internal quality factor due to internal losses in the resonator.

[2] We adopt the quantum convention for wave propagation (i.e. $e^{i(kx-\omega t)}$), which differs by a sign from the one found typically in the microwave textbooks (i.e. $e^{i(\omega t-kx)}$).

An identity can be established relating the round trip frequency $\omega_r/2\pi$, the transmission coefficient, and the energy leakage κ via the ports

$$\kappa = 2 \cdot \omega_r/2\pi \cdot |t|^2 \tag{2.43}$$

The factor 2 stems from the fact that two scattering events occur per round trip. We thus get

$$\boxed{Q_{ext} = \omega_r/\kappa = \pi/\,|t|^2} \tag{2.44}$$

For instance, if the scatterer is a capacitor of capacitance C, we obtain from Eq. (2.39), $|t|^2 \simeq 4C^2\omega_r^2 Z_0^2$ and thus

$$Q_{ext} = \frac{\pi}{4C^2\omega_r^2 Z_0^2} \tag{2.45}$$

In Fig. 2.6a, b, we represented the frequency dependence of the amplitude of the transmitted field $|\tau|$ and of its relative phase. When $Q_{int} = +\infty$ (blue curve), the transmission at $\omega = \omega_r$ is equal to one:

$$|\tau(\omega_r)| = \frac{|t^2|}{|1 - r^2|} \simeq 1 \tag{2.46}$$

The phase shifts by π at resonance. As one increases the ratio Q_{ext}/Q_{int}, the maximum transmission at resonance decreases (red and green curves).

2.1.5.4 λ/4 Resonators with Short Circuit Termination

Another important type of transmission line resonator is the so-called $\lambda/4$ resonator. In this type of resonator, the segment length $L_{res} = \lambda/4$ and is terminated by a short circuit, such that at the resonant frequency $\omega_r = \pi c/(2 * L_{res})$, a phase shift is observed in the reflection of the signal. This results from coherent interference of reflection amplitudes

$$\begin{aligned}\rho &= r - t^2 e^{2ikL_{res}} \sum_{j=0}^{\infty} (-1)^j \left(r e^{2ikL_{res}}\right)^j \\ &= r - e^{\pi i\omega/\omega_r(1+i/Q_{int})} \frac{t^2}{1 + r e^{\pi i\omega/\omega_r(1+i/Q_{int})}}\end{aligned} \tag{2.47}$$

where $2kL_{res} = \frac{\omega}{c}(1 + i/Q_{int}) \times L_{res} = \pi \frac{\omega}{\omega_r}(1 + i/Q_{int})$.

The energy leakage κ via the port is related to the round trip frequency $\omega_r/2\pi$ and to the transmission coefficient by

$$\kappa = \omega_r/2\pi \cdot |t|^2 \tag{2.48}$$

We thus get

$$\boxed{Q_{ext} = \omega_r/\kappa = 2\pi/\,|t|^2} \tag{2.49}$$

Depending on the ratio between Q_{int} and Q_{ext}, we can define three regimes characterized by different behavior of the reflexion coefficent ρ.

The *overcoupled regime* (blue curve) occurs when $Q_{ext} << Q_{int}$. In this regime, $|\rho| \sim 1$ for all frequencies. However, the phase of ρ changes abruptly close to multiples of the resonance frequency and undergoes a 2π shift as shown in Fig. 2.6.

The *critical coupling* (green curve) occurs when $Q_{ext} = Q_{int}$. For this regime, the amplitude reaches almost zero at resonance, while a discontinuity in the phase brings a phase shift of π.

The undercoupled regime (red curve) occurs when $Q_{ext} > Q_{int}$. In this regime, the resonance corresponds to a dip in the amplitude of ρ and a shift $< \pi$ in its phase. The undercoupled resonator is particularly difficult to measure in reflexion since both the amplitude and the phase differs slightly from the out-of-resonance value.

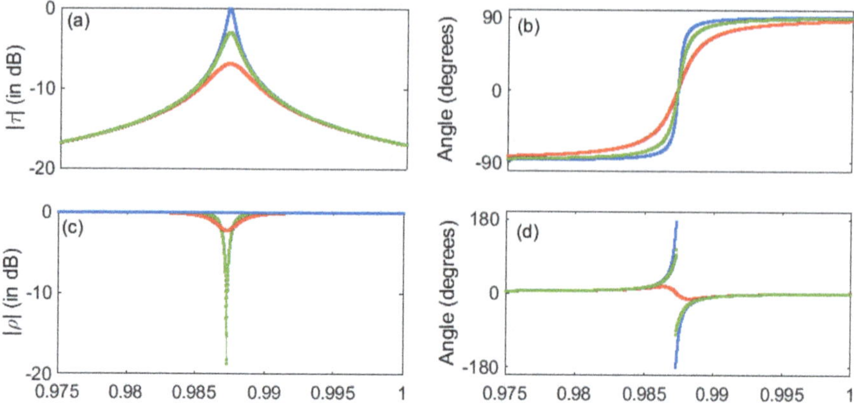

Fig. 2.6 (**a–b**) Modulus of the transmitted field $|\tau|$ and relative phase of the output field via a $\lambda/2$ resonator with symmetrical terminations ($z = 50\,\boldsymbol{i}$, $Q \sim 2000$) assuming no internal losses ($Q_{in} = +\infty$, blue curve) or internal losses of quality factor $Q_{in} = 4000$ (green curve) and $Q_{in} = 1000$ (red curve). (**c–d**) Modulus of the reflected field $|\rho|$ and relative phase of the reflected field on a $\lambda/4$ resonator with short circuit termination ($z = 50\,\boldsymbol{i}$, $Q \sim 4000$) assuming no internal losses ($Q_{in} = +\infty$, blue curve) or internal losses of quality factor $Q_{in} = 4000$ (critical coupling regime, green curve) and $Q_{in} = 1000$ (red curve)

2.1.5.5 Expression for the Local Current and/or Voltage in the Resonator as a Function of Propagation Wave Amplitudes

One can write the current/voltage at a given position in the resonator as a linear response of the propagation wave amplitudes of incoming waves $A_L^{\rightarrow}$ and $A_R^{\leftarrow}$ on both side of the resonator. The obtained linear map can be summarized as

$$I(x,\omega) = \frac{1}{\sqrt{Z_0}}\left(f_{\rightarrow}(\omega,x)\,A_L^{\rightarrow} + f_{\leftarrow}(\omega,x)\,A_R^{\leftarrow}\right) \tag{2.50}$$

$$V(x,\omega) = \sqrt{Z_0}\left(g_{\rightarrow}(\omega,x)\,A_L^{\rightarrow} + g_{\leftarrow}(\omega,x)\,A_R^{\leftarrow}\right) \tag{2.51}$$

The exact numerical value of $f_{\rightarrow/\leftarrow}$ and $g_{\rightarrow/\leftarrow}$ can be calculated by considering the coherent interference from the scattering of all the elements . In the case of a symmetric $\lambda/2$ resonator, we obtain

$$f_{\rightarrow}(\omega,x) = t\frac{e^{ik\mathbf{x}} - re^{ik(2L_{res}-\mathbf{x})}}{1-e^{ikL_{res}}r^2} \tag{2.52}$$

$$f_{\leftarrow}(\omega,x) = t\frac{e^{ik(L_{res}-\mathbf{x})} - re^{ik(L_{res}+\mathbf{x})}}{1-e^{ikL_{res}}r^2}$$

similarly for the voltage, we get

$$g_{\rightarrow}(\omega,x) = t\frac{e^{ik\mathbf{x}} + re^{ik(2L_{res}-\mathbf{x})}}{1-e^{ikL_{res}}r^2}$$

$$g_{\leftarrow}(\omega,x) = t\frac{e^{ik(L_{res}-\mathbf{x})} + re^{ik(L_{res}+\mathbf{x})}}{1-e^{ikL_{res}}r^2}$$

2.1.6 Quantization of Transmission Line Resonators

2.1.6.1 $\lambda/2$ Resonators with Open Circuit Terminations

Let us first consider a transmission line resonator of length L_{res} with open circuit termination on both sides. Contrary to a lumped-element resonator, such a distributed resonator possesses an infinite number of modes. The characteristic impedance of the resonator is given by $Z_0 = \sqrt{\mathscr{L}/\mathscr{C}}$ where $\mathscr{C}$ is the capacitance per unit length and $\mathscr{L}$ the inductance per unit length. We introduce the flux $\Phi(\mathbf{x})$ such that $V = \partial_t\Phi$ at position $\mathbf{x} \in [0, L_{res}]$. The Lagrangian writes as follows:

$$\mathcal{L} = \frac{1}{2}\int_0^{L_{res}}\left(\mathscr{C}V^2 - \mathscr{L}I^2\right)d\mathbf{x} = \frac{1}{2}\int_0^{L_{res}}\left(\mathscr{C}\dot{\Phi}^2 - \frac{1}{\mathscr{L}}(\partial_x\Phi)^2\right)d\mathbf{x} \tag{2.53}$$

Taking into account the boundary condition, it is possible to decompose $V(\mathbf{x})$ and $I(\mathbf{x})$ into an infinite number of stationary wave modes. We thus decompose Φ into infinite stationary modes of mode number j, each verifying the open circuit boundary condition $I = 0$ at $\mathbf{x} = 0$ and $\mathbf{x} = L_{res}$, we thus write

$$\Phi(\mathbf{x}) = \sum_{j=1}^{\infty} \Phi_j \cos\left(\pi j \mathbf{x}/L_{res}\right) \tag{2.54}$$

which we inject into the Lagrangian expression and get

$$\mathcal{L} = \frac{L_{res}}{2} \sum_{j=1}^{\infty} \left(\frac{\mathscr{C}}{2} \dot{\Phi}_j^{\,2} - \frac{1}{2\mathscr{L}} \left(\frac{\pi j}{L_{res}} \Phi_j \right)^2 \right) \tag{2.55}$$

We obtain the Hamiltonian after performing the Legendre transformation

$$\mathcal{H} = \sum_{j=1}^{\infty} \mathcal{H}_j = \frac{1}{L_{res}} \sum_{j=1}^{\infty} \left(\frac{Q_j^2}{\mathscr{C}} + \frac{\pi^2 j^2}{4\mathscr{L}} \Phi_j^2 \right) \tag{2.56}$$

where $Q_j = \partial \mathcal{L}/\partial \dot{\Phi}_j = \mathscr{C} L_{res} \dot{\Phi}_j/2$ is the conjugated variable of Φ_j such that $\left[\Phi_i, Q_j\right] = i\hbar\delta_{ij}$. We can further simplify the Hamiltonian by introducing creation and annihilation operators for each mode

$$\mathcal{H}_j = \hbar\omega_j \left(a_j^{\dagger} a_j + 1/2 \right) \tag{2.57}$$

where $a_j = \sqrt{\frac{j\pi}{4Z_0\hbar}} \Phi_j + i\sqrt{\frac{Z_0}{j\hbar\pi}} Q_j$, $\omega_j = j\omega_r = j\left(\frac{\pi}{L_{res}} c\right)$ and $c = \sqrt{\frac{1}{\mathscr{C}\mathscr{L}}}$ the wave velocity. Using $I(\mathbf{x}) = -\partial_x \Phi(\mathbf{x})/\mathscr{L}$, we get

$$I(\mathbf{x}) = \sum_{j=1}^{\infty} \overbrace{\delta I_0 \sqrt{j} \sin\left(\pi j x/L_{res}\right)}^{\delta I_j(x)} \left(a_j + a_j^{\dagger} \right) \tag{2.58}$$

where $\delta I_0 = \omega_r \sqrt{\frac{\hbar}{\pi Z_0}}$. Similarly, using $V(\mathbf{x}) = \dot{\Phi}(\mathbf{x})$, we get

$$V(\mathbf{x}) = -\boldsymbol{i} \sum_{j=1}^{\infty} \overbrace{\delta V_0 \sqrt{j} \cos\left(\pi j x/L_{res}\right)}^{\delta V_j(x)} \left(a_j - a_j^{\dagger} \right) \tag{2.59}$$

where $\delta V_0 = \omega_r \sqrt{\frac{\hbar Z_0}{\pi}}$.

2.1.6.2 Determining the Current Operator by Filter Function Formalism

Using Eqs. (2.50) and (2.36) we get that

$$\hat{I}(x) = \sqrt{\frac{c}{2\Lambda}} \sum_k \sqrt{\frac{\hbar\omega_k}{Z_0}} \left(f_{\rightarrow}(\omega_k, x)\, a_{L,k}^{\rightarrow} + f_{\leftarrow}(\omega_k, x)\, a_{R,k}^{\leftarrow}\right) + \text{H.c.} \tag{2.60}$$

Let us introduce a new operator $\mathbf{A}$ as a linear combination of $a_{L,k}^{\rightarrow}$ and $a_{R,k}^{\leftarrow}$

$$\mathbf{A} = \frac{\sum_k \sqrt{k}\left(f_{\rightarrow}(\omega_k, x)\, a_{L,k}^{\rightarrow} + f_{\leftarrow}(\omega_k, x)\, a_{R,k}^{\leftarrow}\right)}{\sqrt{\sum_k k\left(|f_{\rightarrow}(\omega_k, x)|^2 + |f_{\leftarrow}(\omega_k, x)|^2\right)}} \tag{2.61}$$

which verifies $\left[\mathbf{A}, \mathbf{A}^{\dagger}\right] = 1$ and can rewrite the current operator under the form

$$\hat{I}(x) = \delta I(x)\left(\mathbf{A} + \mathbf{A}^{\dagger}\right)$$

where

$$\delta I(x) = \sqrt{\sum_k \frac{c}{2\Lambda Z_0} \hbar\omega_k \left(|f_{\rightarrow}(\omega_k, x)|^2 + |f_{\leftarrow}(\omega_k, x)|^2\right)}$$

For convenience, we introduce the density of states

$$\eta(\omega) \equiv \frac{1}{\Delta\omega} = \Lambda/2\pi c \tag{2.62}$$

Injecting the expression of η we obtain an expression independent of Λ

$$\boxed{\delta I(x) = \sqrt{\int_{d\omega} \frac{\hbar\omega}{4\pi Z_0} \left(|f_{\rightarrow}|^2(\omega, x) + |f_{\leftarrow}|^2(\omega, x)\right)}} \tag{2.63}$$

2.2 Superconducting Qubits

2.2.1 Using the Non-linearity of Josephson Junctions

A circuit formed by linear components, such as capacitors and inductors, behaves as an harmonic oscillator and not as a qubit. A non-linear element is therefore essential in order to differentiate the transitions between states $|0\rangle$ and $|1\rangle$ from other higher-lying eigenstates transitions. In superconducting circuits, this non-linearity is obtained by adding to the circuit one or several Josephson junctions. Josephson

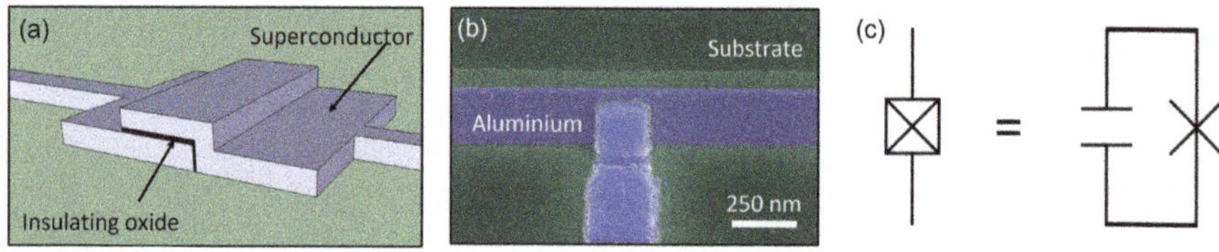

Fig. 2.7 The Josephson junction: a non-linear, non-dissipative element. (**a**) Schematic representation of a Josephson junction showing two superconducting layers separated by a thin insulating oxide layer (**b**) Colored SEM micrograph showing a Josephson junction. (**c**) Circuit diagram representing a Josephson junction, which corresponds to a capacitor of capacitance C_J in parallel with a Josephson (non linear) inductor of inductance $L_J = \frac{\varphi_0}{I_0}$ and represented as a cross

junctions are formed by two superconducting islands separated by a thin insulating layer (see Fig. 2.7) that allows tunneling of Cooper pairs. They are characterized by the so-called Josephson relations:

$$\boxed{\begin{aligned} I &= I_0 \sin\left(\frac{\Phi}{\varphi_0}\right) \\ V &= \dot{\Phi} \end{aligned}} \tag{2.64}$$

where Φ is the flux threading the junction, I_0 is the critical current of the junction and $\varphi_0 = \hbar/2e$ is the reduced magnetic flux quantum.

Josephson junctions are almost non-dissipative. This property allows their use in quantum circuits. The potential energy of the Josephson junction is given by:

$$E = \int_{-\infty}^{t} I(t)V(t)dt = \int_{-\infty}^{t} I_0 \sin\left(\frac{\Phi}{\varphi_0}\right)\dot{\Phi}dt = -E_J \cos(\frac{\Phi}{\varphi_0}) \tag{2.65}$$

where E_J=$I_0\varphi_0$ is called **Josephson energy**.

2.2.2 The Charge Qubit

The simplest version of superconducting qubit, also called Cooper pair box (CPB) consists of two superconducting islands connected by a single Josephson junction of capacitance C_J and Josephson energy E_J. One of the island is electrostatically biased by a voltage source V_g in series with a capacitor C_g. The Cooper pair box was initially developped in 1996 by the Quantronics group at CEA Saclay [6] . In 1999, a team from NEC used the CPB to demonstrate for the first time a coherent superposition of states [7].

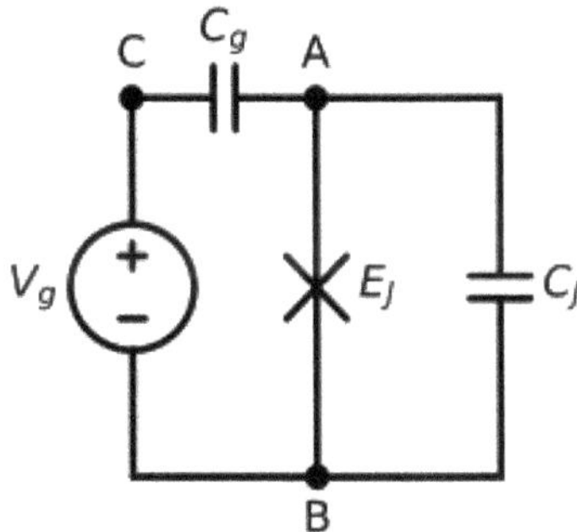

Fig. 2.8 Circuit schematic of a Cooper Pair Box (CPB). The device consists of a Josephson junction with Josephson energy E_J and capacitance C_J, capacitively coupled to a voltage source V_g through a gate capacitor C_g

2.2.2.1 Solving the Cooper Pair Box Hamiltonian

In Fig. 2.8, we present a circuit schematic of a CPB. The points A, B and C are nodes of the circuit and are characterized by electrical potentials V_A,V_B and V_C. We define a spanning tree by choosing the flux Φ connecting nodes A and B.

The inductive energy U is simply given by Eq. (2.65)

$$U = -E_J \cos(\frac{\Phi}{\varphi_0}) \tag{2.66}$$

The capacitive energy K of the system is given by

$$K = \frac{1}{2}C_J(V_A - V_B)^2 + \frac{1}{2}C_g(V_A - V_C)^2$$

Using the definitions $\dot{\Phi} = V_A - V_B$ and $V_g = V_C - V_B$, it is possible to express V_A,V_B and V_C as a function of $\dot{\Phi}$ and V_g, such that

$$K = \frac{1}{2}C_J\dot{\Phi}^2 + \frac{1}{2}C_g\left(\dot{\Phi} - V_g\right)^2 \tag{2.67}$$

It is thus possible to write the Lagrangian $\mathcal{L}$ as

$$\mathcal{L} = \frac{1}{2}C_J\dot{\Phi}^2 + \frac{1}{2}C_g\left(\dot{\Phi} - V_g\right)^2 + E_J \cos(\frac{\Phi}{\varphi_0}) \tag{2.68}$$

The conjugate momentum of our system is given by

$$Q \equiv \frac{\partial \mathcal{L}}{\partial \dot{\Phi}} = (C_J + C_g)\dot{\Phi} - C_g V_g \tag{2.69}$$

and the Hamiltonian is thus given by

$$\mathcal{H} = \dot{\Phi}Q - \mathcal{L} = \frac{1}{2(C_g + C_J)}(Q + C_g V_g)^2 - E_J \cos(\frac{\Phi}{\varphi_0}) - \frac{1}{2}C_g V_g^2 \tag{2.70}$$

By dropping the constant term and introducing the new variables $n = \frac{Q}{2e}$ and $\varphi = \frac{\Phi}{\varphi_0}$ such that $[\varphi, n] = \frac{1}{\hbar}[\Phi, Q] = \boldsymbol{i}$, we can write the Hamiltonian as

$$\boxed{\mathcal{H} = 4E_C(\hat{n} - n_g)^2 - E_J \cos(\hat{\varphi})} \tag{2.71}$$

where $n_g = -\frac{C_g V_g}{2e}$ and $E_C = \frac{e^2}{2(C_J + C_g)}$. In order to find the eigenenergies and corresponding eigenstates, the Hamiltonian can be represented in the basis formed by the eigenstates $|n\rangle$ of operator $\widehat{n}$. Indeed, $[\varphi, n] = \boldsymbol{i}$ implies that $\langle\varphi|n\rangle = e^{\boldsymbol{i}n\varphi}$ and thus, the operator $\cos(\hat{\varphi}) = 1/2(e^{\boldsymbol{i}\hat{\varphi}} + e^{-\boldsymbol{i}\hat{\varphi}})$ can be written in the eigenbasis $|n\rangle$ as

$$\cos(\hat{\varphi})\,|n\rangle = \frac{1}{2}(e^{\boldsymbol{i}\hat{\varphi}} + e^{-\boldsymbol{i}\hat{\varphi}}) \sum |\varphi\rangle\,\langle\varphi|n\rangle = \frac{1}{2}\left(|n+1\rangle + |n-1\rangle\right) \tag{2.72}$$

It is thus easy to represent the Hamiltonian in a truncated charge basis as

$$H = \begin{pmatrix} 4E_C(-2-n_g)^2 & -E_J/2 & 0 & 0 & 0 \\ -E_J/2 & 4E_C(-1-n_g)^2 & -E_J/2 & 0 & 0 \\ 0 & -E_J/2 & 4E_C(0-n_g)^2 & -E_J/2 & 0 \\ 0 & 0 & -E_J/2 & 4E_C(1-n_g)^2 & -E_J/2 \\ 0 & 0 & 0 & -E_J/2 & 4E_C(2-n_g)^2 \end{pmatrix} \tag{2.73}$$

The choice of the truncation size depends on the parameters E_J and E_C and on the precision which is required. The results of such a diagonalization for the ground and first excited states are shown in Fig. 2.9 for different ratios of E_J/E_C. The voltage V_g allows controlling the transition energy of the qubit. As can be

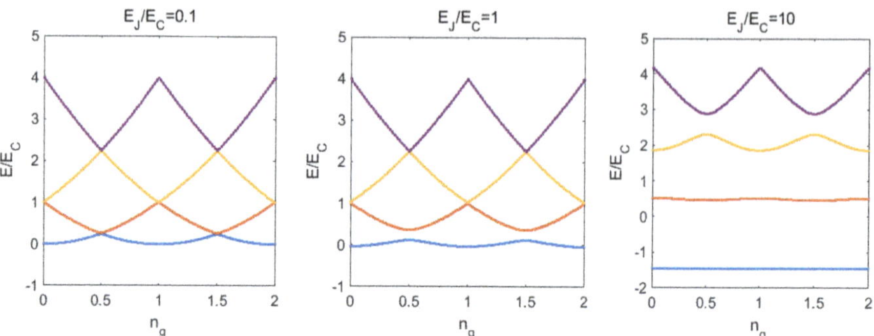

Fig. 2.9 First four energy levels of the Cooper pair box as a function of the reduced gate charge n_g for E_J/E_c ratios equal to 0.1, 1 and 10 (left to right). As can be seen, for $E_J \gg E_c$, the charge dispersion curves becomes more and more flat

seen, the charge-dispersion curve of the first two levels becomes almost flat when $E_J \gg E_c$.

2.2.2.2 The Split Cooper Pair Box

It is possible to get some additionnal control of the qubit energy by replacing the Josephson junction of the CPB by a Superconducting Quantum Interference Device (SQUID) [1]. In the following section we will show how the SQUID allows controlling the qubit transition energy via the magnetic flux Φ_S threading its loop. First, let us write the potential energy U_S of the SQUID, shown in Fig. 2.10:

$$U_S = -\frac{1+d}{2} E_J \cos(\varphi_1) - \frac{1-d}{2} E_J \cos(\varphi_2) \tag{2.74}$$

where d is the asymmetry parameter ,which can get any value in range of [0, 1]. A DC magnetic flux Φ_S is threading the loop of the SQUID such that $\varphi_1 - \varphi_2 = \frac{\Phi_S}{\varphi_0}$ leading to:

$$U_S = \underbrace{-E_J \sqrt{\frac{(1+d^2) + (1-d^2)\cos\left(\frac{\Phi_S}{\varphi_0}\right)}{2}}}_{-E_J(\Phi_S, d)} \cdot \cos\left(\frac{\varphi_1 + \varphi_2}{2} + \arctan\left[-d \cdot \tan\left(\frac{\Phi_S}{2\varphi_0}\right)\right]\right) \tag{2.75}$$

The potential energy of the SQUID is therefore equivalent to the potential energy of a *single* Josephson junction with tunable Josephson energy $E_J(\Phi_S, d)$ that varies between dE_J to E_J. When the asymmetry is large ($d \simeq 1$), the Josephson energy of the SQUID varies slightly and thus is less sensitive than for a symmetric SQUID ($d = 0$).

The remaining part of the quantization process proceeds as before, yielding an Hamiltonian of the split Cooper pair box of the form

$$\mathcal{H} = 4E_C(\hat{n} - n_g)^2 - E_J(\Phi_S, d)\cos\left(\hat{\varphi} + \gamma(\Phi_S, d)\right) \tag{2.76}$$

where $\gamma(\Phi_S, d) = \arctan\left[-d \cdot \tan\left(\frac{\Phi_S}{2\varphi_0}\right)\right]$.

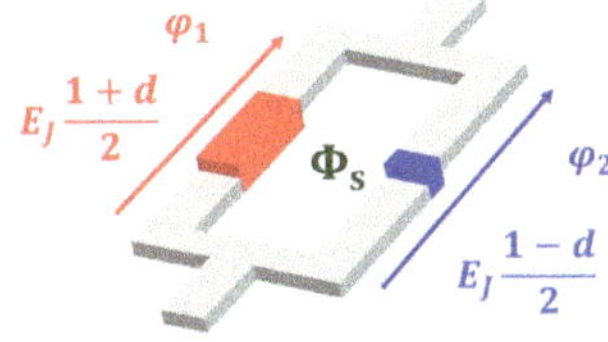

Fig. 2.10 Schematic of an asymmetric SQUID

2.2.2.3 The Transmon Qubit

The transmon qubit has been developed in 2006 in the group of R. Schoelkopf at Yale [8]. It is a CPB whose charging energy is strongly reduced by putting a large capacitance in parallel to the Josephson junction, such that the device is in the regime $E_J \sim 100\ E_C$. As shown in Fig. 2.11b, the charge dispersion of the energy levels of the CPB, $\Delta\omega(n_g)$, becomes extremely weak and the transition energy $\hbar\omega_{01}$ almost insensitive to the value of the gate charge n_g. This reduced sensitivity to charge is highly advantageous in experiments since it makes the qubit almost insensitive to charge noise and thus increases the coherence time. However, when increasing the ratio E_J/E_C one also reduces the anharmonicity $\alpha = (\omega_{12}-\omega_{01})/\omega_{01}$ (see Fig. 2.11a), therefore limiting the speed of gate operations that can be realized with this qubit (Fig. 2.11b).

The Hamiltonian of the system is similar to Eq. (2.71). Yet, the high E_J/E_C ratio reduces strongly the flux fluctuations and one can thus develop the cosine function close to zero as $\cos(\hat{\varphi}) = 1 - \frac{1}{2}\hat{\varphi}^2 + \frac{1}{4!}\hat{\varphi}^4 + O(\hat{\varphi}^6)$.

When taking only into account the terms in $\hat{\varphi}^2$, the system behaves as an harmonic oscillator of frequency $\hbar\omega_r = \sqrt{8E_J E_C}$. It is possible to express the operators $\widehat{n}$ and $\hat{\varphi}$ as a function of the creation and annihilation operator a and a^+

$$\hat{\varphi} = \left(\frac{2E_C}{E_J}\right)^{1/4}(a + a^+)$$

$$\widehat{n} = -\frac{i}{2}\left(\frac{E_J}{2E_C}\right)^{1/4}(a - a^+)$$

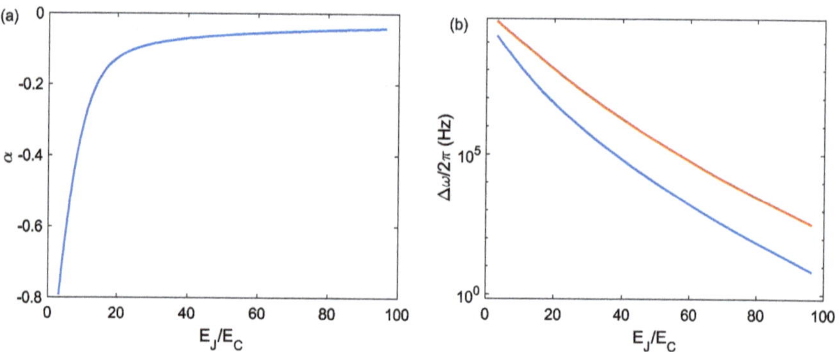

Fig. 2.11 Transmon Properties. (**a**) Anharmonicity $\alpha = (\omega_{12} - \omega_{01})/\omega_{01}$ versus E_J/E_C ratio. (**b**) Amplitude of the charge dispersion $\Delta\omega = \omega(n_g = 0.5) - \omega(n_g = 0)$ versus E_J/E_C for the $0 \rightarrow 1$ transition (in blue) and $1 \rightarrow 2$ transition (in red). The $0 \rightarrow 1$ transition frequency of the qubit is kept around 4.8 GHz

The zero point fluctuations of the phase operator, $\Delta\varphi = \left(\frac{2E_C}{E_J}\right)^{1/4}$, are small when $E_J \gg E_C$, which justifies our approximation a posteriori. Developing the cosine potential to higher order gives

$$\boxed{\mathcal{H} = \hbar\omega_r\left(a^+a + \frac{1}{2}\right) - \frac{E_C}{12}\left(a + a^+\right)^4} \tag{2.77}$$

The second term of the Hamiltonian can be viewed as a Kerr non-linearity. It can be solved perturbatively using first order perturbation theory. The shift of the nth energy level of the harmonic oscillator is given by

$$\begin{aligned}\Delta E_n &= \left\langle n \left| -\frac{E_C}{12}\left(a + a^+\right)^4 \right| n \right\rangle \\ &= -\frac{E_C}{12}\left(6n^2 + 6n + 3\right)\end{aligned} \tag{2.78}$$

and thus $\Delta E_{n+1} - \Delta E_n = -E_C(n+1)$. The Kerr non-linear term modifies the equidistant interlevel spacing and defines a qubit with anharmonicity

$$\alpha = (\omega_{12} - \omega_{01})/\omega_{01} = -\sqrt{E_C/8E_J} \tag{2.79}$$

2.2.2.4 Improving Transmon Design

Typical values for a transmon are $E_C/h = 200$ MHz and $E_J/h = 15$ GHz giving a ratio $E_J/E_C = 75$, a qubit frequency $\omega_{01}/2\pi = 4.7$ GHz and an anharmonicity $\alpha = -0.04$. As mentioned earlier, the low anharmonicity requires that the gate time should be much longer than $h/E_c = 5$ ns.

The typical relaxation times of transmons have been largely improved by the introduction of three dimensional cavity, which reduce the impact of dielectric losses in the circuit [9]. Various works have tried to reduce these losses in order to increase the fidelity of the qubit while keeping a 2D scalable architecture [10, 11]. This can be done by reducing the interface defects between the metal and the substrate [11]. In recent works [12], transmons with typical relaxation time of $T_1 \sim 50$ μs are controlled by $\sim$ 40 ns two qubit gates, leading to two qubit gate fidelity in the range of 99.3–99.8%.

2.2.3 The Superconducting Flux Qubit

The superconducting flux qubit is a superconducting circuit which consists of a micron-size superconducting aluminum loop intersected by three (or more) Josephson junctions, among which one of the junctions is smaller than the others by a factor α (see Fig. 2.12). This qubit was initially developed at Delft University in 1999 [13–17].

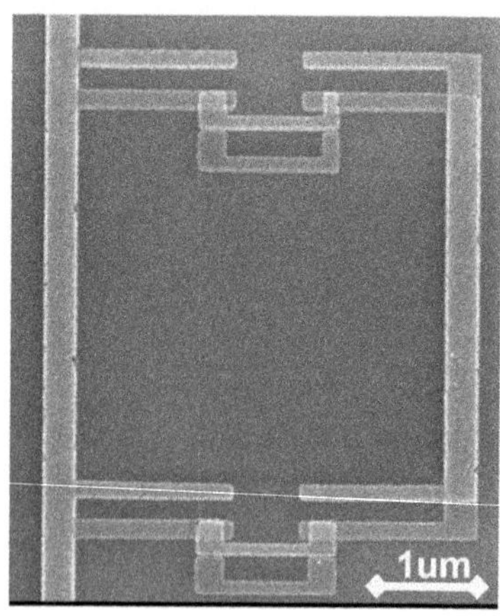

Fig. 2.12 Scanning Electron Microscope (SEM) image of a 4-junction flux qubit

2.2.3.1 Potential Energy of a Flux Qubit

The potential energy of the circuit can be written as a sum of the potential energies of each junction intersecting the loop (see Eq. (2.65)). A DC magnetic flux Φ is threading the loop, therefore due to Faraday law $\sum_{i=1}^{n-1} \varphi_i + \varphi_\alpha = \frac{\Phi}{\varphi_0}$ and thus:

$$U = -E_J \left[\sum_{i=1}^{n-1} \cos(\varphi_i) + \alpha \cos(\frac{\Phi}{\varphi_0} - \sum_{i=1}^{n-1} \varphi_i) \right] \quad (2.80)$$

For a flux qubit with $n = 3$ junctions, the potential energy can be plotted with a pseudo-color plot shown in Fig. 2.13.

When $\Phi/\varphi_0 = \pi$ and for $\alpha > \alpha_{min}$, the potential energy U exhibits two degenerated minima (Fig. 2.13), the potential barrier between these two minima being a function of the parameters of the junctions. The position of the minima are given by solving the partial derivative equations $\partial_{\varphi_i} U = 0$. The two solutions verify the simple equation

$$\sin \varphi^* = \alpha \sin \left((n-1)\varphi^* \right) \quad (2.81)$$

and correspond to two opposite persistent currents $I_P = \pm I_0 \sin \varphi^*$ flowing in the loop.

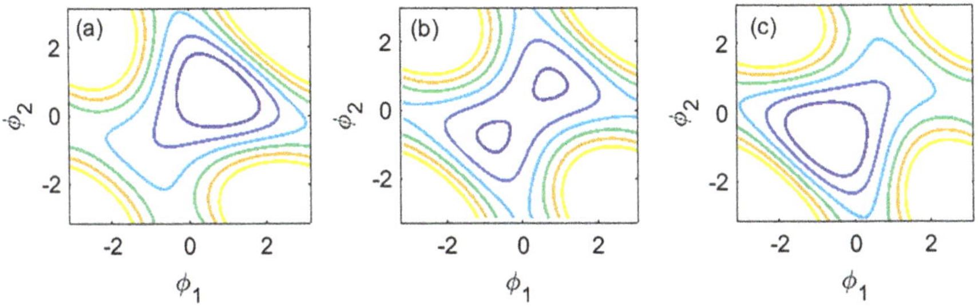

Fig. 2.13 Potential energy landscape of a 3-island flux qubit with parameters $\alpha = 0.7$. (**a**) for $\Phi/\varphi_0 = 0.8\pi$; (**b**) for $\Phi/\varphi_0 = \pi$; (**c**) for $\Phi/\varphi_0 = 1.2\pi$

By inverting this equation, we find that $U_{n-2}(\cos\varphi^*) = \frac{1}{\alpha}$ and

$$\boxed{I_P = \pm I_0\sqrt{1 - \left[U_{n-2}^{-1}\left(\frac{1}{\alpha}\right)\right]^2}} \tag{2.82}$$

where U_{n-2} is the (n-1)-th Chebyshev polynomial of the second kind. The following table summarizes the common values for $n = 3, 4$ or 5.

n	U_{n-2}	$\lvert I_P\rvert$	α_{min}
3	$2X$	$I_0\sqrt{1-\left(\frac{1}{2\alpha}\right)^2}$	$\frac{1}{2}$
4	$4X^2-1$	$I_0\sqrt{\frac{3}{4}-\frac{1}{4\alpha}}$	$\frac{1}{3}$
5	$8X^3-4$	$I_0\sqrt{1-\left(\frac{1}{2}+\frac{1}{8\alpha}\right)^{2/3}}$	$\frac{1}{4}$

2.2.3.2 Kinetic Energy of a Flux Qubit

For a 4-junction qubit, the kinetic energy K of the system is the sum of the capacitive energies of the circuit shown in Fig. 2.14

$$K = \frac{1}{2}\sum_{i\neq j} C_{ij}\left(V_j - V_i\right)^2 \tag{2.83}$$

$$+\frac{1}{2}C_J\left((V_1-V_2)^2+(V_2-V_3)^2+(V_3-V_4)^2+\alpha\left(V_4-V_1\right)^2\right)$$

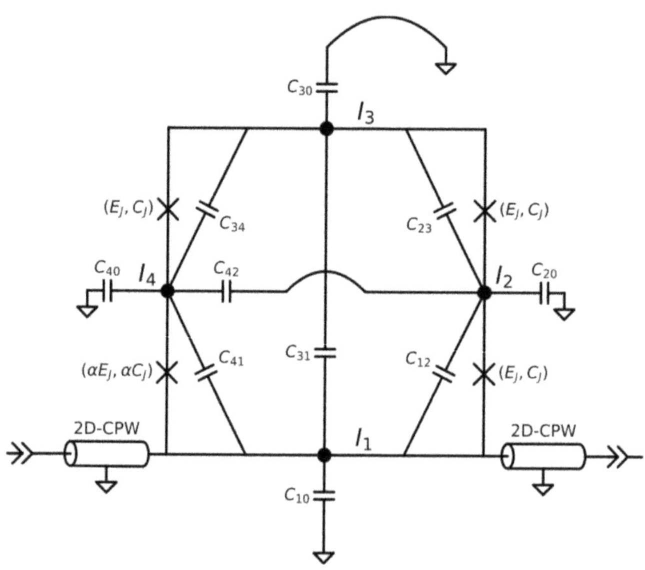

Fig. 2.14 Equivalent circuit diagram of a flux qubit. The Josephson junctions are defined by their Josephson energy E_J and their bare capacitance C_J. The island I_1 is galvanically connected to a coplanar waveguide resonator. Each island is capacitively coupled to its surrounding by geometric capacitances denoted as C_{ij} where $(i, j) \in (0, 1, .., 4)$, the index 0 representing the ground

It is a quadratic form of the island voltages V_i and can thus be written as

$$K = \frac{1}{2}\mathbf{V}^T\mathbf{C}\mathbf{V} \tag{2.84}$$

where $\mathbf{V^T} = \left(V_1\ , V_2\ , V_3,\ V_4\right)$ and $\mathbf{C}$ is a 4×4 matrix which we will refer in the following as the capacitance matrix. The matrix $\mathbf{C}$ can be written as the sum of the Josephson capacitance matrix $\mathbf{C_J}$ and the geometric capacitance matrix $\mathbf{C_{geom}}$.

$$\mathbf{C} = \mathbf{C_J} + \mathbf{C_{geom}} \tag{2.85}$$

where

$$\mathbf{C_J} = C_J \begin{pmatrix} 1+\alpha & -1 & 0 & -\alpha \\ -1 & 2 & -1 & 0 \\ 0 & -1 & 2 & -1 \\ -\alpha & 0 & -1 & 1+\alpha \end{pmatrix} \tag{2.86}$$

and

$$\mathbf{C_{geom}} = \begin{pmatrix} C_{10} & 0 & 0 & 0 \\ 0 & C_{20} & 0 & 0 \\ 0 & 0 & C_{30} & 0 \\ 0 & 0 & 0 & C_{40} \end{pmatrix}$$

$$+ \begin{pmatrix} \sum_{j\neq 1} C_{1j} & -C_{12} & -C_{13} & -C_{14} \\ -C_{21} & \sum_{j\neq 2} C_{2j} & -C_{23} & -C_{24} \\ -C_{31} & -C_{32} & \sum_{j\neq 3} C_{3j} & -C_{34} \\ -C_{41} & -C_{42} & -C_{43} & \sum_{j\neq 4} C_{4j} \end{pmatrix} \tag{2.87}$$

2.2.3.3 Legendre Transformation and Hamiltonian

The Lagrangian of the system is $\mathcal{L} = K - U$. The conjugate momenta of our system are given by

$$n_j \equiv \frac{1}{\hbar}\frac{\partial \mathcal{L}}{\partial \dot{\varphi}_j} \tag{2.88}$$

Since $\frac{\Phi_0}{2\pi}\dot{\varphi}_{j,} = V_{j+1} - V_j$, it is necessary to express the kinetic energy terms in a new basis. Since island I_1 shown in Fig. 2.14 is galvanically connected to the central conductor of a CPW, we can safely assume that $V_1 = 0\,\mathrm{V}$, which simplifies considerably the transformation:

$$\begin{aligned} V_1 &= 0 \\ V_2 &= \cancelto{0}{V_1} + V_{12} \\ V_3 &= \cancelto{0}{V_1} + V_{12} + V_{23} \\ V_4 &= \cancelto{0}{V_1} + V_{12} + V_{23} + V_{34} \end{aligned}$$

where $V_{ij} = V_j - V_i$. The passage matrix P between these two bases can be thus written as

$$\mathbf{P} = \begin{pmatrix} 0\,0\,0 \\ 1\,0\,0 \\ 1\,1\,0 \\ 1\,1\,1 \end{pmatrix} \tag{2.89}$$

The Hamiltonian $\mathcal{H}$ is then obtained by the Legendre transformation $\mathcal{H} = \hbar \sum_{j=1}^{3} \dot{\varphi}_j n_j - \mathcal{L}$ and thus writes

$$\mathcal{H} = \frac{(2e)^2}{2} \boldsymbol{n}^T \left(\mathbf{P}^T \mathbf{C} \mathbf{P}\right)^{-1} \boldsymbol{n} + U \tag{2.90}$$

This Hamiltonian can be expressed in the so-called charge basis $|n_1, n_2, n_3\rangle$, $\forall n_1, n_2, n_3 \in \mathbb{Z}^3$, noting that

$$\cos\varphi_j |n_1, n_2, n_3\rangle = \frac{1}{2}\big(|n_1 + \delta_{j1}, n_2 + \delta_{j2}, n_3 + \delta_{j3}\rangle + |n_1 - \delta_{j1}, n_2 - \delta_{j2}, n_3 - \delta_{j3}\rangle\big) \tag{2.91}$$

In this basis the operator $\frac{(2e)^2}{2}\boldsymbol{n}^T\left(\mathbf{P}^T\mathbf{CP}\right)^{-1}\boldsymbol{n}$ is diagonal while the operator U is sparse. The precision of the eigenvalues and eigenstates depends on the truncation of the n_j bases. With $n_k = -10\ldots 10$, we would need 21^3 coefficients just to describe the wavefunction and another $\left(21^3\right)^2$ to describe the Hamiltonian matrix. Thanks to the the sparsity of the Hamiltonian operator, the number of nonzero entries in this matrix is only $21^3 \times (1 + 4 \times 2)$. This resolution in charge space is computationally feasible both to store and diagonalize matrices efficiently.

2.2.3.4 Pseudo-Hamiltonian

Following the full diagonalization of the Hamiltonian, we obtain the spectrum of the flux qubit by subtracting the energy of the first excited state $|1\rangle$ from the energy of the ground state $|0\rangle$. It can be shown that close to $\Phi/\varphi_0 = \pi$, the system behaves as a two level system and the spectrum can be fully described by two parameters:

- The value of the persistent current I_p, already discussed previously.
- The so-called flux qubit gap, denoted as Δ, which corresponds to the tunneling term between the two potential minima.

The value of the gap can be directly measured by the transition energy at half a flux quantum $\Phi/\varphi_0 = \pi$. This point is known as the *optimal point* of the flux qubit due to its immunity at first order in flux noise, as will be explained in later sections. In the vicinity of the optimal point, the Hamiltonian of the system can be written using perturbation theory as

$$\begin{aligned}\mathcal{H} &= \mathcal{H}_0 - \alpha E_J \partial_\Phi\left(\cos\left(2\pi\tfrac{\Phi}{\Phi_0} - \textstyle\sum_{j=1}^3\varphi_j\right)\right)_{\Phi=\Phi_0/2}\cdot\left(\Phi - \tfrac{\Phi_0}{2}\right)\\ &= \mathcal{H}_0 + \tfrac{1}{\varphi_0}\left[\underbrace{\alpha E_J\sin\left(\varphi_\alpha\right)}_{\hat{I}\cdot\varphi_0}\left(\Phi - \tfrac{\Phi_0}{2}\right)\right] = \mathcal{H}_0 + \hat{I}\cdot\left(\Phi - \tfrac{\Phi_0}{2}\right)\end{aligned} \tag{2.92}$$

When the current operator is projected on the eigenstates $|0\rangle$, $|1\rangle$ of $\mathcal{H}_0$ we get

$$\begin{aligned}\langle 0|\,\hat{I}\,|0\rangle = 0 \;,\; \langle 0|\,\hat{I}\,|1\rangle = I_p\\ \langle 1|\,\hat{I}\,|0\rangle = I_p \;,\; \langle 1|\,\hat{I}\,|1\rangle = 0\end{aligned} \tag{2.93}$$

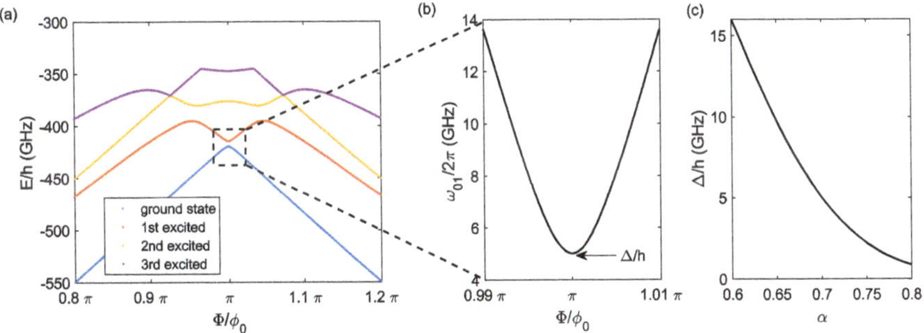

Fig. 2.15 Flux qubit energy levels. (**a**) Calculated eigenenergies of the flux qubit circuit versus applied magnetic field. The calculations were performed using a 3-junction qubit with $E_J/h = 350$ GHz, $E_C/h = 5$ GHz and $\alpha = 0.7$. (**b**) Calculated qubit spectroscopy in the close vicinity of $\Phi/\varphi_0 = \pi$. At precisely $\Phi/\varphi_0 = \pi$, the transition energy of the qubit is minimal and equal to the gap Δ. (**c**) Qubit minimal frequency (Δ/h) versus α parameter

Therefore, the Hamiltonian of the system can be written in this basis as

$$\boxed{\mathcal{H}_{\text{eff}} = \frac{\hbar}{2}\left[\Delta\sigma_z + \varepsilon\sigma_x\right]} \tag{2.94}$$

where $\varepsilon = \frac{2I_p}{\hbar}(\Phi - \pi\varphi_0)$. The frequency of the flux qubit is thus given by

$$\omega_{01} = \sqrt{\Delta^2 + \varepsilon^2} \tag{2.95}$$

2.2.3.5 Improving Flux Qubit Design

As shown in Fig. 2.15b, the flux qubit resonance frequency is strongly dependent on the value of the applied magnetic flux. Away from $\Phi/\varphi_0 = \pi$, the coherence of the qubit will be compromised by the presence of flux noise. We will study this question more in details in Sect. 2.4.2.4. The only point where one can expect to have long coherence time is the so-called "optimal point" where the qubit frequency is minimal and thus immune to flux noise at first order. At that point the flux qubit transition energy is equal to the flux qubit gap Δ. The flux qubit gap is strongly dependent on the design parameters of the junctions (E_J, E_C) and on the value of α as shown in Fig. 2.15c. This means that an extreme precision in the fabrication of the qubit is needed if one wishes to control the value of the gap [18].

Tunable Flux Qubits

It is possible to create a tunable flux qubit by replacing one of the junction by a SQUID as is done for the split Cooper pair box in Sect. 2.2.2.2. This approach brings necessarily a new channel of decoherence to the qubit [19] which should be controlled properly, for instance by using SQUIDS with large assymetry [20].

Fluxonium

The fluxonium has been developed in 2009 at Yale University [21]. The main idea of this design consists of reducing the flux sensitivity of the qubit by increasing the number of junctions intersecting the loop of the circuit. As shown in Sect. 2.2.3.1, the introduction of a large number of junctions (typically ~ 50) reduces dramatically the value of the persistent current flowing in the loop of the qubit and thus its magnetic dipole moment. In addition, this qubit is immune to charge noise and exhibits a large increase of its relaxation time at $\Phi/\varphi_0 = \pi$ due to destructive interference of quasiparticles [22]. A large enhancement of the coherence time compared to flux qubits was indeed observed [23]. The main limitations of this design are its rather low transition frequencies (in the range of few hundreds of MHz at optimal point), which require dynamical initialization of the qubit. Moreover, the tiny magnetic moment of the fluxonium reduces its ability to be easily coupled to other qubits, resonators and/or quantum devices.

Capacitively-Shunted Flux Qubits

In this design developed at MIT in 2016, the flux qubit is connected to a big capacitance which reduces strongly the persistent current and anharmonicity of the qubit [24, 25]. In a recent work [25], a capacitively shunted flux qubit embedded into a three-dimensional cavity has shown relaxation times up to $T_1 \sim 90\ \mu s$ and Ramsey decoherence time of $T_{2R} \sim 18\ \mu s$.

2.3 Coupling Qubits and Resonators

In this chapter, we will describe succinctly how qubits and resonators can be coupled together in order to establish the main ingredients required for the functioning of a quantum processor. Our objective is not to give a comprehensive overview of the field of circuit-QED but rather to focus on the basic principles of qubit readout and manipulation.

2.3.1 Coupling a Qubit with a Resonator

Charge and flux qubit can be coupled to a resonator by capacitive or inductive coupling. The value of this coupling depends on the electric/magnetic zero-field fluctuations of the resonator at the qubit position and on the electric or magnetic moment of the qubit.

2.3.1.1 Transmon Embedded in a Three-Dimensional Cavity

For instance, let us consider a transmon embedded into a three-dimensional rectangular cavity shown in Fig. 2.16. The transmon is composed of two pads separated by a short distance d and connected by a wire intersected by a single Josephson junction (see [9]). The zero-field fluctuations of the electric field in the

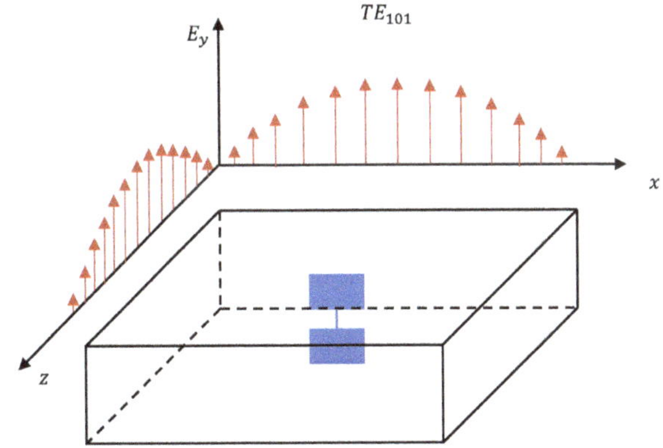

Fig. 2.16 Transmon qubit embedded into a three dimensional cavity. The qubit is coupled to the electric field E_y of the fundamental mode (TE101) of the cavity

cavity can be easily estimated by integrating the electric field energy density over the whole volume V and thus calculating the energy stored in the cavity

$$\int_V \varepsilon_0 \delta E_0^2(x, y, z)\, dx\, dy\, dz = \frac{\hbar\omega}{2} \tag{2.96}$$

The coupling between the fundamental cavity mode (TE101 represented in Fig. 2.16) and a qubit situated in the center of the cavity is thus given by

$$\hbar g = ed\delta E_0 = \frac{ed}{2}\sqrt{\frac{\hbar\omega}{2\varepsilon_0 V}} \tag{2.97}$$

This simple back-of-the-envelope estimation can be applied for instance to the cavity-qubit system described in [9] where $d = 100\,\mu\text{m}$, $\omega = 8\,\text{GHz}$ and $V = 3\,\text{cm}^3$ and gives $\frac{g}{2\pi} = 125\,\text{MHz}$, which is very close to the coupling constant extracted experimentally from spectroscopic measurement in the same publication.

2.3.1.2 Flux Qubit Coupled Inductively to a Lumped Element Resonator

Another interesting example consists of a flux qubit coupled inductively to a lumped element resonator as shown in Fig. 2.17. The flux qubit is placed at a distance d from the resonator inductance and is coupled to the current fluctuations of the resonator.

Assuming the current is flowing in an infinitely-thin wire, it is possible to calculate analytically the magnetic field in the vicinity of the qubit using Biot Savart law. Namely,

$$\delta B_0 = \mu_0 \frac{\delta I_0}{2\pi r} \tag{2.98}$$

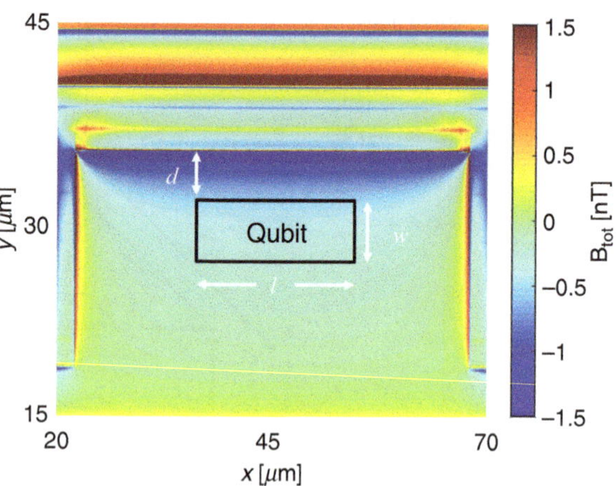

Fig. 2.17 Flux qubit coupled inductively to a resonator. The magnetic field is calculated from the current flowing in the resonator using an electromagnetic simulator (Sonnet)

The coupling between the resonator and the qubit is obtained by integrating the magnetic field threading the loop of the qubit. Using Eq. (2.94),

$$\hbar g = I_p \int \delta B_0 dS = M I_P \delta I_0 \tag{2.99}$$

where $M = \frac{\mu_0 l}{2\pi} \ln\left(1 + \frac{w}{d}\right)$ is the mutual inductance and δI_0 are the current quantum fluctuations of the resonator (see Eq. (2.58)). This simple back-of-the-envelope estimation can be applied for instance to the system shown in Fig. 2.17 where $d = 1$ µm, $w = l = 3$ µm, $I_p = 300$ nA and $\delta I_0 = 40$ nA. We obtain $M = 0.8$ pH and $\frac{g}{2\pi} = 15$ MHz.

The mutual inductance can be increased further by connecting galvanically the qubit loop to the resonator. In this configuration, the mutual inductance per unit length M/l reaches approximately 3 pH/µm for wires of cross section $200 \times 40\,\text{nm}^2$ and thus the coupling can reach $\frac{g}{2\pi} = 170$ MHz, assuming $I_p = 300$ nA and $\delta I_0 =$ 40 nA.

2.3.1.3 Qubit Readout by Dispersive Shift

The Hamiltonian of a qubit coupled to a resonator can be written as

$$\mathcal{H} = \hbar\omega_r a^+ a + \frac{1}{2}\hbar\omega_{01}\sigma_z + V \tag{2.100}$$

where $V = \hbar g \left(\sigma^+ + \sigma^-\right)\left(a + a^+\right)$, g being the coupling calculated in the sections herein above. The expansion of this product of operators involves four terms. The terms proportional to $\sigma^+ a$ and $\sigma^- a^+$correspond to transitions from lower (resp. upper) level of the qubit together with the annihilation (resp. creation) of a photon in the resonator. The two other terms $\sigma^+ a^+$ and $\sigma^- a$ correspond to transitions from lower (resp. upper) level of the qubit together with creation (resp. annihilation) of a photon in the resonator. When the frequency of the qubit and the resonator are sufficiently close, these terms correspond to highly non-resonant processes. Neglecting these anti-resonant terms is a standard approximation in Quantum Electrodynamics called Rotative Wave Approximation (RWA).

For simplicity, we will make this approximation in the following and consider

$$V = \hbar g \left(\sigma^{+} a + \sigma^{-} a^{+}\right) \tag{2.101}$$

Within this approximation, the qubit and the resonator frequencies are assumed to be relatively close. When the qubit and the resonator are detuned and the coupling is sufficiently small ($g \ll |\omega_{01} - \omega_r|$), one can transform the Hamiltonian using a unitary transformation $U = e^S$, where S is an anti-hermitian operator chosen to satisfy (see Sect. 2.5.2.3)

$$\left[S, \hbar\omega_r a^{+} a + \frac{1}{2}\hbar\omega_{01}\sigma_z\right] = -V \tag{2.102}$$

It is straightforward to show that this condition is satisfied by choosing

$$S = \frac{g}{\omega_{01} - \omega_r}\left(\sigma^{+} a - \sigma^{-} a^{+}\right) \tag{2.103}$$

Using this transformation, the Hamiltonian can be described in a perturbative approach (see Sect. 2.5.2.3) as

$$\widetilde{H} \simeq \hbar\omega_r a^{+} a + \frac{1}{2}\hbar\omega_{01}\sigma_z + \frac{g^2}{(\omega_{01} - \omega_r)}\sigma_z\left(a^{+} a + \frac{1}{2}\right) \tag{2.104}$$

This last term corresponds to a Lamb shift effect. The frequency of the qubit is shifted by the presence of photons in the cavity. This will have important consequences on the qubit coherence as will be seen in Sect. 2.4.2.5. The same term viewed from the resonator perspective corresponds to a change of the resonator frequency depending on the state of the qubit. When the qubit is in the excited state, the resonator frequency is offset from ω_r by

$$\boxed{\delta_1\omega_r = \frac{g^2}{(\omega_{01} - \omega_r)}} \tag{2.105}$$

When the qubit is in the ground state, the resonator frequency is offset from its bare frequency $\omega_r$by an opposite value

$$\boxed{\delta_0\omega_r = -\frac{g^2}{(\omega_{01} - \omega_r)}} \tag{2.106}$$

This frequency shift is usually called dispersive shift and allows the detection of the qubit state by looking at the frequency shift of its coupled resonator. In order to be able to detect it easily, this shift should be comparable with the resonator linewidth.

2.3.2 Single Qubit Gates

2.3.2.1 Driving a Qubit by a Classical Drive

Arbitrary single-qubit rotations can be realized by applying a classical drive for a certain time duration. The Hamiltonian of the system can be written as

$$\mathcal{H} = \hbar \frac{\omega_{01}}{2} \sigma_z + \hbar \Omega \sigma_x \cos(\omega_{mw} t + \varphi) \tag{2.107}$$

where ω_{01} is the qubit frequency, ω_{mw} is the drive frequency, φ the phase of the drive and Ω is the Rabi frequency and is proportional to the drive amplitude. The equations of motion which describe the qubit evolution taking into account decoherence are derived in Appendix 2.5.1.7.

Under unitary transformation $U = \exp\left(\boldsymbol{i}\sigma_z \frac{\omega_{mw}}{2} t\right)$, the Hamiltonian becomes (see Appendix 2.5.2.1)

$$\begin{aligned}\widetilde{\mathcal{H}} = \hbar \frac{\delta}{2} \sigma_z + \hbar \frac{\Omega}{2} \left(\sigma^+ \exp(\boldsymbol{i}\omega_{mw} t) + \sigma^- \exp(-\boldsymbol{i}\omega_{mw} t)\right) (\exp(\boldsymbol{i}(\omega_{mw} t + \varphi)) \\ + \exp(-\boldsymbol{i}(\omega_{mw} t + \varphi)))\end{aligned} \tag{2.108}$$

where $\delta = \omega_{01} - \omega_{mw}$ is the detuning between the drive and the qubit frequency. Neglecting the terms rotating at $2\,\omega_{mw}$ (Rotative Wave Approximation), one obtains a time-independent effective Hamiltonian

$$\widetilde{\mathcal{H}} = \hbar \frac{\delta}{2} \sigma_z + \hbar \frac{\Omega}{2} \left(\sigma^+ \exp(-\boldsymbol{i}\varphi) + \sigma^- \exp(+\boldsymbol{i}\varphi)\right) \tag{2.109}$$

For $\delta = 0$, the evolution under such Hamiltonian is relatively simple

$$U(\tau) = \begin{pmatrix} \cos\left(\frac{\Omega\tau}{2}\right) & -\boldsymbol{i}\sin\left(\frac{\Omega\tau}{2}\right) e^{-\boldsymbol{i}\varphi} \\ -\boldsymbol{i}\sin\left(\frac{\Omega\tau}{2}\right) e^{\boldsymbol{i}\varphi} & \cos\left(\frac{\Omega\tau}{2}\right) \end{pmatrix} \tag{2.110}$$

In particular a $\pi/2$ pulse ($\Omega\tau = \pi/2$) will take a qubit in the ground state $|0\rangle$ to an equal superposition of ground and excited:

$$|0\rangle \rightarrow \frac{1}{\sqrt{2}} \left(|0\rangle - \boldsymbol{i}e^{-\boldsymbol{i}\varphi} |1\rangle\right)$$

$$|1\rangle \rightarrow \frac{1}{\sqrt{2}} \left(-\boldsymbol{i}e^{\boldsymbol{i}\varphi} |0\rangle + |1\rangle\right)$$

A π pulse ($\Omega\tau = \pi$) will take a qubit in the ground state $|0\rangle$ to the excited state and vice versa

$$|0\rangle \to -ie^{-i\varphi}|1\rangle$$
$$|1\rangle \to -ie^{i\varphi}|0\rangle$$

2.3.2.2 Driving a Qubit via a Resonator Port

It is possible to drive the qubits that are positioned inside of a resonator by using one of the resonator ports. In the following we calculate the Rabi frequency of a flux qubit coupled inductively to a $\lambda/2$ transmission line resonator (see Sect. 2.1.5.5). According to (2.60)

$$\hat{I}(x) = \frac{1}{\sqrt{Z_0}} \sum_n \left(f_{\rightarrow}(\omega_n, x)\, A^{\rightarrow}_{L,n} + f_{\leftarrow}(\omega_n, x)\, A^{\leftarrow}_{R,n}\right) + \text{H.c.} \tag{2.111}$$

Let us assume that we drive a monochromatic wave ($\omega_n = \omega_{01}$) from the left such that $A^{\leftarrow}_{R,n} = 0, \forall n$. The resulting Rabi frequency (see Eq. (2.94)) is

$$\Omega^{\rightarrow} = \frac{2MI_p}{\hbar}\sqrt{\frac{1}{Z_0}}\left|f_{\rightarrow}(\omega_{01}, x)\, A^{\rightarrow}_{L}(\omega_{01})\right| \tag{2.112}$$

Since $\left|A^{\rightarrow}_{L}(\omega_{01})\right|^2 = \langle P\rangle/2$ (see (2.22)), we get

$$\boxed{(\hbar\Omega^{\rightarrow})^2 / \langle P\rangle = \frac{\left(2MI_p\right)^2}{2Z_0}\left|f_{\rightarrow}(\omega_{01}, x)\right|^2} \tag{2.113}$$

The same can be done for a monochromatic drive from the right, assuming $A^{\rightarrow}_{L,n} = 0, \forall n$. Thus we get

$$\boxed{(\hbar\Omega^{\leftarrow})^2 / \langle P\rangle = \frac{\left(2MI_p\right)^2}{2Z_0}\left|f_{\leftarrow}(\omega_{01}, x)\right|^2} \tag{2.114}$$

2.3.3 Two-Qubit Gates

2.3.3.1 Coupling Two Qubits by Fixed Coupling

One of the simplest two-qubit coupling scheme consists of coupling neighboring qubits with a static coupling. These neighboring qubits are naturally coupled by dipole-dipole interaction. The coupling is mainly electric (capacitive) for charge qubits and magnetic (inductive) for flux qubits. It is possible to increase the coupling strength by using an intermediate lumped element as shown in Fig. 2.18.

In the following, we will illustrate how this coupling is established between two transmon qubits. This kind of coupling has already been described for resonators in Sect. 2.1.2.3.

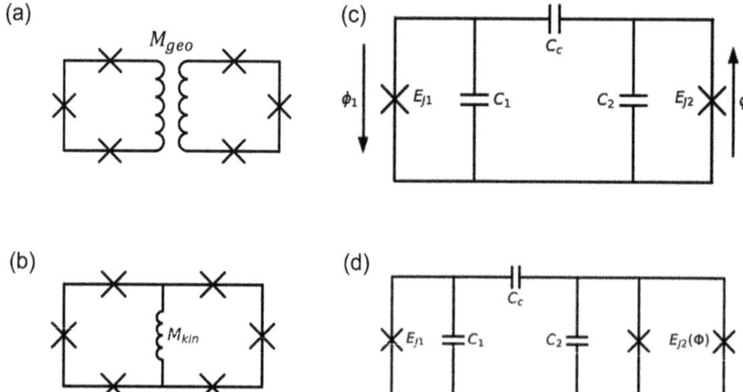

Fig. 2.18 Different coupling schemes (**a**) Inductive coupling by geometric mutual inductance M_{geo} between two flux qubits. (**b**) Inductive coupling by sharing a lumped element inductance M_{kin} between the qubits. (**c**) Capacitive coupling by sharing a coupling capacitance C_c between two transmon qubits. (**d**) Controllable coupling by tuning the frequency of a tunable transmon via a DC flux

The Lagrangian of two capacitively coupled transmon qubits (shown in Fig. 2.18c) is given by

$$\mathcal{L} = T - V = \left[\frac{1}{2}C_1\dot{\Phi_1}^2 + \frac{1}{2}C_2\dot{\Phi_2}^2 + \frac{1}{2}C_C\left(\dot{\Phi_1} - \dot{\Phi_2}\right)^2\right] + \left[E_{J_1}\cos\left(\frac{\Phi_1}{\varphi_0}\right) + E_{J_2}\cos\left(\frac{\Phi_2}{\varphi_0}\right)\right]$$

The conjugate momenta are defined by

$$Q_1 = \frac{\partial\mathcal{L}}{\partial\dot{\Phi_1}} = C_1\dot{\Phi_1} + C_C\left(\dot{\Phi_1} - \dot{\Phi_2}\right)$$

$$Q_2 = \frac{\partial\mathcal{L}}{\partial\dot{\Phi_2}} = C_2\dot{\Phi_2} - C_C\left(\dot{\Phi_1} - \dot{\Phi_2}\right)$$

We then obtain the Hamiltonian of the system $H = Q_1\dot{\Phi_1} + Q_2\dot{\Phi_2} - \mathcal{L}$ and decompose it into three elements $H = H_1 + H_2 + V$ such that

$$H_1 = \frac{1}{2}\beta\left(C_2 + C_C\right)Q_1^2 - E_{J_1}\cos\left(\frac{\Phi_1}{\varphi_0}\right) \sim \frac{1}{2}\hbar\omega_1\sigma_1^z$$

$$H_2 = \frac{1}{2}\beta\left(C_1 + C_C\right)Q_1^2 - E_{J_2}\cos\left(\frac{\Phi_2}{\varphi_0}\right) \sim \frac{1}{2}\hbar\omega_2\sigma_2^z$$

$$\boxed{V = \beta C_C Q_1 Q_2 \sim \hbar g\left(\sigma_1^- - \sigma_1^+\right)\left(\sigma_2^+ - \sigma_2^-\right)} \tag{2.115}$$

where $\beta = \frac{1}{C_C C_1 + C_C C_2 + C_2 C_1}$, $\omega_{1/2} = \frac{\sqrt{E_{J_{1/2}} \beta (C_{2/1} + C_C)}}{\varphi_0}$ and

$$g = -\frac{1}{2} \beta C_C \sqrt{\omega_1 \omega_2} \sqrt{(C_C + C_1)(C_C + C_2)}. \tag{2.116}$$

The product expansion of V involves four terms. Performing a Rotating Wave Approximation (RWA) allows us to neglect the two non-resonant terms, and obtain

$$V = \hbar g \left(\sigma_1^+ \sigma_2^- + \sigma_1^- \sigma_2^+ \right). \tag{2.117}$$

2.3.3.2 iSWAP Gate

This interaction allows us to perform a two qubit gate when the detuning $\delta = \omega_1 - \omega_2$ between the two qubits is smaller than g ($\delta << g$). In order to better understand this point, we calculate the time-evolution matrix of the system

$$U(t) = \begin{pmatrix} 1 & 0 & 0 & 0 \\ 0 & \cos[g_e t] - i \frac{\delta}{g_e} \sin[g_e t] & i \frac{2g}{g_e} \sin[g_e t] & 0 \\ 0 & i \frac{2g}{g_e} \sin[g_e t] & \cos[g_e t] + i \frac{\delta}{g_e} \sin[g_e t] & 0 \\ 0 & 0 & 0 & 1 \end{pmatrix} \tag{2.118}$$

where $g_e = \sqrt{4g^2 + \delta^2}$ is the effective swapping frequency. We can see that in the case where $\delta >> g$, the off-diagonal matrix elements goes to 0, and no energy transfer can be made between the two qubits. However, in the case where $\delta = 0$, the qubits can exchange their excitations. For example, after time $t = \frac{\pi}{4g}$ we can perform the so-called iSWAP gate defined by

$$iSWAP = \begin{pmatrix} 1 & 0 & 0 & 0 \\ 0 & 0 & i & 0 \\ 0 & i & 0 & 0 \\ 0 & 0 & 0 & 1 \end{pmatrix}$$

In practice, it is difficult and undesirable to fabricate qubits at the same resonance frequency. Indeed, if all qubits were at the same frequency, it would be impossible to control them separately. In Fig. 2.18d, we present a slightly different realization of a two-qubit gate dealing with this challenge. The frequency of one qubit is tuned to match the resonance frequency of the other by adding a SQUID which acts as a tunable-inductor (see Sect. 2.2.2.2). The inductance of a SQUID $L_J(\Phi)$ depends on the magnetic flux Φ threading its loop. Therefore, applying DC pulses on the SQUID enables one to tune δ such that $\delta << g$ turning the gate 'on' and 'off' on demand. On one hand, this realization surely shows advantages in controlled gate operations. On the other hand, adding the SQUID to the qubit introduces some decoherence.

2.3.3.3 Controlled-Z Gate

The transmon weak anharmonicity allows implementing a so-called controlled-Z gate. Contrary to the iSWAP gate, this gate is not based on tuning the qubit transition frequencies into resonance with each other but rather exploits the third energy level of the transmon (see Sect. 2.2.2.3).

The idea of this gate consists of tuning the qubits to a point where the $|1, 1\rangle$ and $|0, 2\rangle$ are degenerate in the absence of coupling [12]. In presence of the coupling, the two states can exchange energy and it is thus possible by letting them evolve freely during an appropriate delay time to transfer state $|1, 1\rangle$ to $-|1, 1\rangle$ and thus have

$$CZ = \begin{pmatrix} 1 & 0 & 0 & 0 \\ 0 & 1 & 0 & 0 \\ 0 & 0 & 1 & 0 \\ 0 & 0 & 0 & -1 \end{pmatrix} \tag{2.119}$$

2.3.3.4 Tunable Coupling Mediated by a Resonator or a Qubit

Two-qubit gates can also be mediated by an intermediate resonator [26, 27] or by a coupling qubit [12]. In Fig. 2.19, we present a possible implementation of such a scheme. Two transmon qubits are coupled by capacitors C_{c1} and C_{c2} to a common coupler whose frequency is controllable by a flux Φ threading a SQUID loop. In addition, the two transmons are coupled directly by capacitor C_{12}.

As we will see in the following, the advantage of this implementation is the ability to control directly the coupling between the two qubits without having to detune them out of their optimal working point. In addition, one can also cancel completely the direct coupling due to C_{12} by adjusting the frequency of the coupler and it is thus possible to operate properly each qubit independently with good fidelity.

In the following, we will derive the Hamiltonian of two qubits while considering the coupler as an intermediate resonator. Assuming Rotating Wave Approximation, we have

$$\mathcal{H} = \frac{\hbar\omega_1}{2}\sigma_1^z + \frac{\hbar\omega_2}{2}\sigma_2^z + \hbar\omega_r a^\dagger a + \hbar g_1 \left(\sigma_1^+ a + \sigma_1^- a^\dagger\right) + \hbar g_2 \left(\sigma_2^+ a + \sigma_2^- a^\dagger\right) \tag{2.120}$$

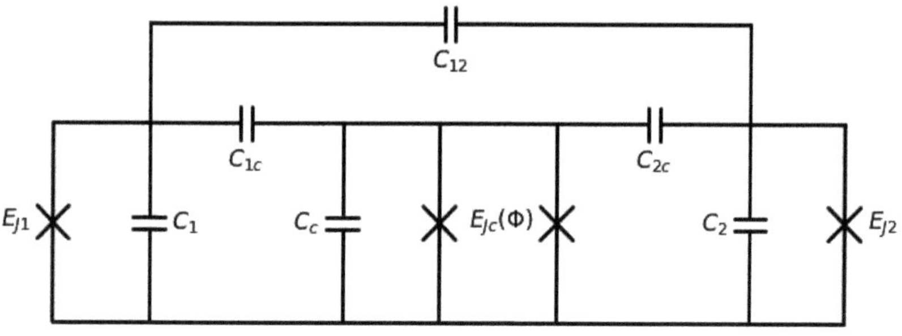

Fig. 2.19 Mediated Coupling between two transmon qubits. The two qubits are coupled to a coupler element by capacitors C_{c1} and C_{c2}. The frequency of the coupler is controllable by a flux threading the SQUID loop

In the case where the detunings are much larger than the coupling constants, it is possible to trace out the degree of freedom of the intermediate coupler by using a Schrieffer Wolff transformation (see Sect. 2.5.2.3)

$$S = \frac{g_1}{\omega_1 - \omega_r}\left(\sigma_1^+ a - \sigma_1^- a^+\right) + \frac{g_2}{\omega_2 - \omega_r}\left(\sigma_2^+ a - \sigma_2^- a^+\right)$$

and get an effective Hamiltonian

$$H_{\text{eff}} = \frac{\hbar\omega_1}{2}\sigma_1^z + \frac{\hbar\omega_2}{2}\sigma_2^z + \hbar\left(\omega_r + \chi_1\sigma_1^z + \chi_2\sigma_2^z\right)a^\dagger a + \hbar g_{\text{eff}}\left(\sigma_1^+\sigma_2^- + \sigma_1^-\sigma_2^+\right) \tag{2.121}$$

where $\chi_i = \frac{g_i^2}{\delta_i}$, and an effective interaction

$$\boxed{g_{\text{e}} = \frac{g_1 g_2\left(\delta_1 + \delta_2\right)}{2\delta_1\delta_2}} \tag{2.122}$$

By tuning properly the frequency of the coupler at $\omega_r = \frac{\omega_1+\omega_2}{2} + \frac{g_1 g_2}{2 g_{12}}\left[1 \pm \sqrt{1 + \left(\frac{g_{12}(\omega_1-\omega_2)}{g_1 g_2}\right)^2}\right]$, it is possible to cancel the small direct coupling term $V_{12} = \hbar g_{12}\left(\sigma_1^+\sigma_2^- + \sigma_1^-\sigma_2^+\right)$ between the qubits. This choice enables to operate single qubit gates with maximum fidelity. When the coupling between two qubits is needed, the coupler frequency is changed quickly by applying DC current on a flux line in the vicinity of the SQUID shown in Fig. 2.19.

2.3.3.5 Microwave Dynamic Coupling

A different approach is to apply a resonant microwave-drive on a qubit in order to dress this qubit in effective resonance with another. The advantage of this coupling scheme is that one can turn on and off the coupling by the application of a microwave tone. In this section, we will try to explain briefly this strategy in a simple case. We consider two qubits which are coupled directly by coupling constant g. One applies a time-dependent resonant Rabi drive on qubit 1. The driven Hamiltonian writes

$$\mathcal{H} = \hbar\frac{\omega_1}{2}\sigma_1^z + \frac{\hbar\omega_2}{2}\sigma_2^z + \hbar g\sigma_1^x\sigma_2^x + \hbar\Omega\sigma_1^x\cos\left(\omega_1 t\right) \tag{2.123}$$

Under unitary transformation $U_1 = \exp\left(\boldsymbol{i}\frac{\omega_1}{2}(\sigma_1^z + \sigma_2^z)t\right)$, the Hamiltonian becomes (see Appendix 2.5.2.1) after rotating wave approximation

$$\widetilde{\mathcal{H}}_1 = \hbar\frac{\delta}{2}\sigma_2^z + \hbar\frac{\Omega}{2}\sigma_1^x + \hbar g\left(\sigma_1^+\sigma_2^- + \sigma_1^-\sigma_2^+\right) \tag{2.124}$$

The eigenstates associated to eigenvalues $\pm\Omega/2$ of $\hbar\Omega/2\sigma_1^x$ are

$$|+\rangle = \frac{|0\rangle + |1\rangle}{\sqrt{2}}$$
$$|-\rangle = \frac{|0\rangle - |1\rangle}{\sqrt{2}}$$

The splitting of these two levels is at the origin of Rabi oscillations. The operators can be rewritten in the basis of $|\mp\rangle$ as

$$\sigma_1^+ = |1\rangle\langle 0| = (|+\rangle - |-\rangle)(\langle +| + \langle -|)/2$$
$$\sigma_1^- = |0\rangle\langle 1| = (|+\rangle + |-\rangle)(\langle +| - \langle -|)/2$$

Under this basis change, the above operators can be replaced by

$$\sigma_1^\pm \to \left(\sigma_1^z \mp i\sigma_1^y\right)/2$$
$$\sigma_1^x \to \sigma_1^z$$

In this basis $\widetilde{\mathcal{H}}$ can be written as

$$\begin{aligned} \widetilde{\mathcal{H}}_1 &= H_0 + V \\ H_0 &= \hbar\Omega\frac{\sigma_1^z}{2} + \hbar\delta\frac{\sigma_2^z}{2} \\ V &= \hbar g\left(\frac{\sigma_1^z - i\sigma_1^y}{2}\sigma_2^- + \frac{\sigma_1^z + i\sigma_1^y}{2}\sigma_2^+\right) \end{aligned} \tag{2.125}$$

The expression of operators $\sigma_1^-, \sigma_1^+, \sigma_2^-, \sigma_2^+$ under unitary transformation $U_2 = \exp\left(i(\frac{\Omega}{2}\sigma_z^1 + \frac{\delta}{2}\sigma_z^2)t\right)$ can be easily estimated using Baker Campbell Hausdorff formula (see Sect. 2.5.2.2)

$$\sigma_1^+ \to \sigma_1^+ e^{+i\Omega t}$$
$$\sigma_1^- \to \sigma_1^- e^{-i\Omega t}$$
$$\sigma_2^+ \to \sigma_2^+ e^{+i\delta t}$$
$$\sigma_2^- \to \sigma_2^- e^{-i\delta t}$$

Therefore under this transformation, the Hamiltonian becomes

$$\widetilde{\mathcal{H}}_2 = U_2 V U_2^+ = \hbar g \left(\frac{\sigma_1^z + \sigma_1^- e^{-i\Omega t} - \sigma_1^+ e^{+i\Omega t}}{2} \sigma_2^- e^{-i\delta t} + \frac{\sigma_1^z + \sigma_1^+ e^{+i\Omega t} - \sigma_1^- e^{-i\Omega t}}{2} \sigma_2^+ e^{+i\delta t} \right) \quad (2.126)$$

If $\Omega \approx \delta$, only two terms of this Hamiltonian will be time independent giving rise to an effective Hamiltonian

$$H_{\Omega=\delta} = -\frac{\hbar g}{2} \left(\sigma_1^+ \sigma_2^- + \sigma_1^- \sigma_2^+ \right)$$

One of the main advantage of the dynamical coupling techniques is that they can be directly generalized to large registers with minimal extra hardware and control lines. Along these lines, several protocols have been proposed and realized in recent years. For instance, one can apply a resonant Rabi frequency drive on both qubits at the same time as suggested in Ref. [28, 29] or one can even drive one qubit at the resonant frequency of the other qubit [30], inducing dynamics in the latter across the connecting resonator. These techniques (FLICFORQ, Cross Resonance,…) are frequently used in recent experiments.

2.4 Relaxation and Decoherence

One of the main limitations of quantum computers is related to the uncontrolled influence of the environment. An efficient quantum processor should have a scalable register of qubits that is easy to initialize, readout and manipulate but that is at the same time well-protected from variations of the parameters of its environment. These variations of parameters can cause uncontrolled changes in the qubits state, reducing our ability to perform well-defined operations. This process is called decoherence and is characterized by two distinct rates. The depolarization rate Γ_1 corresponds to an energy exchange with the environment. The pure dephasing rate Γ_φ is associated to low-frequency noise, which affects the Larmor frequency of the qubit without energy exchange. The two processes of relaxation and pure dephasing combine to the so-called decoherence rate[3]

$$\boxed{\Gamma_2 = \frac{1}{2}\Gamma_1 + \Gamma_\varphi}$$

[3] Note that the definition of Γ_2 as a sum of rates is only strictly valid when the noise spectra are Lorentzians centers around zero frequency and decay functions are exponential.

2.4.1 Relaxation

2.4.1.1 Fermi Golden Rule

The relaxation rate Γ_1 corresponds to energy exchange between the qubit and its environment. In principle, this rate is the sum of excitation and relaxation rates of the qubit with its environment

$$\Gamma_1 = \Gamma^{rel}_{1\to 0} + \Gamma^{exc}_{0\to 1} \tag{2.127}$$

Within the assumption that the qubit is weakly coupled to its environment, $\Gamma^{rel}_{1\to 0}$ and $\Gamma^{exc}_{0\to 1}$ can be estimated using Fermi Golden Rule

$$\Gamma^{rel}_{1\to 0} = \frac{2\pi}{\hbar^2}\sum_{n,m} \rho_{nn} \left|\langle 1, m \left| V_{int} \right| 0, n\rangle\right|^2 \delta\left(\omega_m - \omega_n - \omega_{01}\right) \tag{2.128}$$

$$\Gamma^{exc}_{0\to 1} = \frac{2\pi}{\hbar^2}\sum_{n,m} \rho_{nn} \left|\langle 0, m \left| V_{int} \right| 1, n\rangle\right|^2 \delta\left(\omega_m - \omega_n + \omega_{01}\right) \tag{2.129}$$

where ρ_{nn} are the diagonal elements of the density operator of the surrounding bath. There are no quantum correlations in the bath and thus all non-diagonal elements of the density operator are equal to zero.

2.4.1.2 Link Between the Fermi Golden Rule and the Power Spectrum of a Bath Operator

In the following we assume that the coupling of the qubit to the environment can be written as $V_{int} = \lambda\sigma_x F$ where F is an arbitrary operator acting on the environment degrees of freedom. The correlations of operator $F(t)$ are given by

$$\begin{aligned}
\langle F(t) F(0)\rangle &= \mathrm{Tr}\left(\rho F(t) F(0)\right) = \sum_n \rho_{nn} \langle n \left| F(t)F(0) \right| n\rangle \\
&= \sum_{n,m} \rho_{nn} \langle n \left| F(t) \right| m\rangle \langle m \left| F(0) \right| n\rangle \\
&= \sum_{n,m} \rho_{nn} \left\langle n \left| e^{iHt/\hbar} F(0) e^{-iHt/\hbar} \right| m\right\rangle \langle m \left| F(0) \right| n\rangle \\
&= \sum_{n,m} \rho_{nn} e^{i(\omega_n - \omega_m)t} \left|\langle m \left| F \right| n\rangle\right|^2
\end{aligned}$$

According to Wiener-Khintchine theorem, the power spectrum $S_F(\omega)$ is related to the correlations by

$$\begin{aligned}
S_F(\omega) &= \frac{1}{2\pi}\int_{t\in\mathbb{R}} \langle F(t) F(0)\rangle e^{i\omega t} \\
&= \sum_{n,m} \rho_{nn} \left|\langle m \left| F \right| n\rangle\right|^2 \delta\left(\omega_m - \omega_n - \omega\right)
\end{aligned}$$

It is thus possible to express the relaxations rates as a function of the power spectrum of operator F

$$\boxed{\Gamma_{1\to 0}^{rel} = \frac{2\pi}{\hbar^2} |\lambda|^2 S_F(\omega_{01})} \tag{2.130}$$

$$\boxed{\Gamma_{0\to 1}^{exc} = \frac{2\pi}{\hbar^2} |\lambda|^2 S_F(-\omega_{01})} \tag{2.131}$$

In superconducting qubits, the temperature of the bath is usually much smaller than the qubit frequency and thus $\Gamma_{0\to 1}^{exc}$ is exponentially suppressed by a Boltzmann factor $\exp[-\hbar\omega_{01}/k_B T]$.

2.4.1.3 Wigner Weisskopf Theory of Relaxation

In the following, we will consider a qubit coupled to a bath of harmonic oscillators within the Wigner-Weisskopf theory of relaxation. The system is described by the following Hamiltonian

$$\mathcal{H} = \hbar\frac{\omega_{01}}{2}\sigma_z + \sum \hbar\omega_k a_k^+ a_k + \hbar\sigma_x \sum g_k(a_k^+ + a_k) \tag{2.132}$$

We write the Heisenberg equations for a_k and σ_z operators as

$$\dot{a}_k = \frac{i}{\hbar}\left[\mathcal{H}, a_k\right] = -i\omega_k a_k - i g_k \sigma_x$$

$$\dot{a}_k^+ = \frac{i}{\hbar}\left[\mathcal{H}, a_k^+\right] = i\omega_k a_k^+ + i g_k \sigma_x$$

$$\dot{\sigma}_z = \frac{i}{\hbar}\left[\mathcal{H}, \sigma_z\right] = 2\sigma_y \sum g_k(a_k^+ + a_k)$$

We assume that at time $t = 0$, the spin is in its excited state and thus $\langle\sigma_z(t=0)\rangle = 1$ and the bath is empty $\forall k$ $\langle a_k(t=0)\rangle = 0$. At time $t = +\infty$, the spin is de-excited and $\forall k$ $\langle a_k^+(t=+\infty)\rangle = 0$. Using these boundary conditions, we can easily integrate the first two equations

$$a_k(t) = -i g_k \int_0^t e^{-i\omega_k(t-t')}\sigma_x(t')\,dt'$$

$$a_k^+(t) = -i g_k \int_t^{+\infty} e^{i\omega_k(t-t')}\sigma_x(t')\,dt'$$

By replacing these expression we get

$$\dot{\sigma}_z(t) = -2i \sum g_k^2 \left[\sigma_y(t) \left(\int_t^{+\infty} e^{i\omega_k(t-t')} \sigma_x(t')\, dt' \right) \right.$$
$$\left. + \sigma_y(t) \left(\int_0^t e^{-i\omega_k(t-t')} \sigma_x(t')\, dt' \right) \right] \tag{2.133}$$

We can simplify this expression assuming $g_k = g\ \forall k$ (Markovian approximation) and using

$$\Delta\omega \sum e^{\pm i\omega_k t} \simeq \int_0^{+\infty} e^{\pm i\omega t}\, d\omega = [\pi\delta(t) \pm i P \frac{1}{t}]$$

where P denotes the Cauchy principal value and $\Delta\omega$ the constant spacing between different values of ω_k. Moreover,

$$\int_0^t \sigma_x(t')\delta(t-t')\, dt' = \int_t^{+\infty} \sigma_x(t')\delta(t-t')\, dt' = \sigma_x(t)/2 \tag{2.134}$$

Thus, we get

$$\dot{\sigma}_z(t) = -2\pi g^2 \frac{1}{\Delta\omega} \sigma_z(t) = -\Gamma_1 \sigma_z(t) \tag{2.135}$$

This last expression enables us to extract an expression of the relaxation rate Γ_1 of the qubit

$$\boxed{\Gamma_1 = \frac{2\pi g^2}{\Delta\omega}}$$

2.4.1.4 Relaxation of a Qubit Coupled to a Single Mode Resonator: Purcell Effect

In the following, we will consider a qubit embedded into a single mode resonator. For simplicity, we consider a system described by the so-called Jaynes-Cummings Hamiltonian

$$\mathcal{H} = \hbar \frac{\omega_{01}}{2} \sigma_z + \hbar\omega_r a^+ a + \hbar g(\sigma^+ a + \sigma^- a^+) \tag{2.136}$$

We assume the resonator is coupled to the external environment with a Lindblad jump operator $L = \sqrt{\kappa a}$. The presence of a dissipation channel for the resonator opens an effective dissipation for the qubit due to the presence of the coupling g. This effect is called Purcell effect and should be carefully taken into account when designing a circuit-QED experiment.

When the qubit and the resonator are sufficiently detuned $g \ll |\omega_{01} - \omega_r|$, the Hamiltonian can be replaced by an effective Hamiltonian using a Schrieffer-Wolff transformation $U = e^S$ where $S = \frac{g}{\omega_{01}-\omega_r}\left(\sigma^+ a - \sigma^- a^+\right)$ (see Chap. 2.5.2.4). This transformation enables us to estimate the Purcell effect. Indeed, the equations of motion of the density operator in this new frame $\widetilde{\rho}$ should include Lindblad jump operators expressed in the same frame

$$\widetilde{L} = e^S L e^{-S} \tag{2.137}$$

Using Baker Campbell Hausdorff formula (see Sect. 2.5.2.2), we get

$$\widetilde{L} \sim L + [S, L] = \sqrt{\kappa} a + \frac{g}{\omega_{01} - \omega_r} \sqrt{\kappa} \sigma^- \tag{2.138}$$

We can now write the equation of motion of operator σ_z using Sect. 2.5.1.6

$$\dot{\sigma}_z(t) = -\frac{g^2 \kappa}{(\omega_{01} - \omega_r)^2}(\sigma_z + \mathbb{1}) \tag{2.139}$$

and thus we can extract the Purcell relaxation rate Γ_1^P

$$\boxed{\Gamma_1^P = \frac{g^2 \kappa}{(\omega_{01} - \omega_r)^2}} \tag{2.140}$$

2.4.1.5 Purcell Rate in a Transmission Line Resonator

In the following, we will consider as an illustration the Hamiltonian of a flux qubit coupled inductively to a transmission line resonator

$$H = \hbar \omega_{01} \frac{\sigma_z}{2} + M I_p \hat{I}(x) \sigma_x \tag{2.141}$$

We denote as Γ_1^P the Purcell relaxation rate of the qubit due to its coupling with the resonator. Using Eqs. (2.130) and 2.131, we have

$$\Gamma_1^P = \frac{2\pi}{\hbar^2} M^2 I_p^2 \left[S_I(\omega_{01}) + S_I(-\omega_{01})\right]$$

We recall that according to Eq. (2.50), $I(x, \omega)$ can be expressed as a function of the propagation wave amplitudes $A_L^{\rightarrow}(\omega)$ and $A_R^{\leftarrow}(\omega)$ in the incoming lines as

$$I(x, \omega) = \frac{1}{\sqrt{Z_0}}\left(f_{\rightarrow}(x, \omega) A_L^{\rightarrow}(\omega) + f_{\leftarrow}(x, \omega) A_R^{\leftarrow}(\omega)\right) \tag{2.142}$$

Due to the independence of signals $A_L^{\rightarrow}(\omega)$ and $A_R^{\leftarrow}(\omega)$, we have

$$S_I(\omega) = \frac{1}{Z_0}\left(|f_{\rightarrow}(\omega, x)|^2 S_{A_L^{\rightarrow}}(\omega) + |f_{\leftarrow}(\omega, x)|^2 S_{A_R^{\leftarrow}}(\omega)\right) \tag{2.143}$$

Using $A_k^{\rightarrow} = \sqrt{\frac{c}{2\Lambda}\hbar\omega_k} a_k^{\rightarrow}$ (see Sect. 2.1.4.4) we calculate

$$\begin{aligned} S_{A_L^{\rightarrow}}(\omega) &\equiv \frac{1}{2\pi}\int_{t\in\mathbb{R}} \left\langle A_L^{\rightarrow}(t)\, A_L^{\rightarrow}(0)\right\rangle e^{i\omega t} \\ &= \frac{1}{2\pi}\sum_{n\geq 0}\left(\frac{c}{2\Lambda}\hbar\omega_n\right)\int_{t\in\mathbb{R}}\left(\left\langle \left(a_{L,n}^{\rightarrow}\right)^{\dagger} a_{L,n}^{\rightarrow}\right\rangle e^{+i\omega_n t}\right. \\ &\quad \left. + \left\langle a_{L,n}^{\rightarrow}\left(a_{L,n}^{\rightarrow}\right)^{\dagger}\right\rangle e^{-i\omega_n t}\right) e^{i\omega t} \\ &= \sum_{n\geq 0}\left(\frac{c}{2\Lambda}\hbar\omega_n\right)\left(\left\langle \left(a_{L,n}^{\rightarrow}\right)^{\dagger} a_{L,n}^{\rightarrow}\right\rangle \delta\left(\omega_n+\omega\right) + \left\langle a_{L,n}^{\rightarrow}\left(a_{L,n}^{\rightarrow}\right)^{\dagger}\right\rangle \delta\left(\omega_n-\omega\right)\right) \end{aligned}$$

By replacing the discrete sum $\sum_{n\geq 0}$ by its equivalent integral $\int_{d\omega\in\mathbb{R}_+}\eta(\omega)$, we get

$$\begin{aligned} S_{A_L^{\rightarrow}}(\omega) &= \int_{d\omega'\in\mathbb{R}_+} \frac{\hbar\omega'}{4\pi}\left(\left\langle a^{\dagger}a\right\rangle \delta\left(\omega'+\omega\right) + \left\langle aa^{\dagger}\right\rangle \delta\left(-\omega'+\omega\right)\right) \\ &= \begin{cases} \frac{\hbar\omega}{4\pi}\left\langle aa^{\dagger}\right\rangle & \omega > 0 \\ \frac{\hbar\omega}{4\pi}\left\langle a^{\dagger}a\right\rangle & \omega < 0 \end{cases} \end{aligned} \tag{2.144}$$

By considering that the photon bath is thermalized at a given temperature T,

$$S_{A_L^{\rightarrow}}(\omega) = \begin{cases} \frac{\hbar\omega}{4\pi}\frac{1}{1-e^{-\beta\hbar\omega}} & \omega > 0 \\ \frac{\hbar\omega}{4\pi}\frac{e^{-\beta\hbar\omega}}{1-e^{-\beta\hbar\omega}} & \omega < 0 \end{cases} \tag{2.145}$$

where $\beta = \frac{1}{k_B T}$. Finally, the Purcell decay rate writes

$$\Gamma_1^P = \left(\frac{MI_p}{\hbar}\right)^2 \frac{\hbar\omega_{01}\coth\left(\frac{\hbar\omega_{01}}{2k_B T}\right)\left(\left|f_{\rightarrow}(\omega_{01},x)\right|^2 + \left|f_{\leftarrow}(\omega_{01},x)\right|^2\right)}{2Z_0} \tag{2.146}$$

which we separate into two parts

$$\Gamma_1^P = \Gamma_1^{\rightarrow} + \Gamma_1^{\leftarrow} \tag{2.147}$$

where

$$\Gamma_1^{\rightarrow/\leftarrow} = \left(\frac{MI_p}{\hbar}\right)^2 \frac{\hbar\omega_{01}}{2Z_0}\left|f_{\rightarrow/\leftarrow}(\omega_{01},x)\right|^2 \coth\left(\frac{\beta\hbar\omega_{01}}{2}\right) \tag{2.148}$$

2.4.1.6 Calculating the Purcell Rate from Rabi Frequency

Considering Eqs. 2.148, (2.113), and (2.114), we establish a relationship valid at zero temperature

$$\boxed{\Gamma_1^{\rightarrow/\leftarrow}(0\,K) = \frac{\hbar\omega_{01}\left(\Omega^{\rightarrow/\leftarrow}\right)^2}{4\langle P\rangle}} \tag{2.149}$$

We thus get

$$\Gamma_1^P(0\,K) = \frac{\hbar\omega_{01}\left[(\Omega^{\rightarrow})^2 + (\Omega^{\leftarrow})^2\right]}{4\langle P\rangle} \tag{2.150}$$

The advantage of this formula is that it gives directly the Purcell rate via the measurement of the Rabi frequency for a given average power $\langle P\rangle$ with a precision limited by the uncertainty on $\langle P\rangle$, which is typically ± 1 dB.

2.4.1.7 Dielectric Losses

The typical relaxation times of superconducting qubits have been largely improved by the introduction of three dimensional cavity, which reduce the impact of dielectric losses in the circuit [9, 17, 25]. Various works have tried to reduce these losses in order to increase the fidelity of the qubit [10, 11, 18] while keeping a 2D scalable architecture. This can be done by reducing the interface defects between the metal and the substrate [11].

Dielectric losses take place in the capacitors and can be modeled by a small resistor of resistance R in series with each capacitor. The value of R is determined by the loss tangent of the dielectric material separating each island and is given by

$$RC\omega_{01} = \tan\delta \tag{2.151}$$

As a result, the noise voltage generated by the (lossy) capacitor is given by

$$S_{\delta V}(\omega_{01}) + S_{\delta V}(-\omega_{01}) = \frac{R\hbar\omega_{01}}{\pi}\coth\left(\frac{\beta\hbar\omega_{01}}{2}\right) \approx \frac{\hbar\omega_{01}R}{\pi} \tag{2.152}$$

To calculate the relaxation rate of a qubit, one must determine the transverse term $\sigma_{x/y}$ in the Hamiltonian introduced by a small perturbation δV. For instance, one can calculate the relaxation of a transmon due to dielectric losses in the substrate. The variation of the charge across the capacitor C due to δV is given by $\delta Q = -C\delta V$. At first order in δV, this modifies the kinetic term of the Hamiltonian by $dK = \delta Q\,\hat{V}$. Using Eq. (2.130), we get

$$\Gamma_1^{\mathrm{diel}} = \frac{2\pi}{\hbar^2}C^2\left|\left\langle 1\left|\hat{V}\right|0\right\rangle\right|^2 S_{\delta V}(\omega_{01}) \tag{2.153}$$

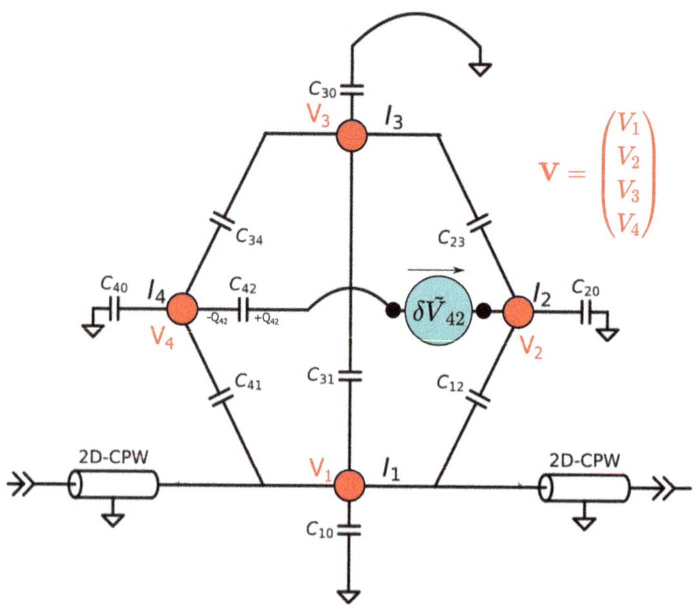

Fig. 2.20 Dielectric losses of a flux qubit. An external bias δV_{ij}(cyan) is applied. Here we show the example of $(i, j) = (4, 2)$, where we write $Q_{42} = C_{42}\left(V_4 - \left(V_2 + \tilde{V}_{42}\right)\right)$

which gives

$$\boxed{\Gamma_1^{\text{diel}} = \frac{16}{\hbar} E_C \left|\langle 1 \left|\hat{n}\right| 0\rangle\right|^2 \tan\delta} \tag{2.154}$$

where $E_C = \frac{e^2}{2C}$, $\hat{n} = \frac{C\hat{V}}{2e}$. For a typical transmon on sapphire substrate $\left(\tan\delta \sim 10^{-6}\right)$ with $E_J/\hbar = 20$ GHz and $E_C/\hbar = 200$ MHz, the value of the transition matrix element is $\left|\langle 1 \left|\hat{n}\right| 0\rangle\right|^2 \sim 1.7$ and thus $\Gamma_1^{\text{diel}} \sim 5$ kHz.

In the following, we will calculate the dielectric losses for a flux qubit with four junctions. As we can see in Fig. 2.20, the variation of charge across the capacitor C_{42} due to δV_{42} is given by

$$\delta Q_{42} = -C_{42}\delta V_{42} \tag{2.155}$$

At first order in δV_{ij}, this modifies the kinetic term by

$$\begin{aligned} dK_{42} &= \delta Q_{42} (V_2 - V_4) \\ &= -C_{42}\delta V_{42}\mathbf{V}^T \begin{pmatrix} 0 \\ 1 \\ 0 \\ -1 \end{pmatrix} \end{aligned}$$

Let us recall that $\mathbf{P}$ from (2.89) is the 4×3 transfer matrix from the junction coordinates to the island phases such that $\mathbf{V} = \mathbf{P}\dot{\mathbf{\Phi}} = \mathbf{P}\left(\mathbf{P}^T\mathbf{C}\mathbf{P}\right)^{-1}\boldsymbol{Q}$ (up to a constant). Injecting and generalizing to all other indices ij, we can write the total perturbation of the Hamiltonian to all perturbations δV_{ij}.

$$
\begin{aligned}
dH &= \sum_{i \neq j} \left(-dK_{ij}\right) \\
&= \sum_{i \neq j} C_{ij}\delta V_{ij}\boldsymbol{Q}^T\left(\mathbf{P}^T\mathbf{C}\mathbf{P}\right)^{-1}\mathbf{P}^T\left(\mathbf{e}_j - \mathbf{e}_i\right)
\end{aligned}
$$

where $\mathbf{e}_i$ is the sparse column vector with 1 only at position i.

Let us define the 3×3 real symmetrical matrix comprised of the quantum overlaps of the charge operators

$$
\mathbf{Q^2} = \left(\langle 1|Q_i|0\rangle\langle 0|Q_j|1\rangle, \forall i, j \in \{1, 2, 3\}\right) \tag{2.156}
$$

Seeing that dH is linear in the $\boldsymbol{Q}$ operators we can write dH under the form

$$
\begin{aligned}
dH &= \delta V_{ij} \cdot \left(\boldsymbol{Q}^T\boldsymbol{L_{ij}}\right) \\
\boldsymbol{L_{ij}} &= C_{ij}\left(\mathbf{P}^T\mathbf{C}\mathbf{P}\right)^{-1}\mathbf{P}^T\left(\mathbf{e}_i - \mathbf{e}_j\right)
\end{aligned}
$$

Subsequently, the loss rate due to δV_{ij} writes

$$
\begin{aligned}
\Gamma_{ij} &= 2\pi\frac{\left[S_{\delta V_{ij}}\left(\omega_{01}\right) + S_{\delta V_{ij}}\left(-\omega_{01}\right)\right]\boldsymbol{L_{ij}}^T\mathbf{Q^2}\boldsymbol{L_{ij}}}{\hbar^2} \\
&= \left(2\omega_{01}R/\hbar\right)\mathrm{Tr}\left(\boldsymbol{L_{ij}}\boldsymbol{L_{ij}}^T\mathbf{Q^2}\right) \\
&= \left(2\tan\delta_{ij}/\hbar C_{ij}\right)\mathrm{Tr}\left(\boldsymbol{L_{ij}}\boldsymbol{L_{ij}}^T\mathbf{Q^2}\right)
\end{aligned}
$$

where we replace the expression of R in the last equality according to (2.151). We get

$$
\Gamma_{ij} = \frac{2}{\hbar}\mathrm{Tr}\left(\left(\mathbf{P}^T\mathbf{C}\mathbf{P}\right)^{-1}\left[\mathbf{P}^T\left(\mathbf{e}_i - \mathbf{e}_j\right)C_{ij}\tan\delta_{ij}\left(\mathbf{e}_i - \mathbf{e}_j\right)^T\mathbf{P}\right]\left(\mathbf{P}^T\mathbf{C}\mathbf{P}\right)^{-1}\mathbf{Q^2}\right) \tag{2.157}
$$

After summation over all island pairs i, j in $[\cdot]$, we get

$$
\boxed{\Gamma_1^{\mathrm{diel}} = \frac{2}{\hbar}\mathrm{Tr}\left(\left(\mathbf{P}^T\mathbf{C}\mathbf{P}\right)^{-1}\left(\mathbf{P}^T\mathbf{P}\right)\left(\mathbf{P}^T\mathbf{C}\mathbf{P}\right)^{-1}\mathbf{Q^2}\right)} \tag{2.158}
$$

where $= \sum_{ij} \left(\mathbf{e}_i - \mathbf{e}_j\right) C_{ij} \tan\delta_{ij} \left(\mathbf{e}_i - \mathbf{e}_j\right)^T$ is the capacitance matrix *weighted* by the $\tan\delta$ of individual capacitive elements (we recover, for the *unweighted* expression, the usual capacitance matrix $\mathbf{C} = \sum_{ij} \left(\mathbf{e}_i - \mathbf{e}_j\right) C_{ij} \left(\mathbf{e}_i - \mathbf{e}_j\right)^T$). In the case where $\tan\delta$ is homogeneous over all the different capacitances , we get

$$\boxed{\Gamma_1^{\mathrm{diel}} = \frac{2}{\hbar} \mathrm{Tr}\left(\left(\mathbf{P}^T \mathbf{C} \mathbf{P}\right)^{-1} \mathbf{Q}^2\right) \tan\delta} \tag{2.159}$$

2.4.2 Pure Dephasing

2.4.2.1 General Framework for the Pure Dephasing of a Qubit

In an ideal system, the decoherence rate Γ_2 is limited by the energy relaxation rate of the qubit and is given by $\Gamma_2 = \Gamma_1/2$. In practice, the decoherence rate of a qubit may be much larger than this theoretical limit due to pure dephasing. The pure dephasing rate Γ_φ is associated to low-frequency noise, which affects the Larmor frequency of the qubit without energy exchange.

In order to estimate this effect more precisely, we write the Heisenberg equations of a qubit in free precession in the frame rotating at the average Larmor frequency of the qubit $\langle\omega_{01}\rangle$

$$\frac{d\sigma^+}{dt} = i\delta(t)\sigma^+$$

$$\frac{d\sigma^-}{dt} = -i\delta(t)\sigma^-$$

$$\frac{d\sigma_z}{dt} = 0$$

where $\delta(t) = \omega_{01}(t) - \langle\omega_{01}\rangle$. These differential equations are decoupled and can be solved straightforwardly

$$\sigma^+(t) = e^{i \int_0^t \delta(t)\,dt} \sigma^+(0)$$

$$\sigma^-(t) = e^{-i \int_0^t \delta(t)\,dt} \sigma^-(0)$$

The phase $\varphi(t) = \int_0^t \delta(t)\,dt$ depends on small fluctuations $\lambda(t)$ which slightly modify the qubit Hamiltonian. At first order, $\delta(t)$ is given by $\delta(t) = \frac{\partial\omega_{01}}{\partial\lambda}\lambda(t)$. The pure dephasing rate of the system corresponds to the decay of the expectation value $\left\langle\sigma^\pm(t)\right\rangle$ and is given by

$$\left\langle\sigma^\pm(t)\right\rangle = \left\langle e^{\pm i\varphi(t)}\right\rangle \sigma^+(0) \tag{2.160}$$

If the fluctuations $\delta\lambda(t)$ are small enough, they can be considered as a random variable with Gaussian distribution [31]. Thus, using Isserlis theorem, we get

$$f_R(t) = \left\langle e^{\pm i\varphi(t)} \right\rangle \simeq e^{-1/2\langle\varphi^2(t)\rangle} \tag{2.161}$$

The expectation value of $\left\langle\sigma^{\pm}(t)\right\rangle$ will therefore decay according to

$$f_R(t) = e^{-1/2\left(\frac{\partial\omega_{01}}{\partial\lambda}\right)^2\left\langle\left(\int_0^t \lambda(t')dt'\right)^2\right\rangle} \tag{2.162}$$

We can write

$$\left\langle\left(\int_0^t \lambda(t')dt'\right)^2\right\rangle = \int_{t\in\mathbb{R}}\int_{u\in\mathbb{R}} dt'\,du\,\left\langle\lambda(t'-u)\lambda(0)\right\rangle H(t')H(u') \tag{2.163}$$

Where $H(u)$ is the boxcar function such that $H(u) = 1$ if $0 \le u \le t$ and zero elsewhere. Using Wiener Khinchine theorem to express the correlations as a function of the power spectrum $S_\lambda(\omega)$

$$\langle\lambda(t)\,\lambda(0)\rangle = \int d\omega e^{-i\omega t}\, S_\lambda(\omega) \tag{2.164}$$

Noting that

$$\int_{u\in\mathbb{R}} du\; H(u)e^{\pm i\omega u} = t\,\mathrm{sinc}\left(\frac{\omega t}{2}\right) \tag{2.165}$$

we get

$$\boxed{f_R(t) = \exp\left(-\frac{t^2}{2}\left(\frac{\partial\omega_{01}}{\partial\lambda}\right)^2 \int_{-\infty}^{\infty} d\omega\, S_\lambda(\omega)\,\mathrm{sinc}^2\left(\frac{\omega t}{2}\right)\right)} \tag{2.166}$$

2.4.2.2 Dynamical Decoupling

Experimentally, the pure dephasing rate of a qubit can be estimated by the so-called *Ramsey sequence*, where two identical $\pi/2$ pulses are played consecutively after a time delay t. It is possible to dynamically decouple the noise responsible for this dephasing by playing a more complex set of pulses. The most popular technique to achieve this is called *Hahn Echo technique* and consists of playing a π-pulse in between the two $\pi/2$ pulses. This π pulse inverses the time evolution and therefore cancels the contribution to dephasing of low frequency noise.

In a Hahn echo sequence, the first $\pi/2$-pulse puts the state of the qubit in a coherent superposition state. During the time t_1, the qubit performs a free evolution and accumulates phase $\varphi_1(t_1) = \int_0^{t_1} \delta_1(t')dt'$. The π—pulse flips the time evolution of the qubit such that during the time t_2 it acquires an opposite phase $\varphi_2(t_2) = -\int_{t_1}^{t_1+t_2} \delta_2(t')dt'$. Assuming that the environment is static during the free evolution, the phase accumulated is completely canceled if $t_1 = t_2 = t/2$.

In practice the noise is not static and the decoherence rate of the qubit—corresponding to the decay $f_E(t) = \langle\sigma^{\pm}(t)\rangle$—is given by

$$f_E(t) = \left\langle e^{\pm i(\delta_1 - \delta_2)} \right\rangle \approx \exp\left(-1/2\left\langle \delta_1^2 + \delta_2^2 - \delta_1\delta_2 - \delta_2\delta_1 \right\rangle\right) \tag{2.167}$$

After a calculation similar to the one performed in previous section [31], one can show that the expectation value of $\langle\sigma^{\pm}(t)\rangle$ will decay according to

$$\boxed{f_E(t) = \exp\left(-\frac{t^2}{2}\left(\frac{\partial\omega_{01}}{\partial\lambda}\right)^2 \int_{-\infty}^{\infty} d\omega\, S_\lambda(\omega) \sin^2(\frac{\omega t}{4}) \operatorname{sinc}^2(\frac{\omega t}{4})\right)} \tag{2.168}$$

White Spectrum

If the power spectrum is white $(S_\lambda(\omega) = S_\lambda(\omega = 0))$, it is possible to calculate analytically the pure dephasing rate. Indeed,

$$\int_{-\infty}^{\infty} d\omega \operatorname{sinc}^2(\frac{\omega t}{2}) = \int_{-\infty}^{\infty} d\omega \sin^2(\frac{\omega t}{4}) \operatorname{sinc}^2(\frac{\omega t}{4}) = \frac{2\pi}{t} \tag{2.169}$$

The dynamical decoupling is therefore not effective for white noise. The Ramsey and Echo sequence give the same *exponential* decay, namely

$$\Gamma_{\varphi R} = \Gamma_{\varphi E} = \pi \left(\frac{\partial\omega_{01}}{\partial\lambda}\right)^2 S_\lambda(\omega = 0) \tag{2.170}$$

This result can be generalized if the power spectrum is regular at $\omega = 0$ on a frequency scale $|\omega| \leq 1/t$. Indeed,

$$\int_{-\infty}^{\infty} d\omega\, S_\lambda(\omega) \operatorname{sinc}^2(\frac{\omega t}{2}) \simeq S_\lambda(0) \int_{-\infty}^{\infty} d\omega \operatorname{sinc}^2(\frac{\omega t}{2}) \tag{2.171}$$

1/f Spectrum

Here we assume that the power spectrum follows a 1/f law in a wide range of frequencies limited by an infrared cutoff ω_{IR} and an ultraviolet cutoff ω_c

$$S_\lambda(\omega) = \frac{A}{|\omega|}, \quad \omega_{IR} \leq |\omega| \leq \omega_c \tag{2.172}$$

The infrared cutoff is usually determined by the total length of the measurement protocol (typically 1Hz) and the ultraviolet cutoff is typically in the range of a few MHz. Using

$$\int_{\omega_{IR}}^{\infty} \frac{d\omega}{|\omega|} \operatorname{sinc}^2(\frac{\omega t}{2}) \simeq 2\ln\left(\frac{1}{\omega_{IR} t}\right) \tag{2.173}$$

$$\int_{-\infty}^{\infty} \frac{d\omega}{|\omega|} \sin^2(\frac{\omega t}{4}) \operatorname{sinc}^2(\frac{\omega t}{4}) = 2\ln(2) \tag{2.174}$$

we get a *Gaussian* decay for both Ramsey and Echo pure dephasing with

$$\Gamma_{\varphi R} = \sqrt{A}\left(\frac{\partial \omega_{01}}{\partial \lambda}\right)\sqrt{\ln\left(\frac{1}{\omega_{IR} t}\right)} \tag{2.175}$$

$$\Gamma_{\varphi E} = \sqrt{A}\left(\frac{\partial \omega_{01}}{\partial \lambda}\right)\sqrt{\ln(2)} \tag{2.176}$$

In particular, $\sqrt{\ln\left(\frac{1}{\omega_{IR} t}\right)} \sim 3.7$ and $\Gamma_{\varphi R}/\Gamma_{\varphi E} \sim 4.5$ for typical cutoff frequencies. This limited echo efficiency is due to the high frequency tail of 1/f noise.

2.4.2.3 Charge Noise in a Transmon Qubit

An important source of pure dephasing for qubits is noise on the charge of each island of the qubit. This noise is due to microscopic charged fluctuators that can be either trapped electrons or ions in defects of the material that move between two or several metastable positions. These charge fluctuators are partly located in the substrate and partly on the oxide layers covering the electrodes of the device.

The charge noise power spectrum follows approximately a 1/f behavior up to frequencies of approximately 1 MHz

$$S_Q(\omega) = \frac{A_Q}{|\omega|} \tag{2.177}$$

The typical amplitude A_Q depends on the parameters of the experiment (temperature, size of the island, screening by other electrodes). A typical value is $\sqrt{A_Q} \sim 10^{-3} e$ where $e = 1.6 \times 10^{-19}$ C is the electron charge.

Several qubit design have been used to reduce the influence of this noise. In particular, a transmon with a large ratio E_J/E_C will have a small charge dispersion (see Fig. 2.11b), which will reduce the dephasing due to charge noise.

Taking a charge qubit with $\omega_{01}/2\pi = 5$ GHz and $E_J/E_C = 10$, we see in Fig. 2.11b that $\frac{\partial\omega_{01}}{\partial n_g} \sim 10^7$ rad/s. Thus, using Eq. (2.176), we get $\Gamma_{\varphi E}^{\text{charge}} \sim 10$ kHz. If we increase further E_J/E_C to the transmon regime, it is possible to cancel almost completely the charge dispersion, reducing therefore the contribution of charge noise to decoherence.

2.4.2.4 Dephasing of a Flux Qubit Away from its Optimal Point

Flux noise is also an important source of dephasing for qubits which posses a superconducting loop, such as flux qubits. The origin of this noise is most likely due to microscopic spins in the vicinity of the qubit which generate a magnetic noise threading the loop of the qubit. The flux noise power spectrum follows approximately a 1/f behavior up to high frequencies [16]

$$S_\Phi(\omega) = \frac{A_\Phi}{|\omega|} \tag{2.178}$$

The typical value of the flux noise amplitude is $\sqrt{A_\Phi} \sim 10^{-6}\Phi_0$ where $\Phi_0 = 2 \times 10^{-15}$ Wb is the magnetic flux quantum.

To illustrate our discussion, we will consider in the following a flux qubit away from its optimal point. As already mentioned in Sect. 2.2.3.5, the high magnetic moment of the flux qubit make its frequency very sensitive to flux

$$\frac{\partial\omega_{01}}{\partial\Phi} = \frac{\partial\varepsilon}{\partial\Phi}.\frac{\partial\omega_{01}}{\partial\varepsilon} = \left(\frac{2I_p}{\hbar}\right)^2 \frac{(\Phi - \Phi_0/2)}{\omega_{01}} \tag{2.179}$$

Taking a flux qubit with $\omega_{01}/2\pi = 5$ GHz, $I_p = 300$ nA, $\Phi - \Phi_0/2 = 50\,\mu\Phi_0$, and using Eq. (2.176), we get $\Gamma_{\varphi E}^{\text{flux}} \sim 170$ kHz.

2.4.2.5 Calculating the Dephasing Rate Due to Photon Noise

Stochastic fluctuations in the number of photons in a resonator coupled to a qubit create random dispersive shifts which translates into dephasing. Using Eq. (2.104) , the frequency of the qubit is given by

$$\omega_{01}(t) = \langle\omega_{01}\rangle + 2\chi\delta n(t) \tag{2.180}$$

where $\chi = g^2/(\omega_{01} - \omega_r)$ and $\delta n(t) = n(t) - \bar{n}$.

Using Eq. (2.162), one can thus determine the decoherence rate due to photon noise

$$f_R(t) = e^{-1/2(2\chi)^2\left\langle\left(\int_0^t \delta n(t')dt'\right)^2\right\rangle} \tag{2.181}$$

In the following, we compute the correlations $\langle\delta n(\tau)\delta n(0)\rangle$ using quantum regression theorem (see Sect. 2.5.1.8) for thermal and coherent states. More complex photon states in the resonator such as squeezed states can be handled with the same formalism [32].

Thermal Photons

In order to estimate the photon correlations in thermal state of a resonator, we write a set of equations of motion for $\boldsymbol{O} = \left(n(t),\ \mathbb{1}\right)^T$ in presence of Lindblad operators $L_1 = \sqrt{\kappa\,(\bar{n}+1)}a$ and $L_2 = \sqrt{\kappa\bar{n}}a^+$. We obtain coupled differential equations that can be written in the form of

$$\frac{d}{dt}\vec{O}\,(t) = G\vec{O}\,(t) \tag{2.182}$$

where

$$G = \begin{pmatrix} -\kappa & \kappa\bar{n} \\ 0 & 0 \end{pmatrix} \tag{2.183}$$

The steady-state expectation values are obtained by the null eigenstate of G, defined by $G\langle\boldsymbol{O}\rangle = 0$, leading to $\langle n(t)\rangle_\infty = \bar{n}$. Similarly, we calculate the steady-state expectation value $\langle n^2(t)\rangle_\infty$ using the kernel of the equation of motion for $\hat{n}^2$ and find

$$\langle n^2(t)\rangle_\infty = \bar{n}(2\bar{n}+1) \tag{2.184}$$

To obtain the two-time correlation $\langle n(t+\tau)n(t)\rangle_\infty$, we use the quantum regression theorem (see Sect. 2.5.1.8), namely

$$\frac{d}{d\tau}\begin{pmatrix} n(t+\tau)n(t) \\ \mathbb{1}\,n(t) \end{pmatrix} = \begin{pmatrix} -\kappa & \kappa\bar{n} \\ 0 & 0 \end{pmatrix}\begin{pmatrix} n(t+\tau)n(t) \\ \mathbb{1}\,n(t) \end{pmatrix} \tag{2.185}$$

Using the initial conditions $\langle n(t+0)n(t)\rangle_\infty = \bar{n}(2\bar{n}+1)$, we solve the linear differential equation and find

$$\langle n(\tau)n(0)\rangle_\infty = \bar{n}^2 + \left(\bar{n}^2+\bar{n}\right)e^{-\kappa|\tau|} \tag{2.186}$$

Using Wiener Khinchine theorem, we obtain the power spectrum

$$S_n\,(\omega) = \frac{1}{2\pi}\int_{t\in\mathbb{R}} \langle n(t)n(0)\rangle\,e^{i\omega t} = \frac{\kappa}{\pi}\left(\bar{n}^2+\bar{n}\right)\frac{1}{\kappa^2+\omega^2} \tag{2.187}$$

Typically, the decay rate κ of the resonator is much higher than the decoherence rate of the qubit. The power spectrum $S_n(\omega)$ is almost constant in the range of interest and thus we obtain

$$\boxed{\Gamma_\varphi^{\text{thermal}} = \pi\left(\frac{\partial\omega_{01}}{\partial n}\right)^2 S_n(\omega=0) = \frac{4\chi^2\left(\bar{n}^2+\bar{n}\right)}{\kappa}} \tag{2.188}$$

Coherent States

When the resonator is driven by a classical drive, the resonator is in a coherent state. In the following, we consider the Hamiltonian of an harmonic oscillator in presence of a well-chosen Lindblad operator

$$\mathcal{H} = \hbar\omega_r a^+ a \tag{2.189}$$

$$L = \sqrt{\kappa}\left(a - \alpha e^{i\omega_r t}\right) \tag{2.190}$$

The steady state of this system corresponds to a pure state $\rho(t) = |\psi(t)\rangle\langle\psi(t)|$, where the coherent state $|\psi(t)\rangle = |\alpha e^{i\tilde{\omega}_r t}\rangle$ satisfies $L|\psi(t)\rangle = 0$. Applying the transformation $U = e^{i\omega_r t a^+ a}$ on the Hamiltonian and the Lindblad operator above we obtain

$$\widetilde{\mathcal{H}} = 0 \tag{2.191}$$

$$\widetilde{L} = \sqrt{\kappa}\,(a - \alpha)\, e^{i\omega_r t}. \tag{2.192}$$

In this new frame, we can write the coupled equations of motion for $\boldsymbol{O}$, where $\boldsymbol{O} = \left(n(t),\ a(t),\ a^\dagger(t),\ \mathbb{1}\right)^T$, in the form of $\frac{d}{dt}\overrightarrow{O}(t) = G\overrightarrow{O}(t)$ with

$$G = \begin{pmatrix} -\kappa & \frac{1}{2}\alpha^*\kappa & \frac{1}{2}\alpha\kappa & 0 \\ 0 & -\frac{1}{2}\kappa & 0 & \frac{1}{2}\alpha\kappa \\ 0 & 0 & -\frac{1}{2}\kappa & \frac{1}{2}\alpha^*\kappa \\ 0 & 0 & 0 & 0 \end{pmatrix} \tag{2.193}$$

The steady state solutions result in

$$\langle \boldsymbol{O} \rangle_\infty = \left(|\alpha|^2,\ \alpha,\ \alpha^*,\ 1\right)^T \tag{2.194}$$

In order to find $\langle n(\tau)n(0)\rangle$ we diagonalize the system and plug in the initial conditions in the steady-state solution for $\tau = 0$. Thus, we obtain

$$\langle n(\tau)n(0)\rangle_\infty = |\alpha|^4 + |\alpha|^2 e^{-\frac{1}{2}\kappa|\tau|} \tag{2.195}$$

Using Wiener Khinchine theorem, we obtain the power spectrum

$$S_n(\omega) = \frac{1}{2\pi}\int_{t\in\mathbb{R}} \langle n(t)n(0)\rangle\, e^{i\omega t} = \frac{\kappa}{2\pi}\bar{n}^2 \frac{1}{(\kappa/2)^2 + \omega^2} \tag{2.196}$$

We obtain therefore the dephasing rate due to photon noise in a coherent-state

$$\boxed{\Gamma_\varphi^{\text{coherent}} = \pi\left(\frac{\partial\omega_{01}}{\partial n}\right)^2 S_n(\omega = 0) = \frac{8\chi^2\bar{n}}{\kappa}} \tag{2.197}$$

2.4.3 Decoherence Under Driven Evolution

The decoherence rate of a driven qubit involves the noise spectral density at its Rabi frequency. To demonstrate this, we will consider in the following a qubit under driven evolution with a drive tuned at the average frequency of the qubit $\langle\omega_{01}\rangle$ assuming that some external fluctuators may modify its instantaneous frequency $\omega_{01}(t)$. The Hamiltonian of the system can be written as

$$\mathcal{H} = \hbar\frac{\omega_{01}(t)}{2}\sigma_z + \hbar\Omega\sigma_x\cos\left(\langle\omega_{01}\rangle t\right) \tag{2.198}$$

We introduce a Lindblad jump operator $L_1 = \sqrt{\Gamma_1}\sigma^-$ to describe the relaxation processes. In the following, we assume that $\Gamma_1$is a constant and does not depend on the frequency of the qubit.

We first transform this Hamiltonian with a unitary transformation $U_1(t) = e^{i\frac{\langle\omega_{01}\rangle}{2}\sigma_z t}$ such that the Hamiltonian becomes after rotating wave approximation

$$\widetilde{\mathcal{H}} = \hbar\frac{\delta(t)}{2}\sigma_z + \frac{1}{2}\hbar\Omega\sigma_x \tag{2.199}$$

The eigenstates associated to eigenvalues $\pm\Omega/2$ of $\hbar\Omega/2\sigma_x$ are

$$|+\rangle = \frac{|0\rangle + |1\rangle}{\sqrt{2}}$$

$$|-\rangle = \frac{|0\rangle - |1\rangle}{\sqrt{2}}$$

The Hamiltonian can be rewritten in the basis of $|\mp\rangle$. Under this basis change, the above operators can be replaced by

$$\sigma_z \rightarrow \sigma_x$$

$$\sigma_x \rightarrow \sigma_z$$

In this basis, the Hamiltonian becomes

$$\widetilde{\mathcal{H}} = \frac{1}{2}\hbar\Omega\sigma_z + \hbar\frac{\delta(t)}{2}\sigma_x \tag{2.200}$$

Using Fermi-Golden rule (see Eq. (2.130)), we get

$$\boxed{\Gamma_\varphi = \Gamma_{+-} = \frac{\pi}{2}\left(\frac{\partial\omega_{01}}{\partial\lambda}\right)^2 \left(S_\lambda\left(+\Omega\right)\right)} \tag{2.201}$$

The equations of motion of the density operator in this new frame $\widetilde{\rho}$ should also include Lindblad jump operators expressed in the same frame. Under this transformation, the Lindblad jump operator becomes $\widetilde{L}_1 = \sqrt{\Gamma_1}\left(\sigma_z + i\sigma_y\right)/2$. Introducing this transformed operator into the equations of motion for σ^+operator (Eq. (2.218)), we get

$$\boxed{\Gamma_2 = \frac{3}{4}\Gamma_1 + \Gamma_\varphi} \tag{2.202}$$

This rate corresponds to the decay rate of the so-called Rabi oscillations.

2.5 Appendix

2.5.1 Master Equation Formalism

The quantum state of a qubit register is fragile and evolve in a non-unitary way, making it impossible to model its evolution using Schrodinger equation alone. The Master Equation formalism allows to treat the qubit register as an open system which interacts with its environment [33, 34].

2.5.1.1 Density Matrix Representation

The density matrix ρ for a system is a positive semi-definite Hermitian operator of trace one acting on the Hilbert space of the system. It is a generalization of the more usual state vectors or wavefunctions: while those can only represent pure states, density matrices can also represent mixed states, i.e. states where the physical system under study is entangled with its environment.

The general description of a density operator is

$$\rho = \sum_j p_j \left|\psi_j\right\rangle\left\langle\psi_j\right| \tag{2.203}$$

where $\left|\psi_j\right\rangle$ is a pure state of the system and p_j its probability. A density operator represents a pure state ($\rho = |\psi\rangle\langle\psi|$) if and only if $\mathrm{tr}(\rho^2) = 1$.

Interestingly, the expectation value of an operator O is given by

$$\boxed{\langle O(t)\rangle = \mathrm{Tr}\left[O\rho(t)\right]} \tag{2.204}$$

2.5.1.2 Density Matrix of a Qubit

For a single qubit, the density operator is a 2×2 matrix and can be written as

$$\rho = \frac{1}{2}\left(\mathbb{1} + \vec{P}.\vec{\sigma}\right) \tag{2.205}$$

where $\vec{\sigma} = (\sigma_x, \sigma_y, \sigma_z)$ represents a vector of Pauli matrices, and $\vec{P}$ represents a vector of the so-called Bloch sphere.

It is important to emphasize the difference between a probabilistic mixture of quantum states and their superposition. For instance, if the qubit is prepared as a statistical mixture of eigenstates $|0\rangle$ and $|1\rangle$ with equal probability, it can be described by the density matrix operator

$$\rho = \frac{1}{2}\begin{pmatrix} 1 & 0 \\ 0 & 1 \end{pmatrix} \tag{2.206}$$

On the other hand, a quantum superposition of these two states with equal probability amplitudes results in the pure state $|\psi\rangle = \frac{1}{\sqrt{2}}(|0\rangle + |1\rangle)$ and is associated with density matrix operator

$$\rho = \frac{1}{2}\begin{pmatrix} 1 & 1 \\ 1 & 1 \end{pmatrix} \tag{2.207}$$

In a general manner, the state of a qubit is a pure state if $\left|\vec{P}\right| = 1$ and is entangled with the environment if $\left|\vec{P}\right| < 1$.

2.5.1.3 Liouville-von Neumann Equation

The time evolution of the density operator is directly obtained from Schrodinger equation

$$\begin{aligned}\dot{\rho} &= \sum_j p_j \left(|\partial_t \psi_j\rangle\langle\psi_j| + |\psi_j\rangle\langle\partial_t \psi_j|\right) = \frac{1}{i\hbar}\sum_j p_j \left(\mathcal{H}|\psi_j\rangle\langle\psi_j| - |\psi_j\rangle\langle\psi_j|\mathcal{H}\right) \\ &= \frac{1}{i\hbar}\left[\mathcal{H}, \rho\right]\end{aligned}$$

The evolution of the density operator according to Louville-von Neumann equation is unitary. However, a quantum system interacts with its environment and thus some unavoidable non-unitary evolution will happen.

2.5.1.4 Krauss Theorem

In general, any evolution of a quantum system can be described by a quantum map [33, 34] . Such quantum map is a linear super-operator that transforms the density operator ρ into a new operator $\mathcal{L}(\rho)$. To be valid, the quantum map must fulfill several conditions:

- **Linearity**—The super-operator must be linear $\mathcal{L}(\alpha\rho + \beta\rho') = \alpha\mathcal{L}(\rho) + \beta\mathcal{L}(\rho')$ with $\alpha + \beta = 1$.
- **Preservation of the trace**—The super-operator must conserve the trace of the density matrix $\mathrm{tr}(\mathscr{L}(\rho)) = 1$

- **Complete positivity**—$\mathcal{L}(\rho)$ must be positive semi-definite for any composite quantum system including the system and parts of its environment.

Under these conditions, the Krauss theorem states that any quantum map can be expressed as a sum of operators

$$\mathscr{L}(\rho) = \sum_{\mu=0}^{K-1} M_\mu \rho M_\mu^+ \tag{2.208}$$

where one can choose $K - 1 < d^2$, d being the dimension of the Hilbert space of the system and K is the Krauss number. In order to conserve the trace, it is straightforward to show that the Krauss operators M_μ must satisfy the normalization condition

$$\sum_{\mu=0}^{K-1} M_\mu^+ M_\mu = \mathbb{1} \tag{2.209}$$

2.5.1.5 Lindblad Equation

Assuming the system is Markovian (no memory), we define the derivative of the density operator as

$$\dot{\rho} = \frac{\mathscr{L}_\tau(\rho) - \rho}{\tau} \tag{2.210}$$

In order to be defined for infinitesimal τ, one of the Krauss operators M_0 should satisfy the condition $\lim_{\tau\to 0} M_0 = \mathbb{1}$. At first order in τ, we can thus write it as

$$M_0 = \mathbb{1} - \boldsymbol{i} L_0 \tau \tag{2.211}$$

We write the operator L_0 as a sum of hermitian and antihermitian operators $L_0 = \frac{1}{2}\left(L_0 + L_0^+\right) + \frac{1}{2}\left(L_0 - L_0^+\right)$ and introduce $H = \frac{\hbar}{2}\left(L_0 + L_0^+\right)$ and $J = \frac{i}{2}\left(L_0 - L_0^+\right)$. Up to first order in τ, we have

$$M_0 \rho M_0^+ = \rho - \boldsymbol{i}\tau/\hbar\,[H, \rho] - \tau\,(J\rho + \rho J) \tag{2.212}$$

All other Krauss operators ($\mu \neq 0$) will contribute at first order only and thus can be written as

$$M_\mu = \sqrt{\tau} L_\mu \tag{2.213}$$

Using the normalization condition, we get

$$\sum_{\mu=0}^{K-1} M_\mu^+ M_\mu = \mathbb{1} - 2J\tau + \tau \sum_{\mu\neq 0} L_\mu^+ L_\mu = \mathbb{1} \tag{2.214}$$

we can thus express J as a function of L_μ as

$$J = \frac{1}{2} \sum_{\mu \neq 0} L_\mu^+ L_\mu \tag{2.215}$$

Reorganizing the terms of the equation, we get the Lindblad equation

$$\boxed{\frac{d}{dt}\rho = -\frac{i}{\hbar}[H, \rho] + \sum_{\mu \neq 0} L_\mu \rho L_\mu^\dagger - \tfrac{1}{2}\left(L_\mu^\dagger L_\mu \rho + \rho L_\mu^\dagger L_\mu\right)} \tag{2.216}$$

2.5.1.6 Equations of Motion

For a system described by the Lindblad master-equation, the time evolution of an arbitrary operator $O(t)$ is given by

$$\frac{d}{dt}\langle O(t)\rangle = \frac{d}{dt}\mathrm{Tr}\left[O\rho(t)\right] = \mathrm{Tr}\left[O\frac{d}{dt}\rho(t)\right] \tag{2.217}$$

Using Eq. (2.216), and noting that

$$\begin{aligned}
\mathrm{Tr}\left[O\left[H, \rho\right]\right] &= \mathrm{Tr}\left[\left[O, H\right]\rho\right] = \langle[O(t), H]\rangle \\
\mathrm{Tr}\left[O L_\mu \rho L_\mu^\dagger\right] &= \mathrm{Tr}\left[L_\mu^\dagger O L_\mu \rho\right] = \left\langle L_\mu^\dagger O(t) L_\mu\right\rangle \\
\mathrm{Tr}\left[O \rho L_\mu^\dagger L_\mu\right] &= \mathrm{Tr}\left[L_\mu^\dagger L_\mu O \rho\right] = \left\langle L_\mu^\dagger L_\mu O(t)\right\rangle \\
\mathrm{Tr}\left[O L_\mu^\dagger L_\mu \rho\right] &= \left\langle O(t) L_\mu^\dagger L_\mu\right\rangle
\end{aligned}$$

We get

$$\boxed{\frac{d}{dt}\langle O(t)\rangle = \frac{i}{\hbar}\langle[H, O(t)]\rangle + \sum_{\mu \neq 0}\left(\left\langle L_\mu^\dagger O(t) L_\mu\right\rangle - \frac{1}{2}\left\langle L_\mu^\dagger L_\mu O(t)\right\rangle - \frac{1}{2}\left\langle O(t) L_\mu^\dagger L_\mu\right\rangle\right)} \tag{2.218}$$

2.5.1.7 Some Simple Examples

As a matter of illustration, we will consider in the following two simple examples.

- **Cavity with damping**

In this first example, we consider the equation of motion of an harmonic oscillator driven by a classical field assuming an Hamiltonian $\mathcal{H} = \hbar\omega_r a^+ a + \hbar\varepsilon(a + a^+)$

in presence of a Lindblad jump operator $L = \sqrt{\kappa}a$, where ω_r is the resonance frequency of the resonator and ε the strength of the field. Using Eq. (2.218), we get

$$\frac{da}{dt} = -i\omega_r a - i\varepsilon - \frac{\kappa}{2}a \tag{2.219}$$

from which we obtain that

$$a[\omega] = \frac{-i\varepsilon[\omega]}{i(\omega_r - \omega) + \kappa/2} \tag{2.220}$$

- **Qubit evolution under classical drive**

In this second example, we consider the equations of motion of a qubit driven by a classical field assuming an Hamiltonian $\mathcal{H} = \frac{1}{2}\hbar\delta\sigma_z + \frac{1}{2}\hbar\Omega\sigma_x$ in presence of Lindblad jump operators $L_1 = \sqrt{\Gamma_1}\sigma^-$ and $L_\varphi = \sqrt{\frac{\Gamma_\varphi}{2}}\sigma_z$ where $\delta = \omega_{01} - \omega_P$ is the detuning between the pump and the resonance frequency of the qubit and Ω is the Rabi frequency. Using Eq. (2.218), we get

$$\begin{aligned}
\frac{d\sigma_x}{dt} &= -\delta\sigma_y - \left(\frac{\Gamma_1}{2} + \Gamma_\varphi\right)\sigma_x \\
\frac{d\sigma_y}{dt} &= \delta\sigma_x - \Omega\sigma_z - \left(\frac{\Gamma_1}{2} + \Gamma_\varphi\right)\sigma_y \\
\frac{d\sigma_z}{dt} &= \Omega\sigma_y - \Gamma_1\left(\sigma_z + \mathbb{1}\right) \\
\frac{d\mathbb{1}}{dt} &= 0
\end{aligned}$$

This system of equations can be written in a matrix for with $\vec{O} = \left(\sigma_x(t), \sigma_y(t), \sigma_z(t)\ \mathbb{1}\right)^T$ as $\frac{d}{dt}\vec{O}(t) = G\vec{O}(t)$ with

$$G = \begin{pmatrix} -\left(\frac{\Gamma_1}{2} + \Gamma_\varphi\right) & -\delta & 0 & 0 \\ \delta & -\left(\frac{\Gamma_1}{2} + \Gamma_\varphi\right) & -\Omega & 0 \\ 0 & \Omega & -\Gamma_1 & -\Gamma_1 \\ 0 & 0 & 0 & 0 \end{pmatrix} \tag{2.221}$$

The steady-state expectation values are obtained by the null eigenstate of G, defined by $G\langle \boldsymbol{O} \rangle = 0$, leading to

$$\langle\sigma_x\rangle_\infty = \frac{\delta\Omega T_2^2}{1 + \delta^2 T_2^2 + \Omega^2 T_1 T_2} w_0 \tag{2.222}$$

$$\langle \sigma_y \rangle_\infty = \frac{\Omega T_2}{1 + \delta^2 T_2^2 + \Omega^2 T_1 T_2} w_0 \tag{2.223}$$

$$\langle \sigma_z \rangle_\infty = \frac{1 + \delta^2 T_2^2}{1 + \delta^2 T_2^2 + \Omega^2 T_1 T_2} w_0 \tag{2.224}$$

where $\frac{1}{T_2} = \frac{\Gamma_1}{2} + \Gamma_\varphi$ and w_0 is the expectation value of σ_z when the system is not driven ($\Omega = 0$).

2.5.1.8 Quantum Regression Theorem

As shown in the example herein above, it is often possible to write an equation of motion for an operator $A(t)$ as a linear combination of a set of system's operators B_j, namely

$$\frac{d}{dt} \langle A(t) \rangle = \sum_j G_j \langle B_j(t) \rangle \tag{2.225}$$

Using Lindblad master-equation $\frac{d\rho}{dt} = \mathscr{L}\rho$, we can thus write

$$\mathrm{Tr}\,[A \mathscr{L} \rho] = \mathrm{Tr}\left[\sum_j G_j B_j \rho\right] \tag{2.226}$$

This equation being satisfied for any $\rho(t)$, we obtain that

$$A\mathcal{L} = \sum_j G_j B_j \tag{2.227}$$

This result allows us to compute the time derivative

$$\frac{d}{d\tau} \langle A(t+\tau) O(t) \rangle = \frac{d}{d\tau} \mathrm{Tr}\left[A e^{\mathcal{L}\tau} O \rho(t)\right] \tag{2.228}$$

$$= \mathrm{Tr}\left[A \mathcal{L} e^{\mathcal{L}\tau} O \rho(t)\right]. \tag{2.229}$$

Using Eq. (2.227) we obtain

$$\frac{d}{d\tau} \langle A(t+\tau) O(t) \rangle = \mathrm{Tr}\left[\sum_j G_j B_j e^{\mathcal{L}\tau} O \rho(t)\right] \tag{2.230}$$

$$= \sum_j G_j \mathrm{Tr}\left[B_j e^{\mathcal{L}\tau} O\rho(t)\right] \tag{2.231}$$

$$= \sum_j G_j \langle B_j(t+\tau) O(t)\rangle \tag{2.232}$$

Equation (2.232) is the quantum regression theorem [32].

2.5.2 Schrieffer Wolff Transformation

2.5.2.1 Unitary Transformation

Let's consider a unitary transformation $U(t)$ acting on the Hilbert space $\mathcal{E}_H$ and characterized by the relation $U^+U = \mathbb{1}$. Under this transformation, the state of the system $|\Psi\rangle$ becomes $\left|\widetilde{\Psi}\right\rangle = U(t)\,|\Psi\rangle$. Writing the Schrodinger equation for $\left|\widetilde{\Psi}\right\rangle$ gives

$$\begin{aligned} i\hbar\partial_t \left|\widetilde{\Psi}\right\rangle &= i\hbar\partial_t\left(U(t)\,|\Psi\rangle\right) = i\hbar\partial_t U(t)\,|\Psi\rangle + i\hbar U(t)\partial_t\,|\Psi\rangle \\ &= i\hbar\partial_t U(t)U^+(t)\left|\widetilde{\Psi}\right\rangle + U(t)H\,|\Psi\rangle \\ &= \left(i\hbar\partial_t U(t)U^+(t) + U(t)HU^+(t)\right)\left|\widetilde{\Psi}\right\rangle \end{aligned}$$

Thus, under this unitary transformation the Hamiltonian becomes

$$\boxed{\widetilde{H} = U(t)HU^+(t) + i\hbar\dot{U}(t)U^+(t)} \tag{2.233}$$

In particular, if the unitary transformation does not depend on time, $\widetilde{H} = UHU^+$. If the transformation is time dependent, a new term will be added in the Hamiltonian. This term is the quantum equivalent of fictitious forces that appear in a non-inertial frame of reference in classical physics.

2.5.2.2 Baker Campbell Hausdorff Formula

For two operators A and B, we consider the expression $\widetilde{B}(\lambda) = e^{\lambda A}Be^{-\lambda A}$ where $\lambda \in \mathbb{C}$. The Baker Campbell Hausdorff formula states that it is possible to express $\widetilde{B}(\lambda)$ as a formal series of operators A and B and iterated commutators thereof, namely

$$\begin{aligned} \widetilde{B}(\lambda) = e^{\lambda A}Be^{-\lambda A} &= B + \lambda\,[A,B] + \frac{\lambda^2}{2}\,[A,[A,B]] \\ &+ \ldots + \frac{\lambda^n}{n!}\,[A,[A,\ldots\,[A,B]]]]]_n \end{aligned} \tag{2.234}$$

The demonstration of this formula is extremely simple. For $\lambda = 0$, the formula is true. Moreover, it is straightforward to show that the two sides of the equality are solutions of the same linear first order differential equation

$$\frac{\partial \widetilde{B}(\lambda)}{\partial \lambda} = \left[A, \widetilde{B}(\lambda)\right] \tag{2.235}$$

Thus, according to the Cauchy Lipschitz theorem, the two sides of the equation must be identical.

2.5.2.3 Schrieffer Wolff Transformation

In quantum mechanics, the Schrieffer–Wolff transformation is a unitary transformation used to simplify the Hamiltonian of a system $H = H_0 + V$ to second order in the interaction V. Under a unitary transformation, the Hamiltonian of the system becomes

$$\widetilde{H} = e^S H e^{-S} \tag{2.236}$$

where S is an anti-hermitian operator. Using Baker Campbell Hausdorff formula

$$\begin{aligned} \widetilde{H} = H_0 + V + [S, H_0 + V] + \frac{1}{2}[S, [S, H_0 + V]] \\ + \ldots + \frac{1}{n!}[S, [S, \ldots [S, H_0 + V]]]]]_n \end{aligned} \tag{2.237}$$

If one chooses properly S such that

$$[S, H_0] = -V \tag{2.238}$$

$$\widetilde{H} = H_0 + V - V + [S, V] - \frac{1}{2}[S, V] + \frac{1}{2}[S, [S, V]] + \ldots$$

Thus, the transformed Hamiltonian can be written up to second order in V as

$$\boxed{\widetilde{H} = H_0 + \frac{1}{2}[S, V] + O(V^3)} \tag{2.239}$$

2.5.2.4 A Simple Application of Schrieffer Wolff Transformation

As an illustration, we consider we consider the Hamiltonian of a spin coupled non-resonantly to a resonator (assuming for simplicity rotating wave approximation). The Hamiltonian of the system can be written as

$$\mathcal{H} = \hbar\omega_r a^+ a + \frac{1}{2}\hbar\omega_{01}\sigma_z + \hbar g\left(\sigma^+ a + \sigma^- a^+\right) \tag{2.240}$$

We consider the coupling term $V = \hbar g\left(\sigma^+ a + \sigma^- a^+\right)$ as a small perturbation and introduce the operator $S = \frac{g}{\omega_{01}-\omega_r}\left(\sigma^+ a - \sigma^- a^+\right)$. It is easy to check that S is anti-hermitian $\left(S = -S^+\right)$ and that

$$\frac{g}{\omega_{01}-\omega_r}\left[\left(\sigma^+ a - \sigma^- a^+\right), \omega_r a^+ a + \frac{1}{2}\omega_{01}\sigma_z\right] = -g\left(\sigma^+ a + \sigma^- a^+\right) = -V \tag{2.241}$$

The commutator $[S, V]$ can be calculated straightforwardly

$$[S, V] = \frac{\hbar g^2}{\omega_{01}-\omega_r}\left[\sigma^+ a - \sigma^- a^+, \sigma^+ a + \sigma^- a^+\right] = \frac{2\hbar g^2}{\omega_{01}-\omega_r}\left[\sigma^+ a, \sigma^- a^+\right] \tag{2.242}$$

Thus, the transformed Hamiltonian can be written as

$$\boxed{\widetilde{H} \simeq \hbar\omega_r a^+ a + \frac{1}{2}\hbar\omega_{01}\sigma_z + \frac{\hbar g^2}{(\omega_{01}-\omega_r)}\sigma_z\left(a^+ a + \frac{1}{2}\right)} \tag{2.243}$$

This Hamiltonian exhibits the so-called dispersive shift of the resonator. The resonance frequency of the resonator is slightly shifted from its bare resonance due to the presence of the spin. The sign of this shift depends on the state of the spin and thus is used frequently in circuit QED for reading out a qubit state. Inversely, the number of photons in the resonator will give rise to a shift in the transition frequency of the spin (Lamb shift). This reverse effect is in general detrimental for the spin coherence. Shot noise of photons in the resonator (photon noise) will give rise to noise in the resonance frequency of the spin which will lead to loss of coherence.

Acknowledgments The material of this book chapter is based on lectures on superconducting quantum circuits given at Bar Ilan University (Ramat Gan, Israel) from 2016 to 2022, at GradNet Quantum Computing workshop (April 2021) and at the International Training course in the Physics of Strongly Correlated Systems (October 2021). I wish to thank my wife, Myriam, for her constant encouragement and support over the years, inspiring me to write these notes and bringing this book chapter to life. We are indebted to all the colleagues, students, postdocs and visitors who have worked with us over the years. In particular, I wish to thank D. Esteve, P. Bertet, D. Vion, M. Devoret, E. Dalla Torre, E. Ginossar, G. Catelani, T. Chang, I. Holzman, P. Brookes, I. Shani and T. Cohen for numerous discussions and collaborative works that have led to this manuscript.

References

1. M. Tinkham, *Introduction to Superconductivity*, 2nd edn. (Dover Publications, Mineola, 2004). http://www.worldcat.org/isbn/0486435032 64, 85
2. J. Millman, Proc. IRE **28**(9), 413 (1940). https://doi.org/10.1109/JRPROC.1940.225885 66, 71
3. D.M. Pozar, *Microwave Engineering*, 4th edn. (Wiley, New Delhi, 2012). Includes index 67

4. A. Caldeira, A. Leggett, Ann. Phys. **149**(2), 374 (1983). https://doi.org/10.1016/0003-4916(83)90202-6 70
5. B. Yurke, J.S. Denker, Phys. Rev. A **29**, 1419 (1984). https://doi.org/10.1103/PhysRevA.29.1419 70
6. V. Bouchiat, D. Vion, P. Joyez, D. Esteve, M.H. Devoret, Phys. Scripta **1998**(T76), 165 (1998). https://doi.org/10.1238/Physica.Topical.076a00165 82
7. Y. Nakamura, Y.A. Pashkin, J.S. Tsai, Nature **398**(6730), 786 (1999). https://doi.org/10.1038/19718 82
8. J. Koch, T.M. Yu, J. Gambetta, A.A. Houck, D.I. Schuster, J. Majer, A. Blais, M.H. Devoret, S.M. Girvin, R.J. Schoelkopf, Phys. Rev. A **76**, 042319 (2007). https://doi.org/10.1103/PhysRevA.76.042319 86
9. H. Paik, D.I. Schuster, L.S. Bishop, G. Kirchmair, G. Catelani, A.P. Sears, B.R. Johnson, M.J. Reagor, L. Frunzio, L.I. Glazman, S.M. Girvin, M.H. Devoret, R.J. Schoelkopf, Phys. Rev. Lett. **107**, 240501 (2011). https://doi.org/10.1103/PhysRevLett.107.240501 87, 94, 95, 111
10. R. Barends, J. Kelly, A. Megrant, D. Sank, E. Jeffrey, Y. Chen, Y. Yin, B. Chiaro, J. Mutus, C. Neill, P. O'Malley, P. Roushan, J. Wenner, T.C. White, A.N. Cleland, J.M. Martinis, Phys. Rev. Lett. **111**, 080502 (2013). https://doi.org/10.1103/PhysRevLett.111.080502 87, 111
11. A.P.M. Place, L.V.H. Rodgers, P. Mundada, B.M. Smitham, M. Fitzpatrick, Z. Leng, A. Premkumar, J. Bryon, A. Vrajitoarea, S. Sussman, G. Cheng, T. Madhavan, H.K. Babla, X.H. Le, Y. Gang, B. Jäck, A. Gyenis, N. Yao, R.J. Cava, N.P. de Leon, A.A. Houck, Nat. Commun. **12**(1) (2021). https://doi.org/10.1038/s41467-021-22030-5 87, 111
12. Y. Sung, L. Ding, J. Braumüller, A. Vepsäläinen, B. Kannan, M. Kjaergaard, A. Greene, G.O. Samach, C. McNally, D. Kim, A. Melville, B.M. Niedzielski, M.E. Schwartz, J.L. Yoder, T.P. Orlando, S. Gustavsson, W.D. Oliver, Phys. Rev. X **11**, 021058 (2021). https://doi.org/10.1103/PhysRevX.11.021058 87, 102
13. J.E. Mooij, T.P. Orlando, L. Levitov, L. Tian, C.H. van der Wal, S. Lloyd, Science **285**(5430), 1036 (1999). https://doi.org/10.1126/science.285.5430.1036 87
14. I. Chiorescu, Y. Nakamura, C.J.P.M. Harmans, J.E. Mooij, Science **299**(5614), 1869 (2003). https://doi.org/10.1126/science.1081045
15. F. Yoshihara, K. Harrabi, A.O. Niskanen, Y. Nakamura, J.S. Tsai, Phys. Rev. Lett. **97**, 167001 (2006). https://doi.org/10.1103/PhysRevLett.97.167001
16. J. Bylander, S. Gustavsson, F. Yan, F. Yoshihara, K. Harrabi, G. Fitch, D.G. Cory, Y. Nakamura, J.S. Tsai, W.D. Oliver, Nat. Phys. **7**(7), 565 (2011). https://doi.org/10.1038/nphys1994 118
17. M. Stern, G. Catelani, Y. Kubo, C. Grezes, A. Bienfait, D. Vion, D. Esteve, P. Bertet, Phys. Rev. Lett. **113**, 123601 (2014). https://doi.org/10.1103/PhysRevLett.113.123601 87, 111
18. T. Chang, I. Holzman, T. Cohen, B.C. Johnson, D.N. Jamieson, M. Stern, Phys. Rev. Appl. **18**, 064062 (2022). https://doi.org/10.1103/PhysRevApplied.18.064062 93, 111
19. F.G. Paauw, A. Fedorov, C.J.P.M. Harmans, J.E. Mooij, Phys. Rev. Lett. **102**, 090501 (2009). https://doi.org/10.1103/PhysRevLett.102.090501 93
20. T. Chang, T. Cohen, I. Holzman, G. Catelani, M. Stern, Phys. Rev. Appl. **19**, 024066 (2023). https://doi.org/10.1103/PhysRevApplied.19.024066 93
21. V.E. Manucharyan, J. Koch, L.I. Glazman, M.H. Devoret, Science **326**(5949), 113 (2009). https://doi.org/10.1126/science.1175552 94
22. I.M. Pop, K. Geerlings, G. Catelani, R.J. Schoelkopf, L.I. Glazman, M.H. Devoret, Nature **508**(7496), 369 (2014). https://doi.org/10.1038/nature13017 94
23. L.B. Nguyen, Y.H. Lin, A. Somoroff, R. Mencia, N. Grabon, V.E. Manucharyan, Phys. Rev. X **9**, 041041 (2019). https://doi.org/10.1103/PhysRevX.9.041041 94
24. F. Yan, S. Gustavsson, A. Kamal, J. Birenbaum, A.P. Sears, D. Hover, T.J. Gudmundsen, D. Rosenberg, G. Samach, S. Weber, J.L. Yoder, T.P. Orlando, J. Clarke, A.J. Kerman, W.D. Oliver, Nat. Commun. **7**(1), 12964 (2016). https://doi.org/10.1038/ncomms12964 94
25. L.V. Abdurakhimov, I. Mahboob, H. Toida, K. Kakuyanagi, S. Saito, Appl. Phys. Lett. **115**(26), 262601 (2019). https://doi.org/10.1063/1.5136262 94, 111
26. A. Blais, A.M. van den Brink, A.M. Zagoskin, Phys. Rev. Lett. **90**, 127901 (2003). https://doi.org/10.1103/PhysRevLett.90.127901 102

27. J. Majer, J.M. Chow, J.M. Gambetta, J. Koch, B.R. Johnson, J.A. Schreier, L. Frunzio, D.I. Schuster, A.A. Houck, A. Wallraff, A. Blais, M.H. Devoret, S.M. Girvin, R.J. Schoelkopf, Nature **449**(7161), 443 (2007). https://doi.org/10.1038/nature06184 102
28. C. Rigetti, A. Blais, M. Devoret, Phys. Rev. Lett. **94**, 240502 (2005). https://doi.org/10.1103/PhysRevLett.94.240502 105
29. C. Rigetti, M. Devoret, Phys. Rev. B **81**, 134507 (2010). https://doi.org/10.1103/PhysRevB.81.134507 105
30. S. Kirchhoff, T. Keßler, P.J. Liebermann, E. Assémat, S. Machnes, F. Motzoi, F.K. Wilhelm, Phys. Rev. A **97**, 042348 (2018). https://doi.org/10.1103/PhysRevA.97.042348 105
31. G. Ithier, E. Collin, P. Joyez, P.J. Meeson, D. Vion, D. Esteve, F. Chiarello, A. Shnirman, Y. Makhlin, J. Schriefl, G. Schön, Phys. Rev. B **72**, 134519 (2005). https://doi.org/10.1103/PhysRevB.72.134519 115, 116
32. I. Shani, E.G. Dalla Torre, M. Stern, Phys. Rev. A **105**, 022617 (2022). https://doi.org/10.1103/PhysRevA.105.022617 118, 128
33. S. Haroche, J.M. Raimond, *Exploring the Quantum* (Oxford University Press, Oxford, 2006). https://doi.org/10.1093/acprof:oso/9780198509141.001.0001 122, 123
34. H.M. Wiseman, G.J. Milburn, *Quantum Measurement and Control* (Cambridge University Press, Cambridge, 2009) 122, 123

3 Subgap States in Semiconductor-Superconductor Devices for Quantum Technologies: Andreev Qubits and Minimal Majorana Chains

Rubén Seoane Souto and Ramón Aguado

Abstract

In recent years, experimental advances have made it possible to achieve an unprecedented degree of control over the properties of subgap bound states in hybrid nanoscale superconducting structures. This research has been driven by the promise of engineering subgap states for quantum applications, including Majorana zero modes predicted to appear at the interface of superconductor and other materials, like topological insulators or semiconductors. In this chapter, we revise the status of the field towards the engineering of quantum devices in controllable semiconductor-superconductor heterostructures. We begin the chapter with a brief introduction about subgap states, focusing on their mathematical formulation. After introducing topological superconductivity using the Kitaev model, we discuss the advances in the search for Majorana states over the last few years, highlighting the difficulties of unambiguously distinguish these states from nontopological subgap states. In recent years, the precise engineering of bound states by a bottom-up approach using quantum dots has led to unprecedented experimental advances, including experimental demonstrations of Andreev qubits based on a quantum dot Josephson junctions and a minimal Kitaev chains based on two quantum dots coherently coupled by the bound states of an intermediate superconducting segment. These experimental advances have revitalized the field and helped to understand that, far from being a disadvantage, the presence of subgap bound states can be exploited for new qubit designs and quantum coherence experiments, including Majorana-based qubits.

R. Seoane Souto · R. Aguado (✉)
Instituto de Ciencia de Materiales de Madrid (ICMM), Consejo Superior de Investigaciones Científicas (CSIC), Madrid, Spain
e-mail: ruben.seoane@csic.es; ramon.aguado@csic.es

R. Aguado et al. (eds.), *New Trends and Platforms for Quantum Technologies*,
Lecture Notes in Physics 1025, https://doi.org/10.1007/978-3-031-55657-9_3

Why Hybrid Semiconductor-Superconductor Qubits Based on Subgap States?

Despite the amazing progress of the last twenty years in qubit research [1], going beyond the actual sizes, with a few tens of noisy qubits [2], to scalable quantum processors with thousands of qubits is far from trivial and represents an engineering challenge [3]. If we focus on solid state implementations, transmon qubits in superconducting circuits [4, 5] and spin qubits [6–8] in semiconductors are currently two of the most promising platforms for quantum computing, but both have their pros and cons. Spin qubits, on one hand, are promising from a scalability standpoint and compatibility with industrial semiconductor processing. However, realizing spin-spin interactions over extended distances still represents a challenge. Transmon-based circuits, on the other hand, can be readily controlled, read and coupled over long distances using circuit quantum electrodynamics (QED) techniques. However, they have a limited speed of qubit operations and they are relatively large, making them prone to unwanted cross-coupling with other circuit elements.

Hybrid combinations of materials provide an interesting change of paradigm, trying to benefit from combining superconductors with semiconductors. This platform can be controlled by voltages that are less susceptible to heating and crosstalk. The first demonstration of this concept was the so-called gatemon qubit: a superconducting transmon qubit based on Josephson junction with gate-tunable semiconducting elements [9–11]. Following this demonstration, new routes have been opened for the study of various forms of hybrid semiconductor–superconductor qubits. Among them, one interesting option is to exploit fermionic-like Bogoliubov de Gennes quasiparticle degrees of freedom in the Josephson junction, which manifest as subgap states generically known as Andreev bound states. The combination of such fermionic degrees of freedom with the bosonic ones associated to the superconducting circuit results in a plethora of novel physics and functionalities.

Using this general concept, various designs have been proposed. One interesting option is to encode a qubit in the spin degree of freedom of a quasiparticle occupying a so-called Andreev bound state in the junction, leading to an effective spin 1/2 degree of freedom where information can be encoded. This qubit is the superconducting counterpart of conventional spin qubits, realized in quantum dots [8], and is hence dubbed superconducting spin qubit [12] or Andreev spin qubit. An important point that must be stressed is that the qubit uses true spin states in the odd fermionic parity sector of the junction as opposed to the Andreev pair qubit [13, 14] which uses the ground and excited states of a standard Andreev level (even parity) as a qubit. Importantly, the intrinsic spin-supercurrent coupling of the Andreev spin qubit facilitates fast, high-fidelity readout and monitoring of the state of the spin qubit using circuit quantum electrodynamics techniques [15–18]. Recent

(continued)

developments that consolidate this architecture as an interesting alternative to other more mature platforms include the demonstration of strong coherent qubit-qubit coupling between a transmon and an Andreev spin qubit [19], as well as strong tunable coupling between two distant qubits [20].

The generalization of this idea to other types of bound states in superconductors and even to "pseudo-spin" degrees of freedom based on topological states with Majorana zero modes Majorana zero modes (MZM)s, undoubtedly makes the field of semiconductor-superconductor hybrid qubits one of the most vibrant in recent years. In particular, a large part of the interest in hybrids is motivated by the recent experimental demonstration [21] of bottom-up engineering of a minimal Kitaev chain [22]. The degree of control on these devices is exquisite. This has allowed the technical challenge of inducing triplet-like equal spin superconducting electron pairing, both in nanowires [23, 24] and in two-dimensional electron gases [25]. Remarkably, already a minimal Kitaev chain with only two quantum dots separated by a superconductor can host Majorana bound states (MBS) [26]. These experiments have given new impetus to a field that seemed somewhat stagnant after more than ten years trying to detect Majorana bound states in hybrid semiconducting-superconducting nanowires [27].

Using such minimal Kitaev chains, the field seems ripe for the demonstration of the basic steps toward a minimal Majorana qubit. In the long run, however, two key steps must be completed before Majorana qubits can be used in topological quantum computation [28]. First, researchers must unambiguously establish that fully nonlocal Majoranas can be fabricated in the laboratory in longer quantum dot chains, with the necessary topological protection. More importantly, unambiguous demonstration of non-abelian braiding of Majoranas would represent an unprecedented breakthrough for fundamental physics.

We end this introductory part by mentioning that, although this chapter will focus exclusively on InAs or InSb-based hybrid superconducting-semiconducting platforms, other recent interesting material combinations include, for example, PbTe nanowires and planar Josephson junctions [29,30], which are currently the subject of active research due to their reduced disorder. From a completely different perspective, other material platforms with promising properties include oxide heterostructures owing to their potential for novel Josephson physics and Majoranas [31,32] or van der Waals heterostructures for superconducting qubit applications [33].

3.1 Subgap States in Superconductors

Discrete states may appear below the superconducting gap for various reasons, such as the existence of normal (non-superconducting) regions within the superconductor, the presence of magnetic fields or magnetic impurities [34], which develop so-called Yu-Shiba-Rusinov (YSR) states [35–37], disorder, etc. All these subgap states are generically known as Andreev bound states (ABS) [38–40]. In the last decades, the discovery of topological materials has allowed for a new twist in the possibilities to engineer subgap states in superconductors. Inspired by notions of topology such as topological invariants and bound states at topological defects, several authors have predicted the existence of a new phase of matter known as the topological superconducting state. This topological phase is characterized by the emergence of a rather special type of bound states occurring at precisely zero energy, where an electron and a hole contribute with exactly the same amplitude to form a quasiparticle. Such a zero energy quasiparticle is equal to its own antiparticle and has therefore Majorana character. As opposed to standard ABSs that can be pushed out of the gap by continuous gap-preserving deformations of the Hamiltonian, well-separated MBSs cannot be removed from zero energy by any local perturbation or noise. This is possible owing to the bulk-boundary correspondence principle of band topology, that predicts the onset of protected edge states at the boundaries between topological and non-topological materials. These states are protected against perturbations by the topology of the bulk material. Quite remarkably, MBSs do not follow fermion statistics, unlike the original particles predicted by Ettore Majorana, but rather possess non-Abelian exchange statistics which, together with their topological protection against local noise, holds promise for applications in fault-tolerant quantum computing. Majorana quasiparticles were first predicted to occur in so-called one-dimensional p-wave superconductors, which are characterized by rare triplet-like pairing instead of the singlet-like pairing of standard s-wave superconductors. These ideas date back to almost two decades, with various seminal papers including the prediction of topological superconductivity in one-dimensional p-wave superconductors by Alexei Kitaev [22].

After a short review about the basic properties of Bogolioubov de Gennes (BdG) quasiparticle excitations in superconductors (Sect. 3.1.1), ABSs (Sect. 3.1.2) and other types of subgap states, including YSR in quantum dot junctions (Sect. 3.1.3), Majorana zero modes will be explained in some detail in the context of the Kitaev model (Sect. 3.2). After a short discussion about the state-of-the-art in Majorana detection in hybrid semiconducting-superconducting nanostructures (Sect. 3.3), we will describe how the Kitaev model, apart from its pedagogical value, has recent direct applicability in current experiments demonstrating a minimal Kitaev chain in hybrid semiconductor-superconductor nanostructures based on quantum dots (Sect. 3.4). In Sect. 3.5, we discuss recent advances in hybrid semiconductor-superconductor qubits based on subgap states, including the Andreev spin qubit. We finish the chapter with a short review in Sect. 3.6 on various theoretical ideas for

Table 3.1 For the shake of clarity and readability, we present here a list of acronyms used along the chapter

Acronym	Meaning
ABS	Andreev bound state
CAR	Crossed Andreeev reflection
ECT	Elastic cotunneling
MBS	Majorana bound state
MP	Majorana polarization
MZM	Majorana zero mode
QD	Quantum dot
PMM	Poor man's Majorana
YSR	Yu-Shiba-Rusinov

coherent operating and manipulation of Majorana states (Fig. 3.1). For the shake of readability, we present in Table 3.1 a list of acronyms used along the chapter.

3.1.1 Bogoliubov de Gennes Quasiparticles in Superconductors

In the so-called Bogoliubov de Gennes formulation of the mean-field BCS Hamiltonian for superconductivity, the excitation spectrum is symmetric with respect to the superconductor's Fermi level (taken as zero energy as a reference), with $\pm E$ energy levels corresponding to opposite occupancies of the same fermion electronic state. This makes the Hamiltonian intrinsically particle-hole symmetric, at the price of a doubling of the Hamiltonian dimension. This can be understood by considering a generic Hamiltonian with BCS mean-field pairing:

$$\mathcal{H} = \sum_{\sigma,\sigma'} H_0^{\sigma,\sigma'} c_\sigma^\dagger c_{\sigma'} + (\Delta c_\uparrow^\dagger c_\downarrow^\dagger + H.c.). \tag{3.1}$$

Here $H_0^{\sigma,\sigma'}$ is a generic spin-dependent quadratic Hamiltonian (kinetic + single electron potentials), Δ is the superconducting pair potential and $c_\sigma^\dagger$ creates an electron with spin $\sigma = \uparrow, \downarrow$ (an implicit an momentum dependence is assumed in this notation). By using the four-component Nambu spinor

$$\hat{\Psi} = \begin{pmatrix} c_\uparrow \\ c_\downarrow \\ c_\downarrow^\dagger \\ -c_\uparrow^\dagger \end{pmatrix}, \tag{3.2}$$

the BCS Hamiltonian in Eq. (3.1) can be cast in the so-called BdG form

$$\mathcal{H} = \frac{1}{2} \hat{\Psi}^\dagger H_{BdG} \hat{\Psi}, \tag{3.3}$$

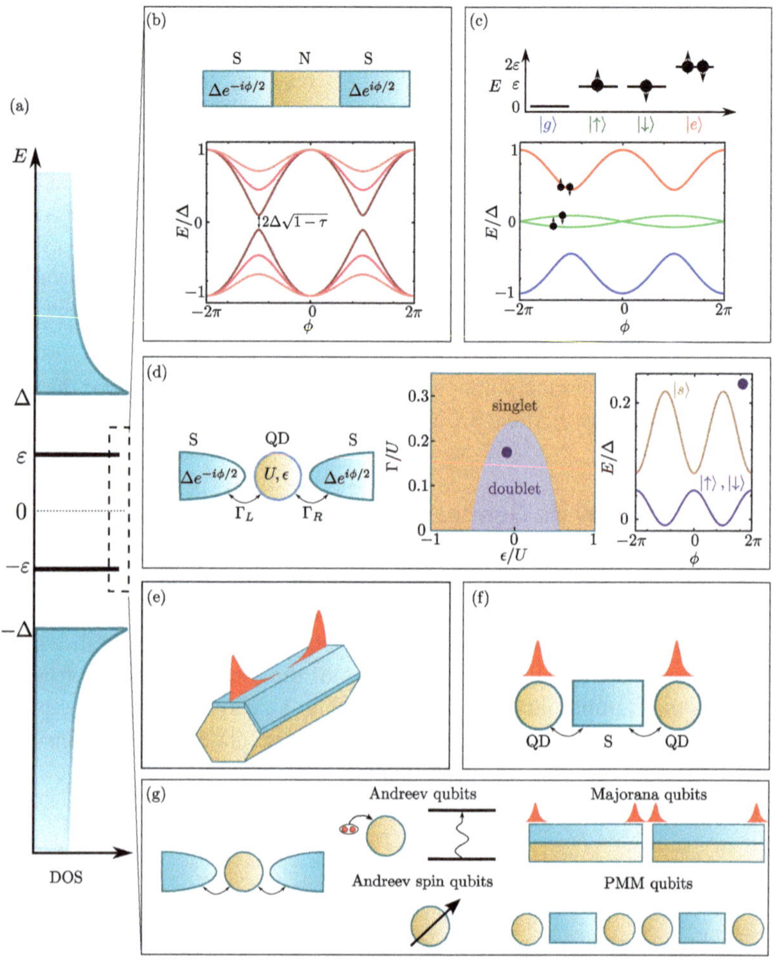

Fig. 3.1 (**a**) Sketch of the density of states of a superconducting system with an energy spectrum containing subgap states (black lines), separated from the continuum of states (blue). (**b**) Sketch (top) and subgap spectrum as a function of the superconducting phase difference, ϕ (bottom) of a short SNS junction of transparency τ (different colours denote decreasing transparency as the energies get closer to the gap edges at $\pm\Delta$). When $\tau \approx 1$ and near $\phi \approx \pi$ the Andreev subgap levels are well detached from the continuum of excitations near Δ and can be used as a qubit system. (**c**) Many-body eigenenergies (top), that the include the four possible occupations of subgap states. The intermediate states $|\uparrow\rangle$ and $|\downarrow\rangle$ are correspond to odd fermionic parity (doublet) when only a single quasiparticle populates the subgap state. Bottom panel shows the two possible qubits that can be realized depending on fermionic parity. Blue and red lines correspond to standard Andreev excitations in the even manifold and constitute a so-called Andreev pair qubit. In the odd fermionic sector, a spin splitting leads to spin-resolved subgap states (green lines) that can be used as another type of qubit (Andreev spin qubit). (**d**) Quantum dot junction, showing the sketch of the junction. Middle panel shows the phase diagram as a function of the coupling to the superconducting leads (Γ) and the dot level position (ϵ), showing two possible ground states: singlet and doublet. Right panel: energy spectrum at a place where the ground state is a doublet (black dot in the middle panel). (**e**) Sketch of a Majorana wire. (**f**) Sketch of a Poor man's Majorana system. (**g**) Different semiconductor-superconductor qubits summarized in this chapter

with

$$H_{BdG} = \begin{pmatrix} H_0 & \Delta \\ \Delta^* & -\sigma^y H_0^* \sigma^y \end{pmatrix}. \tag{3.4}$$

Crucially, the term $-\sigma^y H_0^* \sigma^y$ is the time-reversal of H_0 and appears since holes are the time-reversed version of electrons. The problem defined by Eqs. (3.3) and (3.4) can be solved by seeking for eigenvalues fulfilling the equation

$$H_{BdG}\Phi_n = E_n\Phi_n, \tag{3.5}$$

with $\Phi_n = [u_{n\uparrow}, u_{n\downarrow}, v_{n\uparrow}, v_{n\downarrow}]^T$, such that the diagonalised Hamiltonian becomes

$$\mathcal{H} = \frac{1}{2}\sum_n E_n \gamma_n^\dagger \gamma_n, \tag{3.6}$$

with BdG quasiparticle operators defined as

$$\begin{aligned} \gamma_n &= \int d\mathbf{r}\, \Phi_n^\dagger(\mathbf{r})\hat{\Psi}(\mathbf{r}) \\ &= \int d\mathbf{r}[u_{n,\uparrow}^*(\mathbf{r})c_{\mathbf{r},\uparrow} + u_{n,\downarrow}^*(\mathbf{r})c_{\mathbf{r},\downarrow} - v_{n,\uparrow}^*(\mathbf{r})c_{\mathbf{r},\downarrow}^\dagger + v_{n,\downarrow}^*(\mathbf{r})c_{\mathbf{r},\uparrow}^\dagger], \end{aligned} \tag{3.7}$$

where we have restored the spatial dependence for clarity.

Note that, since we have explicitly included the hole states, thus doubling the dimension of the Hamiltonian, the BdG description is redundant. Therefore, there must necessarily be some symmetry constraint between eigenstates that fixes the number of independent solutions. This constraint reads:

$$P H_{BdG}(\mathbf{r}) P^\dagger = -H_{BdG}(\mathbf{r}), \tag{3.8}$$

where the operator $P \equiv CK$, defined in terms of the charge conjugation operator $C = \tau^y \otimes \sigma^y$ and the complex conjugation operator K, reflects electron-hole symmetry. This means that if there is a solution $\Phi_n(\mathbf{r})$ at positive energy E_n, then there is also a solution $\Phi_m(\mathbf{r})$ at $E_m \equiv -E_n$ fulfilling

$$\Phi_m(\mathbf{r}) = P\Phi_n(\mathbf{r}), \tag{3.9}$$

with

$$E_m \equiv -E_n. \tag{3.10}$$

(continued)

Equivalently, $\gamma_n^\dagger = \gamma_m$ which, in other words, means that creating a Bogoliubov quasiparticle with energy E or removing one with energy $-E$ are identical operations.

3.1.2 Andreev Bound States

At high transparent normal-superconducting (NS) interfaces, a process known as Andreev reflection may take place, whereby electrons are coherently retro-reflected back as holes with inverted spin and momentum, while transferring a Cooper pair into the superconductor [41]. In superconductor-normal-superconductor (SNS) junctions, constructive interference between Andreev processes at both NS interfaces leads to a coherent electron-hole superposition, known as ABSs. Such superpositions result in standing waves with quantized energy leading to subgap states at energies below the superconducting gap. Intuitively, they can be viewed as the electron-hole counterparts of particle-in-a-box states of a quantum well where the boundaries are replaced by superconducting walls. The specific energies of this particular case of BdG excitations depend not only on the superconducting phase difference between both superconductors but also on specific properties of the normal region, such as the transmission probability of electron transport across the normal link, and on the ratio between the length of the normal segment L_N and the superconducting coherence length ξ. For a ballistic system, the latter reads $\xi = \hbar v_F/\Delta$, written in terms of the Fermi velocity v_F of the quasiparticles within the normal region and the superconducting gap Δ.

Short Junction Limit

The simplest limit is the so-called short junction limit ($L_N \ll \xi$) at zero magnetic field and in the Andreev approximation (valid for high density systems, where the Fermi energy is much greater that Δ). In this case, two ABSs appear as subgap BdG excitations in the junction at energies below the gap:

$$E_A^{\pm}(\phi) = \pm\Delta\sqrt{1-\tau\sin^2(\phi/2)}, \tag{3.11}$$

where τ is the normal transmission probability of the conducting channel in the junction. The electrodynamics of such Josephson weak link in a circuit depends not only on the ABSs energy but also on their occupation.

Importantly, in the absence of spin-splitting, ABSs are spin degenerate. Specifically, there is a manifold of even fermionic parity spanned by the ground state and an excited state with two BdG quasiparticles of opposite spin and total energy $2E_A^+(\phi)$. This Hilbert space can be used to define a so-called Andreev pair qubit. There is another possibility corresponding to odd fermionic parity in which only one spinful BdG quasiparticle occupies the junction. Using the spin degrees of freedom of such single quasiparticle one can define a spin qubit inside the superconducting junction. A full discussion of how to use the different Hilbert spaces defined by the fermionic parity of an SNS junction in order to create different Andreev qubit versions will be presented in Sect. 3.5.

3.1.3 Subgap Levels in Quantum Dot Junctions

In the previous discussion about ABSs, we have just assumed that the channel connecting the superconducting electrodes allows for coherent transport through ballistic segments. In contrast, its is quite common that quantum dots (QDs) form in semiconductors near depletion in SNS junctions by e.g. inducing barriers with electrostatic gates. In this situation, charges localize in the channel and the effects of confinement and electrostatic charging become important.

The Superconducting Anderson Model
The standard theoretical model that describes a Josephson junction containing a QD is the single impurity Anderson model coupled to two superconducting reservoirs, the so-called superconducting Anderson model. Despite its apparent simplicity, this model correctly captures the interplay between different physical mechanisms that appear owing to the competition between electron repulsion, originated from Coulomb interactions, and electron pairing, coming from the coupling to the superconductor. The Hamiltonian of the superconducting Anderson model reads

$$H = H_{QD} + H_S^L + H_S^R + H_T^S\,. \tag{3.12}$$

Here, H_{QD} refers to the decoupled dot Hamiltonian given by

$$H_{QD} = \sum_{\sigma} \epsilon_\sigma d_\sigma^\dagger d_\sigma + U n_\uparrow n_\downarrow, \tag{3.13}$$

where d_σ ($d_\sigma^\dagger$) annihilates (creates) an electron with spin $\sigma = \{\uparrow, \downarrow\}$ and energy $\epsilon_\sigma = \epsilon_0 \mp V_Z$, where $V_Z = g\mu_B B/2$ refers to the Zeeman energy owing to the external magnetic field B (with g and μ_B being, respectively, the gyromagnetic factor and the Bohr's magneton). Electron-electron correlations

(continued)

in the dot occupations $n_\sigma = d_\sigma^\dagger d_\sigma$ are included by means of the charging energy U.

The BCS pairing Hamiltonian at each lead is given by

$$H_S^{L/R} = \sum_{k_S\sigma} \varepsilon_{k_S}^{L/R} c_{k_S\sigma}^\dagger c_{k_S\sigma} + \sum_{k_S} \Delta^{L/R} (c_{k_S\uparrow}^\dagger c_{-k_S\downarrow}^\dagger + h.c.), \tag{3.14}$$

where $c_{k_S}^\dagger$ (c_{k_S}) denotes the creation (destruction) operator in the lead with momentum k_S, spin σ and energy ε_{k_S}, which is measured with respect to the chemical potential of the superconductor. The complex superconducting order parameter on each lead is $\Delta^{L/R} = |\Delta^{L/R}|e^{i\phi_{L/R}}$, where $\phi_{L/R}$ is the superconducting phase of the left/right lead. The remaining two terms in the Hamiltonian are the couplings between the quantum dot and the left/right superconducting lead:

$$H_T^S = \sum_{i\in L,R} \sum_{k_S\sigma} \left(V_{k_S}^i c_{k_S\sigma}^\dagger d_\sigma + h.c. \right), \tag{3.15}$$

These two tunneling terms define the two tunneling rates: $\Gamma_{L/R} = \pi \sum_{k_S} |V_{k_S}^{L/R}|^2 \delta(\omega - \epsilon_{k_S}) = \pi |V_S^{L/R}|^2 \rho_S^{L/R}$, with $\rho_S^{L/R}$ being the normal density of states of the S leads evaluated at the Fermi energy. The total coupling between the QD and the superconducting electrodes is $\Gamma = \Gamma_L + \Gamma_R$ and the superconducting phase bias is $\phi = \phi_R - \phi_L$.

3.1.3.1 Non-interacting Limit: Resonant Level Model

For negligible charging effects $U = 0$, the resonant state inside the QD dominates the physical properties of the system. In the single resonant level limit, Eq. (3.11) becomes

$$\varepsilon_A(\phi) = \pm\tilde{\Delta}\sqrt{1 - \tau \sin^2(\phi/2)}, \tag{3.16}$$

where $\tau = \frac{\Gamma^2}{\epsilon_0^2 + \Gamma^2}$ is now the transmission trough a Breit-Wigner resonance of energy ϵ_0 and width Γ corresponding to the total tunneling rate between the two superconducting electrodes and the resonant level. The energy scale $\tilde{\Delta}$ corresponds to the position of the subgap states at $\phi = 0, 2\pi$ and can be obtained from the transcendental equation:

$$\tilde{\Delta}^2 \left[1 + \tau \frac{\Delta^2 - \tilde{\Delta}^2}{\Gamma^2} + 2\tau \sqrt{\frac{\Delta^2 - \tilde{\Delta}^2}{\Gamma^2}} \right] = \Delta^2, \tag{3.17}$$

which defines various interesting limits. Specifically, it goes from $\tilde{\Delta} \approx \Delta$ for $\Gamma \gg \Delta$ and $\tilde{\Delta} \approx \Gamma$ in the opposite $\Gamma \ll \Delta$ limit. For $|\epsilon_0| \ll \Delta$, Eq. (3.16) can be conveniently rewritten as [42, 43]

$$\varepsilon_A(\phi) = \pm\frac{\Delta}{\Delta+\Gamma}\sqrt{\epsilon_0^2 + \Gamma^2\cos^2(\phi/2) + (\Gamma_L - \Gamma_R)^2\sin^2(\phi/2)}, \tag{3.18}$$

which is valid for $\Gamma \ll \Delta$ at any ϕ or $\Gamma \gtrsim \Delta$ provided that $|\pi - \phi| \ll 1$. The prefactor $\frac{\Delta}{\Delta+\Gamma}$ characterizes the extent to which the wave function of the ABS is localized at the quantum dot. If the tunneling between the superconducting leads and the dot is weak, $\Gamma \ll \Delta$, the wave function is predominantly localized at the dot and $\frac{\Delta}{\Delta+\Gamma} \approx 1$. In the opposite limit, $\Gamma \gg \Delta$, the support of the wave function is mostly in the leads. Interestingly, Eq. (3.18) can be written as an ABS in a short junction (see Eq. (3.11)) by noticing that it can be rewritten as [42, 44]

$$\varepsilon_A(\phi) = \pm\Gamma_A\frac{\Delta}{\Delta+\Gamma}\sqrt{1 - \tau\sin^2(\phi/2)}, \tag{3.19}$$

with $\Gamma_A^2 = \epsilon_0^2 + \Gamma^2$ and $\tau = 1 - |r|^2$ the transparency of the junction controlled by the reflection coefficient $r = (\epsilon_0 + i\delta\Gamma)/\Gamma_A$, which depends on both the level position ϵ_0 and the rate asymmetry $\delta\Gamma = (\Gamma_L - \Gamma_R)$. At perfect transparency $\tau = 1$, which is achieved for $\epsilon_0 = \delta\Gamma = 0$, the subgap levels appear at

$$\varepsilon_A(\phi) = \pm\Gamma\frac{\Delta}{\Delta+\Gamma}\cos(\phi/2), \tag{3.20}$$

defining 4π-periodic branches with a zero-energy level crossing at $\phi = \pi$. Note that for $\Gamma \ll \Delta$ the subgap levels lie very deep inside the gap of the superconductor even at phases $\phi = 0, 2\pi$.

3.1.3.2 Interacting Quantum Dots: Some Limits

The physics of QDs coupled to superconductors is richer if interactions are included. Electron-electron interactions tend to fix the number of electrons in the QD to a value that can be tuned using electrostatic gates. On the other hand, superconducting correlations coming from the superconductor tend to pair electrons into singlet Cooper pair. This leads to an interesting competition between these two effects.

Therefore, in the interacting case, $U \neq 0$, two ground states are possible: a spin doublet, $|D\rangle$, with spin $\frac{1}{2}$, and a spin singlet, $|S\rangle$, with spin 0. Transitions between the ground state and the first excited state of the system, i.e., between a doublet and a singlet state or vice-versa, are manifested as a subgap resonance at energies $\pm\varepsilon_A$. Changes in the parity of the ground state of the system appear as points in parameter space where ε_A changes sign (signaled by zero energy crossing of the subgap states). Whether the system is a doublet or a singlet state is determined by the nontrivial interplay between interactions and quantum fluctuations induced by the coupling to the superconductors Γ_S: Coulomb repulsion enforces a one by one electron filling, favoring odd electron occupations with a doublet ground state. The coupling to the superconductor, Γ_S, on the other hand, privileges a singlet ground state. Its physical nature crucially depends on the ratio between the induced pair correlations and interaction strength, Δ/U, as we discuss now.

Large Gap Limit $\Delta \gg U$: BCS-Like Charge Singlets

In the large Δ/U limit, the coupling to the superconductor mainly induces local superconducting correlations in the QD, which acquires local superconducting pairing terms owing to the proximity effect. Such local superconducting pairing leads to Bogoliubov-type singlets, which are BCS-like superpositions of the empty $|0\rangle$ and doubly-occupied $|\uparrow\downarrow\rangle$ states in the QD. In this so-called atomic limit [44–46], the quasiparticles of the superconducting electrodes are removed from the problem and the coupling to the superconductor Γ_S mainly induces local superconducting correlations in the quantum dot, resulting in the Hamiltonian:

$$H_{QD}^{S} = \sum_{\sigma} \epsilon_\sigma d_\sigma^\dagger d_\sigma + U n_\uparrow n_\downarrow + \Delta_\phi d_\uparrow^\dagger d_\downarrow^\dagger + h.c, \tag{3.21}$$

with $\Delta_\phi = \Gamma_L e^{i\phi_L} + \Gamma_R e^{i\phi_R}$, which for $\Gamma_L = \Gamma_R$ can be written in a compact form as $\Delta_\phi = 2\Gamma \cos\phi/2 \equiv \Gamma_S \cos\phi/2$.

Without quasiparticles, the Hilbert space is just four dimensional spanned by the local states in the QD: the empty and doubly occupied states $|0\rangle$, $|2\rangle$ (even fermionic parity) and the doublet $|\uparrow\rangle$, $|\downarrow\rangle$ (odd fermionic parity). Owing to the Pauli exclusion principle, the doubly occupied level in the QD must be a singlet $|2\rangle = 1/\sqrt{2}(|\uparrow\downarrow\rangle - |\downarrow\uparrow\rangle)$. Using this four dimensional Hilbert space, the Hamiltonian in Eq. (3.21) can be written in a compact form as:

$$H = \begin{pmatrix} 0 & \Delta_\phi & 0 & 0 \\ \Delta_\phi & \epsilon_\uparrow + \epsilon_\downarrow + U & 0 & 0 \\ 0 & 0 & \epsilon_\uparrow & 0 \\ 0 & 0 & 0 & \epsilon_\downarrow \end{pmatrix}, \tag{3.22}$$

which is block-diagonalized into even and odd fermion parity subspaces. The eigenvalues of the odd subspace are obviously the same as the spin doublet states, where we allow to have spin-degeneracy breaking by e.g. a Zeeman term $\epsilon_{\uparrow,\downarrow} = \epsilon_0 \pm E_Z$. The even parity eigenvalues correspond to bonding and anti-bonding combinations of empty and doubly occupied states in the QD, namely $|S_\pm\rangle = u|0\rangle \pm v|2\rangle$, with energies

$$E_{e\pm} = \frac{(\epsilon_\uparrow + \epsilon_\downarrow + U)}{2} \pm \sqrt{(\epsilon_\uparrow + \epsilon_\downarrow + U)^2/4 + \Delta_\phi^2}. \tag{3.23}$$

(continued)

The u and v are BCS coherence factors are

$$u^2 = \frac{1}{2}\left[1 + \frac{(\epsilon_\uparrow + \epsilon_\downarrow + U)}{2\sqrt{(\epsilon_\uparrow + \epsilon_\downarrow + U)^2/4 + \Delta_\phi^2}}\right],$$

$$v^2 = \frac{1}{2}\left[1 - \frac{(\epsilon_\uparrow + \epsilon_\downarrow + U)}{2\sqrt{(\epsilon_\uparrow + \epsilon_\downarrow + U)^2/4 + \Delta_\phi^2}}\right], \tag{3.24}$$

In this atomic limit, the singlet-doublet (SD) boundary can be easily obtained as the degeneracy condition between even and odd states. This defining a parity crossing at $\xi_\downarrow = \epsilon_\downarrow - E_{e_-} = 0$, resulting in the boundary

$$\frac{(\epsilon_\downarrow - \epsilon_\uparrow)}{2} + \sqrt{(\epsilon_\uparrow + \epsilon_\downarrow + U)^2/4 + \Delta_\phi^2} - \frac{U}{2} = 0. \tag{3.25}$$

This parity crossing occurs as a quantum phase transition and can be experimentally tuned and detected by means of the Andreev conductance in QDs coupled to superconductors [47–49].

For $\phi = 0$, this parity crossing condition reads [50]

$$E_Z + \frac{U}{2} = \sqrt{(\epsilon_0 + \frac{U}{2})^2 + \Gamma_S^2} \tag{3.26}$$

Interestingly, this condition for a Zeeman-induced zero-energy crossing in a QD is essentially the same as the condition for obtaining MZMs in a semiconducting nanowire with proximity-induced superconductivity [51–54]. This expression illustrates the profound connection between fermionic parity crossings and MZMs, even in this zero-dimensional example.

Finally, we end this part by just mentioning that finite Δ corrections to the above atomic limit boundary are captured by the so-called generalized atomic limit where the original parameters of the atomic limit are rescaled by the quantity $1/(1 + \Gamma_S/\Delta)$ [55–57].

Large Charging Limit $U \gg \Delta$: Yu-Shiba-Rusinov Singlets

In the opposite limit, $U \gg \Delta$, the QD is in the so-called Coulomb blockade regime with a well defined number of electrons. The emergence of subgap states can be understood by considering just one spinful quantum dot level coupled to a superconducting electrode. Conceptually, this scenario is

(continued)

identical to having an isolated magnetic impurity in a superconducting host, a problem well known since the 60s for classical spin impurities resulting in so-called YSR subgap states [35–37]. For QDs the problem is richer owing to quantum fluctuations induced by Γ_S. Quantum fluctuations lead to a very complex scenario since exchange is mediated by Kondo processes, as the unpaired spin in the QD couples to the BdG quasiparticles outside the gap of the SC, with an exchange interaction $J \sim 2\Gamma_S/U$. This exchange interaction creates YSR singlets well beyond the original classical limit since they originate from Kondo-like correlations [50]. The main difference with the standard Kondo effect is that in the superconducting electrode, no quasiparticles are available below the gap, hence Kondo screening is incomplete. There are two main physical consequences of being in the YSR regime with $U \gg \Delta$ as compared to the BCS-like regime with $U \ll \Delta$: (i) instead of the fully *local* BCS-like superpositions of zero and doubly occupied states in the QD, the singlets in this limit are *nonlocal*, corresponding to Kondo-like superpositions between the spin doublet and Bogoliubov quasiparticles in the superconductor. In the presence of phase bias $\phi \neq 0$, the degree of deslocalization of the screening quasiparticles depends on the phase; (ii) the boundary of the singlet-doublet transition in Eq. (3.25) gets replaced by the condition $T_K \sim 0.3\Delta$, where T_K is the Kondo temperature of the problem. Early work on hybrid QDs indicated the importance of Kondo-like correlations [58, 59], while more recent experimental work has provided precise boundaries for the transition [48] and many other aspects [60–63].

While a full solution of this intricate problem needs sophisticated numerics, such as e.g. the numerical renormalization group method [50, 64], deep in the Kondo limit, with $U \to \infty$, one can use renormalised parameters using e.g. slave boson theory and write an effective resonant level model as

$$\varepsilon_A \approx \pm\tilde{\Delta}\sqrt{1 - \tilde{\tau} sin^2(\phi/2)}, \tag{3.27}$$

with $\tilde{\tau} = \frac{\tilde{\Gamma}_S^2}{\tilde{\epsilon}_0^2+\tilde{\Gamma}_S^2}$ and renormalised parameters defined through the constraint $\sqrt{\tilde{\epsilon}_0^2 + \tilde{\Gamma}_S^2} = T_K \approx De^{-\frac{\pi|\epsilon_0|}{\Gamma_S}}$, with D a high-energy bandwidth cutoff [65, 66]. In the unitary limit $\tilde{\epsilon}_0 \to 0$ and $\tilde{\Gamma}_S \to T_K$, we can write the ABSs as [66–68]:

$$\varepsilon_A \approx \pm\Delta\left[1 - 2\left(\frac{\Delta}{T_K}\right)^2\right]\cos(\phi/2), \tag{3.28}$$

which is essentially Eq. (3.20) for a resonant level model written in terms of Kondo parameters.

3.1.3.3 Competition Between Pairing and Charging Energy

The interacting problem becomes analytically unsolvable in the limit where there is not a clear separation between the different energy scales. In this limit, only a numerical treatment can provide an exact solution to the Hamiltonian in Eq. (3.12). In equilibrium conditions, the numerical renormalization group method lead to an exact description of the low-energy properties of the system. The method is based on a logarithmic discretization of the superconductor's band that has an exponential resolution at low energies, close to the superconductor's Fermi level. Precise results on the energy of the ABSs, the phase diagram, and the supercurrent have been reported, see for example Refs. [69–79]. Quantum Monte Carlo [80–83] and functional renormalization [75,84] group are alternative numerical methods to describe the system. The system under non-equilibrium conditions has also been studied by means of exact numerical techniques, like time-dependent density matrix renormalization group [85].

Apart from the previously mentioned numerical methods, there exists a series of approximations that allow for a simpler treatment of the problem. These methods are based on reducing the number of relevant degrees of freedom, so the system can be diagonalized exactly, or perturbation expansion son some of the parameters. One of these approximations is the infinite gap approximation, given in Eq. (3.22), that only considers the induced gap in the superconductor. Below we describe other approximations that have been used to understand the properties of the system.

The Zero Bandwidth Approximation

The zero-bandwidth approximation is based on reducing the infinite number of degrees of freedom of the superconductor to one spin-degenerate state with pairing amplitude Δ. Under this approximation, the Hamiltonian of the superconducting leads, Eq. (3.14), transforms into

$$H_S^{L/R} = \varepsilon^{L/R} c_\sigma^\dagger c_\sigma + \Delta^{L/R} c_\uparrow^\dagger c_\downarrow^\dagger + h.c., \tag{3.29}$$

that describes two quasiparticle states at energies $\pm\Delta$. Consistently, the hoping between the QD and the superconductors has the same form as in Eq. (3.15).

The model provides a good qualitative description of the system properties in a broad range of parameters, including the $0 - \pi$ transition induced by the many-body interactions at the QD [86–88]. However, the absence of a real continuum in the superconductor imposes constrains on the validity range of the approximation. For instance, the approximations fails at describing the regime where the QD is strongly coupled to the superconductor, predicting an nonphysical anti-crossing between the QD and the states at $\pm\Delta$. It also fails at describing non-equilibrium effects, as they rely on the properties of

(continued)

the continuum. Extensions of this model including more states in the leads have been demonstrated to improve results from the zero-bandwidth model, curing some of these shortcomings, see [89] for a discussion. Reproducing the subgap spectrum requires a small number of suitably chosen states.

Perturbation Theory

Perturbation theory provides another way to treat interactions in the system. At the lowest-order expansion in U, the electron-electron interaction renormalizes self-consistently the level and the induced pairing amplitude at the dot as [90]

$$\begin{aligned} \tilde{\varepsilon}_\sigma &= \varepsilon_\sigma + U \langle n_{\bar{\sigma}} \rangle , \\ \tilde{\Delta} &= -U \langle d_\uparrow d_\downarrow \rangle . \end{aligned} \tag{3.30}$$

This approximation reproduces the energy of the ABSs and the supercurrent through the system in the small-U limit [77]. However, it also predicts an spontaneous breaking of the spin-degeneracy that is unphysical. Higher orders in the expansion with U/Γ_S include quantum fluctuations, essential to accurately describe the properties of the system. In fact, second-order expansions have been shown to correctly describe properties of the system up to a moderate values of U/Γ_S [55, 56, 85].

3.2 Kitaev Chain: Basics

3.2.1 Majorana Zero Modes

There is an interesting connection between the BdG description of superconductivity and Majoranas. Most importantly, the four-component Nambu spinor in Eq. (3.2) is nothing but an operator version of the Majorana wave functions. Indeed, by particle-hole symmetry, the Nambu spinor fulfils the so-called Majorana pseudo-reality condition [91], $\hat{\Psi}(\mathbf{r}) = P\hat{\Psi}(\mathbf{r}) = C\hat{\Psi}^*(\mathbf{r})$, where P and C operations are defined around Eq. (3.8). Therefore, the BdG theory for quasiparticle excitations in a superconductor already possesses all the key properties of Majorana fermions. Note that this Majorana character is satisfied by the entire (time-dependent) quantum field but not by the eigenmodes of well defined energy. This can be, in principle, a serious problem towards the observability of the Majorana character of BdG quasiparticles. To avoid the full time dependence of the Nambu spinor, one can focus instead on zero-energy BdG quasiparticles.

The profound meaning of zero-energy BdG quasiparticles can be understood by analyzing the properties of the eigenstates of Eq. (3.5) that are self-conjugate, namely $\gamma_n = \gamma_n^\dagger$. According to Eqs. (3.9) and (3.10), this can only happen at $E = 0$, so that a stationary solution with Majorana character fulfils necessarily,

$$H_{BdG}(\mathbf{r})\Phi_0(\mathbf{r}) = 0. \tag{3.31}$$

The corresponding real space spinor satisfies $\Phi_0(\mathbf{r}) = P\Phi_0(\mathbf{r})$, which implies

$$\gamma_0 = i\int d\mathbf{r}[u_{0,\uparrow}^*(\mathbf{r})c_{\mathbf{r},\uparrow} + u_{0,\downarrow}^*(\mathbf{r})c_{\mathbf{r},\downarrow} - u_{0,\downarrow}(\mathbf{r})c_{\mathbf{r},\downarrow}^\dagger - u_{0,\uparrow}(\mathbf{r})c_{\mathbf{r},\uparrow}^\dagger], \tag{3.32}$$

which is clearly self-conjugate. Such exotic self-conjugate BdG excitations receive the name of MZMs or MBSs.

3.2.2 The Kitaev Model

In the previous subsection, we have just assumed that the MBS exists but that this is a rather peculiar situation. Indeed, the general symmetry obeyed by BdG eigenstates is $\Phi_m(\mathbf{r}) = P\Phi_n(\mathbf{r})$ with $E_m = -E_n$, see Eqs. (3.9), (3.10), and not $\Phi_n(\mathbf{r}) = P\Phi_n(\mathbf{r})$ with $E_n = -E_n = 0$. More importantly, if one of these zero modes exists, it *cannot* acquire a nonzero energy E by any smooth deformation of the Hamiltonian since *finite energy BdG excitations always occur in pairs*. Namely, symmetry would require *another mode* to appear at energy $-E$, in violation of unitarity. There is, however, a way out to the previous conundrum: if the gap separating the zero mode from other quasiparticle excitations *closes*, nothing prevents the system to host pairs of standard BdG excitations as the gap reopens. Such closing and re-opening of the superconducting gap is an instance of a *topological transition*: broadly speaking, a transition that separates two phases characterised by the value of a topological invariant (instead of a broken symmetry). In this particular case, the topological invariant counts the number of MBSs. The superconductors hosting such exotic zero modes with Majorana character are called topological superconductors and the fact that the zero mode cannot acquire a nonzero energy without closing the gap is called topological protection. In this section, we explain how this MBSs appear in, arguably, the simplest model exhibiting topological superconductivity: the Kitaev model [22].

The Kitaev Model as a Tight-Binding Lattice

The Kitaev model is essentially a lattice model that describes a one-dimensional p-wave superconductor, see Fig. 3.2a. Owing to its simplicity, the model grasps the main property we are seeking: a topological phase with emergent MBSs, in a rather intuitive fashion. Specifically, the model describes a chain with N sites of spinless fermions with long-range p-wave superconductivity

$$H = -\mu \sum_{j=1}^{N} \left(c_j^{\dagger} c_j - \frac{1}{2} \right) + \sum_{j=1}^{N-1} \left[-t \left(c_j^{\dagger} c_{j+1} + c_{j+1}^{\dagger} c_j \right) + \Delta \left(c_j c_{j+1} + c_{j+1}^{\dagger} c_j^{\dagger} \right) \right], \quad (3.33)$$

where μ represents the onsite energy, t is the nearest-neighbor hopping amplitude and Δ is the is the p-wave pairing amplitude (assumed real for the moment). This model is quite simple but already contains all the relevant ingredients for topological superconductivity. First note that time-reversal symmetry is broken (since electrons do not have spin degeneracy) Furthermore, the superconducting pairing is rather non-standard (it couples electrons with the same spin in contrast to standard s-wave pairing). Note also that electrons on adjacent sites are paired.

In order to reveal the nontrivial properties of the model, let us first consider a chain with open boundary conditions, and then write the fermionic operators in terms of two new operators γ_j^A and γ_j^B as

$$c_j = \frac{1}{2}\left(\gamma_j^A + i\gamma_j^B\right), \quad c_j^{\dagger} = \frac{1}{2}\left(\gamma_j^A - i\gamma_j^B\right). \quad (3.34)$$

Using standard fermionic anticommutation algebra for c_j, it is very easy to verify that the new operators satisfy the following algebra

$$\{\gamma_i^A, \gamma_j^B\} = 2\delta_{ij}\delta_{AB}, \quad \gamma_j = \gamma_j^{\dagger}, \quad \gamma_j^2 = \gamma_j^{\dagger 2} = 1. \quad (3.35)$$

Thus γ_j^A and γ_j^B are Majorana operators.

Equation (3.34) can be understood as the decomposition of a complex Dirac fermion into real and imaginary parts that correspond to Majorana fermions. The inverse transformation gives us the Majorana operators,

$$\gamma_j^A = c_j + c_j^{\dagger}, \quad \gamma_j^B = i(c^{\dagger} - c_j). \quad (3.36)$$

(continued)

In terms of these new operators, the Hamiltonian in Eq. (3.33) reads

$$H = -\frac{i\mu}{2}\sum_{j=1}^{N}\gamma_j^A\gamma_j^B + \frac{i}{2}\sum_{j=1}^{N-1}\left[\omega_+\gamma_j^B\gamma_{j+1}^A + \omega_-\gamma_j^A\gamma_{j+1}^B\right], \tag{3.37}$$

where $\omega_- = \Delta - t$ and $\omega_+ = \Delta + t$ represent hopping amplitudes between MBSs in neighbouring sites, while μ pairs MBSs in the same site. Let us consider now different possibilities depending on the value of ω_- and ω_+. For $t = \Delta = 0$, the Hamiltonian is trivial, coupling Majorana operators in the same site, Fig. 3.2b, and given by

$$H = -\frac{i\mu}{2}\sum_{j=1}^{N}\gamma_j^A\gamma_j^B. \tag{3.38}$$

This Hamiltonian just expresses the fact that Majorana operators from the same physical site are paired together to form a standard fermion. A less obvious case occurs for $t = \Delta$, namely $\omega_- = 0$ and $\mu = 0$. In this case, Eq. (3.37) becomes,

$$H = it\sum_{j=1}^{N-1}\gamma_j^B\gamma_{j+1}^A. \tag{3.39}$$

Despite its innocent-looking form, Eq. (3.39) is rather nontrivial. First, notice that Majorana operators on the same site are now decoupled (remember that they originally represented a single fermionic degree of freedom!). Furthermore, long range coupling is established since Majorana operators on neighbouring sites are now coupled, Fig. 3.2c. Finally, the Majorana operators at the end of chain γ_1^A and γ_N^B seem to have disappeared from the problem. In order to reveal the deep meaning of all these features in full, let us rewrite the Hamiltonian by defining a new set of fermionic operators

$$d_j = \frac{1}{2}\left(\gamma_j^B + i\gamma_{j+1}^A\right), \quad d_j^\dagger = \frac{1}{2}\left(\gamma_j^B - i\gamma_{j+1}^A\right). \tag{3.40}$$

In terms of these new operators, Eq. (3.37) reads

$$H = 2t\sum_{j=1}^{N-1}\left(d_j^\dagger d_j - \frac{1}{2}\right). \tag{3.41}$$

(continued)

The fermionic operators d_j diagonalise the superconducting problem and therefore describe Bogoliubov quasiparticles with energy $t = \Delta$. Importantly, the diagonalised Hamiltonian contains $N - 1$ quasiparticle operators while the original problem has N sites. The missing fermionic degree of freedom is hiding in the highly delocalised combination

$$f = \frac{1}{2}\left(\gamma_1^A + i\gamma_N^B\right), \quad f^\dagger = \frac{1}{2}\left(\gamma_1^A - i\gamma_N^B\right). \tag{3.42}$$

This fermionic operator does not appear in the Hamiltonian and thus has zero energy. This is an obvious consequence of the fact that the Majorana operators at the ends of the chain commute with the Hamiltonian $[H, \gamma_1^A] = [H, \gamma_N^B] = 0$. Furthermore, this is a standard fermion operator which, as usual, can be empty or occupied. However, this fermionic state is special since both the empty and occupied configurations are degenerate, owing to their zero energy. Note that this ground state degeneracy is rather peculiar since both states differ in fermion parity. This is very different from standard superconductors where, despite the breakdown of particle number conservation, fermion parity is conserved (with even being the parity of the ground state). In contrast, the ground state degeneracy found here, corresponding to different fermionic parities, is unique to topological superconductors and has profound consequences as we shall discuss in the next subsection.

In general, for a small but non-zero μ, the Majorana bound states are not really localised at the ends of the wire, but their wave-functions exhibit an exponential decay into the bulk of the wire. The non-zero spatial overlap of the two Majorana wave-functions results in a non-zero energy splitting between the two Majorana states. Of course, for long wire's lengths, the splitting can be so small that the two Majorana states can be considered to be degenerate. Moreover, the Majoranas can also split when the higher-energy states in the bulk come very close to zero energy, hence the Majorana modes are protected as long as the bulk energy gap is finite. This follows from the particle-hole symmetry involved in the problem, where the spectrum has to be symmetric around zero energy. Therefore, trying to move the Majorana zero modes from zero energy individually is impossible, as it would violate particle-hole symmetry.

The Kitaev Model in Momentum Space

To make connection with the theory of topological bands, nontrivial topological numbers and gap closings in terms of bulk properties, we now consider

(continued)

periodic boundary conditions which, using translational invariance $c_j = \frac{1}{\sqrt{N}} \sum_p e^{ipj} c_p$, allows to write Eq. (3.33) in momentum space p as,

$$H = \sum_p \xi_p \left(c_p^\dagger c_p - \frac{1}{2} \right) - \sum_p t \cos pa + \Delta \sum_p (c_p^\dagger c_{-p}^\dagger e^{ipa} + c_k c_{-p} e^{-ipa}), \tag{3.43}$$

with $\xi_p = -(\mu + 2t \cos pa)$ and a the lattice spacing. Dropping the unimportant constant, the above Hamiltonian can be written in BdG form as

$$H = \frac{1}{2} \sum_p \psi_p^\dagger H_{BdG} \psi_p, \quad \psi_p = \begin{pmatrix} c_p \\ c_{-p}^\dagger \end{pmatrix}, \tag{3.44}$$

with $H_{BdG} = \xi_p \tau_z + \Delta_p \tau_y = \mathbf{h} \cdot \boldsymbol{\tau}$, where $\mathbf{h} = (0, \Delta_p, \xi_p)$, $\Delta_p = -2\Delta \sin p * a$ and $\boldsymbol{\tau} = (\tau_x, \tau_y, \tau_z)$ the Pauli matrices in electron-hole space. The excitation spectrum of H_{BdG} is then given by

$$E_{p,\pm} = \pm \sqrt{(\mu + 2t \cos p * a)^2 + 4\Delta^2 \sin^2 p * a}\,. \tag{3.45}$$

This spectrum is mostly gapped except in special cases: for $\Delta \neq 0$, the energy gap closes when both elements inside the square root vanish simultaneously.

(continued)

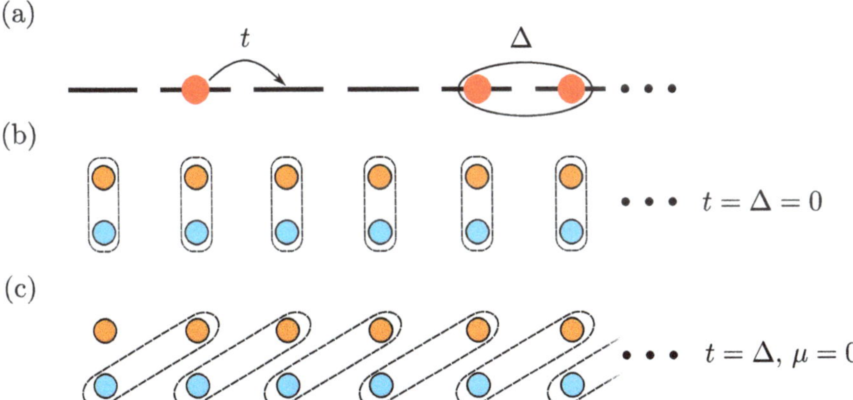

Fig. 3.2 (**a**) Sketch of the Kitaev chain, composed by spinless fermions that are allowed to tunnel (t) and pair (Δ) between neighboring sites. (**b**, **c**) Pictorial decomposition of the local fermionic operators into Majorana operators (orange and blue dots). Panel (**b**) shows the paradigmatic trivial situation with $t = \Delta$, where Majorana operators couple onsite. (**c**) Topological scenario where Majoranas couple between nearest neighbor sites, leaving two unpaired MBSs at the ends

The normal-state dispersion ξ_p vanishes at $\pm p_F$, where the Fermi wavevector is determined by the condition $\mu + 2t \cos p_F a = 0$. The pairing term, on the other hand, vanishes at $p = 0$ and $p = \pm\pi/a$, which is a direct consequence of its p-wave nature. Thus the system is gapless only when the Fermi wavevector equals 0 or $p = \pm\pi/a$. This happens when the chemical potential is at the edges of the normal state dispersion, namely $\mu = -2t$ (for $p_F = 0$) or $\mu = 2t$ (for $p_F = \pm\pi/a$). The lines $\mu = \pm 2t$ define phase boundaries corresponding to two distinct topological phases. These phases are distinguished by the presence or absence of unpaired MBSs at the ends in the geometry with open boundary conditions (bulk-boundary correspondence). These phases are characterised by a $\mathbb{Z}_2$ topological invariant, the Majorana number $M = (-1)^\nu$, where ν represents the number of pairs of Fermi points. The topological superconducting phase occurs for an odd number of pairs of Fermi points, namely $M = -1$, while an even number corresponds to the trivial one. Owing to the bulk-boundary correspondence, one thus expects unpaired MBSs for the open chain when $M = -1$. Indeed, the special point $\mu = 0$ and $\Delta = t$ discussed above for the open chain is well within the topological phase.

3.2.3 Majorana Non-Abelian Properties

One of the most fascinating aspect of Majorana states appearing in one-dimensional topological superconductors are their non-Abelian properties. In 3 dimensions, particles obey either the Fermi-Dirac or the Bose-Einstein distribution. The distribution determines the sign of the wavefunction after the exchange of two identical particles, $|\ldots, \Psi_N, \ldots \Psi_{N'}, \ldots\rangle = \pm |\ldots, \Psi_{N'}, \ldots \Psi_N, \ldots\rangle$, where the $\pm$ sign denotes the result for bosons/fermions. Anyons in low-dimensional systems allow to generalize these commutation relations to more complex ones. The simplest extension to the commutation relation is the one where the wavefunction acquires an arbitrary phase after the exchange of two particles, i.e. $|\ldots, \Psi_N, \ldots \Psi_{N'}, \ldots\rangle = e^{i\theta} |\ldots, \Psi_{N'}, \ldots \Psi_N, \ldots\rangle$. Even more interesting is the case where the ground state of the system is degenerate. In this case, the anyon exchange can transform one ground state into another via an unitary transformation. This transformation can depend on the way the anyons are exchanged, but not on the exact path they follow and timescales, as long as they are sufficiently slow for the system to remain in the ground state, $|\ldots, \Psi_N, \ldots \Psi_{N'}, \ldots\rangle = U(A \rightarrow B) |\ldots, \Psi_{N'}, \ldots \Psi_N, \ldots\rangle$. For these reasons, these anyons are often referred to as non-Abelian, as exchange operations do not commute.

Majorana bound states are quasiparticles with non-Abelian exchange properties. To understand this property, we introduce the concept of fermion parity, related to the number of electrons being even or odd in a given system or part of the system.

In a superconducting system, the total fermion parity is a conserved quantity. In the ground states, the parity of the topological superconductor is determined by the Majorana parity, defined as

$$P_{12} = 1 - 2n = 1 - 2f^{\dagger}f = i\gamma_1^A\gamma_2^B\,, \tag{3.46}$$

where we have used Eq. (3.42) for the definition of the non-local fermion operators as a function of MBS operators. The eigenvalues of the parity operator, P, are ± 1, corresponding to even/odd fermion parity. In case of having additional MBS pairs, the total fermion parity is the sum of the one of each pair. The ground state defined by a pair of MBSs is twofold degenerate, as even and odd fermion parity have the same energy. In general, the ground state of m MBS pairs is 2^m-fold degenerate, corresponding to every pair having even/odd fermion parity. This ground state degeneracy is a requirement for the Majorana non-Abelian exchange properties.

To understand the origin of non-Abelian properties, we can start with the simple picture of moving MBSs around each other, originally conceived by D. Ivanov for vortices in superconductors [92]. Encircling a vortex implies a change of 2π in the phase, which is mathematically taken into account via the branchcut associated to every vortex. During the exchange, one of the MBSs will cross a branchcut, while the other will not, resulting in the exchange

$$\gamma_1 \to -\gamma_2, \qquad \gamma_2 \to \gamma_1\,. \tag{3.47}$$

The physical operation that implements this exchange is given by

$$U_{12} = \frac{1}{\sqrt{2}}(1 + \gamma_1\gamma_2) = e^{-i\gamma_1\gamma_2\pi/4}\,. \tag{3.48}$$

The consequences of braiding two MBSs become clearer when considering the fermion parity of a pair of MBSs, defined in Eq. (3.46). If we consider the two MBSs are in an initial parity state $|n\rangle$, where $n = 0, 1$ for even/odd state, braiding two MBSs transforms the state as

$$U_{12}\,|n\rangle = e^{\pi/4(1-2n)}\,|n\rangle\,. \tag{3.49}$$

this corresponds to a global phase added to the wavefunction of the system.

The braid operation becomes more interesting in the situation with more than a pair of MBSs. The simplest non-trivial scenario is the one with two pairs of MBSs, whose state can be described by the wavefunction $|n_{12}n_{34}\rangle$. In this case, exchanging MBSs that belong to the same pair leads to global phases in the wavefunction,

$$U_{12}\,|n_{12}, n_{34}\rangle = e^{\pi/4(1-2n_{12})}\,|n_{12}, n_{34}\rangle\,, \tag{3.50}$$

$$U_{34}\,|n_{12}, n_{34}\rangle = e^{\pi/4(1-2n_{34})}\,|n_{12}, n_{34}\rangle\,. \tag{3.51}$$

The implications of braiding become more evident when exchanging MBSs that do not belong to the same pair

$$U_{23}\,|n_{12},n_{34}\rangle = \frac{1}{\sqrt{2}}\left[|n_{12},n_{34}\rangle + i(-1)^{n_{12}}\,|1-n_{12},1-n_{34}\rangle\right]. \quad (3.52)$$

In this case, the exchange of two MBSs leads to a non-trivial rotation of the ground state. As expected, the total parity of the system is preserved. However, the local parity of the two MBS pairs can be reversed by splitting/recombining a Cooper pair, where the two electrons flip the parity of the two non-local states, defined by each pair of MBSs. Of course, the choice of two MBSs belonging or not to the same pair is arbitrary. Therefore, fact that a braid looks like a trivial phase or a non-trivial rotation of the state depends on the choice of the basis. This implies that the state $|n_{12},n_{34}\rangle$ is going to look as an equal superposition between the two states with the same total parity in the basis $|n_{13},n_{24}\rangle$ or $|n_{14},n_{23}\rangle$. It means that a system that is initialize in a given basis can look as a superposition between two states in another one. This exotic property is also known as Majorana fusion rules, and related to the possible outcomes when coupling an even number of MBSs, that depends solely on the system's state.

Protocols to demonstrate Majorana fusion and braid statistics are discussed in Sects. 3.6.2 and 3.6.3, respectively.

3.3 Top Down Approach Toward Topological Superconductivity

In 2008 Fu and Kane put forward the conceptual breakthrough of effectively creating p-wave superconductivity and MBSs out of standard s-wave superconductors by virtue of the proximity effect acting onto the helical edge states of topological insulators (propagating edge states with spin-momentum locking) [93]. The possibility of combining different materials to engineer the topological superconducting state has spurred immense interest in the physics of Majorana states in hybrid systems. Fu and Kane's idea was soon extended to other materials with helical states, but different from topological insulators. In 2010 Lutchyn and Oreg, together with their coworkers, described how to bring this proposal to reality by combining well-studied semiconducting materials [51–54]. They proposed the use of hybrid superconducting-semiconducting nanowires subjected to an external magnetic field B. To induce a topological transition (namely, the closing and reopening of an energy gap), the idea is to exploit the competition among three different effects. First, the s-wave superconducting proximity effect, with an induced gap Δ_{ind}: the generation of spin-singlet Cooper pairs in a semiconductor. Second, the Zeeman energy $E_Z = 1/2g\mu_B B$, with g being the nanowire's Landé factor and μ_B the Bohr magneton, tends to break Cooper pairs by aligning their electron spins and closing the superconducting gap. Third, spin-orbit coupling, negates the external magnetic field by preventing the spins from reaching full alignment. The competition between the second and third effects creates regions in parameter space where the gap closes and reopens again. Such phase transition occurs when

$$E_Z = \sqrt{\mu^2 + \Delta_{\text{ind}}^2}, \tag{3.53}$$

where μ is the chemical potential of the semiconductor taken from the bottom of the band. Interestingly, above this critical Zeeman field the proximitized semiconductor effectively behaves as a topological spinless p-wave superconductor, the conceptual model for one-dimensional p-wave superconductivity proposed by Kitaev [22], see Sect. 3.2.

This idea has been extensively analyzed both theoretically and experimentally in the last two decades. In this platform, the semiconductor plays an important role as a booster of the Zeeman field in the semiconductor that allows crossing the topological phase transition without suppressing superconductivity (the Landé factor of the parent metallic superconductor is usually much smaller than the one of the semiconductor $g_S \ll g$). Also it is crucial to be able to gate the semiconducting nanowire to a low-density regime. Specifically, a small μ minimizes the value of the magnetic field needed to enter the topological regime, Eq. (3.53), making it clear why a semiconductor with a low-carrier concentration is advantageous. Also, the Zeeman field cannot overcome the Chandrasekhar-Clogston limit in the superconductor, where superconductivity is suppressed, $\frac{1}{2} g_S \mu_B B < \Delta_S/\sqrt{2}$, where Δ_S is the gap of the parent superconductor. Therefore, the desirable situation involves a material combination with a very large g-factor imbalance, $|g/g_S| \gg 1$ and small μ. Although entering the topological regime is easier for small Δ_{ind}, the topological gap, which sets the topological protection, depends on this parameter. For this reason, a large induced gap is desirable. We note that the spin-orbit coupling does not appear explicitly in the topological criterion, Eq. (3.53). The topological gap, however, depends on the value of the spin-orbit coupling, which has to be large to ensure a good energy separation between the MBSs and the continuum of states.

The profound connection between fermion parity crossings in QDs, which could be considered as the zero-dimensional limit of a proximitized semiconductor, and the above topological criterion is evident by just comparing Eq. (3.53) with Eq. (3.26). It just expresses that parity crossings occur when the effective Zeeman effect in the QD overcomes the superconducting pairing induced by the proximity effect.

3.3.1 The Semiconductor-Superconductor Platform

From the experimental side, the Majorana field took over in 2012 with the first observation of robust zero-bias conductance peak in spectroscopy, signal-

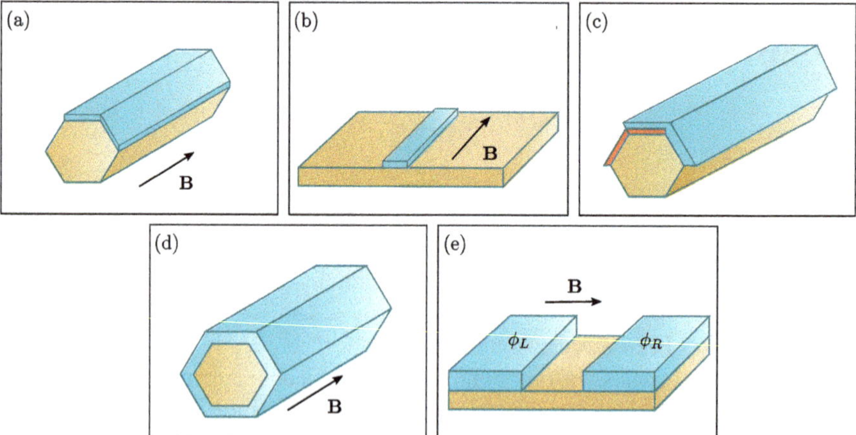

Fig. 3.3 Sketch of some of the most studied semiconductor-superconductor platforms for Majorana physics. (**a**) Semiconductor nanowire (orange) partially covered by a superconducting shell. (**b**) Illustration of a superconducting stripe on top of a semiconductor two-dimensional electron gas. In this case, one-dimensional channels can be achieved by electrostatic confinement. (**c**) Half-shell nanowire partially covered by a ferromagnetic insulator shell (red). (**d**) Full shell wire, where the superconductor wraps around the semiconductor nanowire. (**e**) Planar Josephson junction, where the phase difference between the superconductors ($\phi_L - \phi_R$) can tune the topological transition

ing zero-energy states at the end of semiconductor nanowires partially covered with a superconductor [94], see Fig. 3.3a for an sketch. This observation was followed by similar measurements obtained in other laboratories using different material combinations [95–98]. A major breakthrough appeared with the advent of epitaxial semiconductor-superconductor heterostructures that allowed for clean interfaces that result in hard superconducting gaps, i.e. without undesired states inside the gap [99–102]. In parallel to these developments in half-shell wires, other semiconductor-superconductor platforms have been studied, see Fig. 3.3 for sketches of the most common ones and Refs. [103, 104] for reviews. In the following we mention the most important ones.

Two-dimensional electron gases are versatile platforms for topological superconductivity as they allow to define superconducting one-dimensional channels in different orientations of the plane. This can be done by electrostatic confinement using gate electrodes. A superconducting stripe can be used to induce the proximity pairing needed to get MBSs, see Fig. 3.3b for a sketch. The superconductor screens the electric field coming from the gates, allowing a regime where a quasi one-dimensional channel develops below the superconductor. Epitaxial growth has been also developed for this platform [105, 106]. Recent experiments have explored local and non-local spectroscopy in this platform [107–109]. Some of these experiments reported low-energy states appearing under sufficiently big parallel magnetic fields, consistent with theory predictions for Majorana states [110–112].

Orbital effects can destroy superconductivity well-below the theoretical limit imposed by the Chandrasekhar-Clogston limit. Orbital effects can be minimized by using thin superconductors and magnetic fields that align to their smallest cross section, imposing strong constrains on the geometries that can be studied using nanowires. Ferromagnetic materials offer the possibility to overcome these two issues. In order to avoid undesired quasiparticles in the system, the ferromagnet should be insulating. The platform combining semiconductor-superconductor-ferromagnet in epitaxial growth was developed In Ref. [113], see the sketch of the system in Fig. 3.3c. Early experiments in these systems reported low-energy states at the end of the wire [114] that are spin-polarized [115] and supercurrent reversal due to a large exchange field [116]. The theory of these devices showed that the exchange field induced in the semiconductor is essential to reach the topological regime [117–122]. Recently, the platform has been extended to two-dimension electron gas [123, 124].

Full-shell nanowires, i.e. semiconducting nanowires entirely surrounded by a superconductor, are another way to decrease the required external field to reach the topological regime. The narrow superconducting cylinder feature the so-called Little-Parks effect [125]: a set of superconducting lobes that close when the magnetic field is half a flux quantum across the nanowire section, with a maximal gap when the flux is an integer number of the flux quantum [126–129]. Signatures of low-energy states were found in the first lobe, where the phase of the superconducting parameter winds 2π and describes a fluxoid [126], consistent with theory predictions [130, 131]. Later experiments found that near-zero energy subgap states of non-topological origin could explain the measured low-energy features in the data [132, 133]. Apart from the low-energy state, additional subgap states appear in the first lobe, analogous to the Caroli-de Gennes-Matricon states observed in type II superconductors [134].

The superconducting phase difference between superconductors offers a way to engineer topological phases without large magnetic fields, see Ref. [135] for a review. The simplest proposal in terms of superconductors is composed by a planar semiconductor proximitized by two superconductors in a Josephson junction geometry [136–141]. This proposal exploits the gap closure in transparent bulk Josephson junctions close to phase difference of π. At this point, any small Zeeman field can induce a topological phase transition. Planar Josephson junctions have been also investigated experimentally in Refs. [142–146], finding low-energy states. Other alternatives not discussed here that are being actively investigated are time-driven phase transitions and unconventional superconductivity [147].

? Have Majoranas Been Detected?

Several experiments revealed signatures consistent with the presence of MBSs in semiconductor-superconductor heterostructures, see a recent review of the field in https://arxiv.org/abs/2406.17568. Apart from the aforementioned zero-bias conductance peak reproduced in different platforms, researchers studied the temperature dependence of the peak height, finding a behavior that is consistent with the theory prediction for MBSs [110]. Other experiments studied the hybridization of the low-energy states in semiconductor-superconductor heterostructures with a

QD [107, 148, 149], that provides information about the localization of Majorana modes [150, 151].

Simultaneous transport measurements at both sides of the nanowire provides additional information about the system, including the presence of subgap states at both ends. Additionally, the non-local conductance, i.e. conductance measurements after applying a voltage at the other side of the system, contains additional information about the localization of states inside the gap and their BCS charge [108, 152–154]. These signatures were compiled in a protocol to identify the topological regime based on the simultaneous detection of states at the two ends of the wire and the closing and reopening of the induced gap in the nanowire [155]. Recently, researchers at Microsoft claimed that some nanowires have passed their topological criterion, based on a combination of local and non-local conductances [112].

Majorana bound states can also appear in floating wires, where the thin superconductor is not connected to a big reservoir of electrons. In this configuration, charging energy becomes important, providing an effective coupling between the two MBSs that splits the ground state degeneracy. This splitting is tunable using an external gate that controls the charge occupation of the island. Early proposals studied electron transport through Majorana islands, demonstrating long-range coherent transport between the two ends of the wire, so-called electron teleportation [156, 157]. Experiments in semiconductor-superconductor islands revealed an exponential trend of the energy splitting with the length of the island [126, 158]. Also, the predicted periodicity after the addition of 1 electron in Majorana islands has been demonstrated [159–161], including interferometry features [158]. Finally, cotunneling features in the transport through the island allowed to determine the "spinless" nature of low energy states in ferromagnetic wires [115].

Despite all the evidence, a definitive demonstration of the topological origin of the measured zero-energy states would require measuring Majorana's non-local properties, Sect. 3.6 for details. The reason is that robust zero-energy states of non-topological origin can appear at zero energy in the presence of disorder or smooth confinement potentials in semiconductor-superconductor wires, as discussed in, for example, Refs. [162–169]. In fact, spatially overlapping MBSs at one of the ends of the nanowire, sometimes referred to quasi- or pseudo-Majoranas [165], can mimic the local properties of topological MBSs. Interestingly, this Andreev versus Majorana controversy, and why topologically trivial zero modes often mimic Majorana properties, can be clarified when framed in the language of non-Hermitian topology. This change of paradigm allows one to understand topological transitions and the emergence of zero modes in more general systems than can be described by band topology. This is achieved by studying exceptional point bifurcations in the complex spectrum of the system's non-Hermitian Hamiltonian. Within this broader topological classification, Majoranas from both conventional band topology and a large subset of Andreev levels at zero energy are in fact topologically equivalent, which explains why they cannot be distinguished [168].

Finally, and very importantly, unintentional QD formation in tunneling experiments is a common source of zero bias peaks in conductance owing to the YSR physics that we discussed in Sect. 3.1.3.2. Specifically, as the charge state, and thereby the fermion parity, of the QD is tuned against some external parameter (e.g gate voltage or magnetic field), the ground state switches between the singlet and doublet states [47, 48, 132]. Such parity crossings appear in tunneling conductance as zero-bias peaks [50].

The presence of nontopological subgap states in semiconductor-superconductor hybrid devices is often considered a drawback that complicates the interpretation of experiments. However, as we now discuss, this view is perhaps *too short-sighted*: recent advances in this field are, in fact, strongly driven by the recent demonstration that topologically trivial ABS can also be used to tune artificial Kitaev chains in which topological MBSs can be designed, see Sect. 3.4. In addition, qubits based on Andreev and YSR states offer another interesting alternative in which hybrid qubit research is progressing very rapidly, see Sect. 3.5. In the following sections, we review all these fascinating advances based on subgap states.

3.4 Bottom-Up Approach: Minimal Kitaev Chains

The Kitaev chain, based on spinless fermions with p-wave pairing is the simplest toy model to understand topological superconductivity, see in Sect. 3.3 for details [22]. The model predicts a topological phase transition characterized by the onset of Majorana states at the ends of the system, which are non-local states with non-Abelian exchange properties. The Kitaev chain has a very simple solution when the nearest-neighbor hoping (t) and the p-wave pair amplitude (Δ) are equal, and the chemical potentials of the sites (ε_i) are set to zero, see Sect. 3.2. For these parameters, the chain hosts two MBSs perfectly localized at the two ends of the system. The perfect localization implies that the Majorana states should persist independently from the chain's length. This is true even for minimal chains, composed by two sites. As explained below, Majorana states in minimal Kitaev chains are not topologically protected, as deviations from the ideal parameters will hybridize them, lifting the ground state degeneracy. For this reason, they are usually referred to as poor man's Majoranas (PMMs). Nevertheless, PMMs share all the properties with their topological counterparts, including non-Abelian statistics, as we shall discuss in Sect. 3.6.4.

The recent controversy on the topological nature of the measure subgap states in different platforms has set PMMs as one of the most promising routes to demonstrate non-Abelian quasiparticles in condensed matter physics. In this respect, the robustness of the platform against disorder and the ability of tuning the system in and out of the regime where PMMs appear might be seen as an advantage rather than a disadvantage.

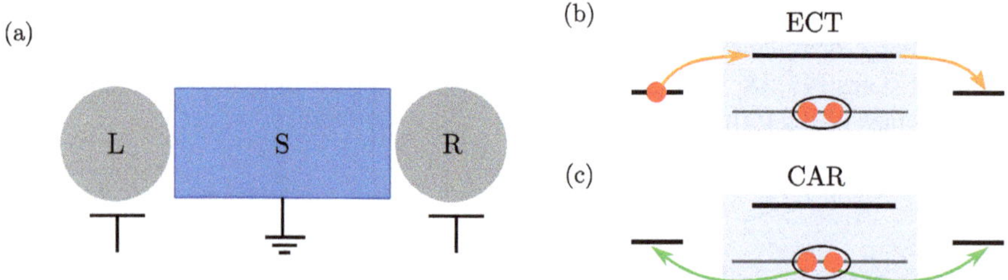

Fig. 3.4 (**a**) Sketch of a Poor man's Majorana (PMM) setup, composed by a narrow superconductor that mediates the coupling between two QDs. The superconductor is grounded and can host subgap states that allow for an efficient tuning of the coupling between the QDs [175]. (**b**) Schematic representations for the elastic cotunneling (ECT) process, where an electron (red circle) is transferred between the left and right dots occupying a virtual excited state in the superconductor (thick line in the middle segment). (**c**) Crossed Andreev reflection (CAR), where a Cooper pair splits (recombines) with one electron going (coming) from each QD

3.4.1 Majorana States in Double Quantum Dots

The first theory proposal for a minimal Kitaev chain appeared in 2012 by two groups independently, Refs. [26, 170]. Their setup is similar to the one used in Cooper pair splitter geometries, see for example Refs. [171–174], with two QDs coupled at the two sides of a superconducting segment, Fig. 3.4. The QDs' properties can be tuned using electrostatic gates, including their energies and the tunnel coupling to the central superconductor. The superconductor should be smaller than the superconducting coherence length to allow for crossed Andreev reflection (CAR) processes, i.e. the splitting of a Cooper pair, where each one of the electrons ends up in a different QD. CAR plays the role of the p-wave pairing, Δ, in the Kitaev chain. Elastic cotunneling (ECT) allows for an electron to jump between the dots, occupying a virtual excited state at the superconductor, which plays the role of the nearest-neighbor hoping, t, in the Kitaev model.

To get Majorana states in the minimal setup sketched in Fig. 3.4, the modulus of t and Δ has to be the same and the system subject to a strong magnetic field to polarize the QDs. In the absence of spin-mixing terms, either Δ or t are zero, depending on whether equal or different spin QD orbitals align. Therefore, spin-mixing terms are required to engineer MBSs, which can be introduced by different mechanisms, including spin-orbit coupling or different magnetization angles between the QDs. Irrespectively from the mechanism, the minimal model for the PMM system is given by [26]

$$H = \sum_{\alpha=L,R} \varepsilon_\alpha d_\alpha^\dagger d_\alpha + \left(t d_L^\dagger d_R + \Delta d_L d_R + \text{H.c.} \right), \tag{3.54}$$

where d_α ($d_\alpha^\dagger$) creates (annihilates) an electron in dot $\alpha = L, R$, ε_α is the dot's energy measured from the superconductor's chemical potential, and t and Δ the ECT and CAR amplitudes. In this equation, we have integrated the effect of the central superconductor and assumed that the system is subject to a strong magnetic field that polarizes the QDs. Therefore, one can ignore one of the spin-spices of the QDs. Extensions of this model are described below.

The Single Particle Formulation

In the single particle basis, the Hamiltonian of the system is given by

$$H = \Psi^\dagger \begin{pmatrix} \varepsilon_L & t & 0 & \Delta \\ t & \varepsilon_R & -\Delta & 0 \\ 0 & -\Delta & -\varepsilon_L & -t \\ \Delta & 0 & -t & -\varepsilon_R \end{pmatrix} \Psi \,, \tag{3.55}$$

where we are using the Nambu spinor $\Psi = (d_L, d_R, d_L^\dagger, d_R^\dagger)$ and t and Δ are chosen real. The model has a sweet spot for $t = \Delta$ and $\varepsilon_L = \varepsilon_R = 0$, where the system has two eigenstates with energy $E = 0$ and wavefunctions given by

$$\psi_1 = \frac{1}{\sqrt{2}}(1, 0, 1, 0)^T \,, \qquad \psi_2 = \frac{i}{\sqrt{2}}(0, 1, 0, -1)^T \,. \tag{3.56}$$

In second quantization, the operators associated with these two states are given by $\gamma_1 = (d_L + d_L^\dagger)/\sqrt{2}$ and $\gamma_2 = i(d_R - d_R^\dagger)/\sqrt{2}$, which are Hermitian operators $[(\gamma_{1,2})^\dagger = \gamma_{1,2}]$ and describe Majorana states completely localized in the dots. The same result can be found in the many-body basis, which reveals the even and odd fermion parity structure, see discussion around Eq. (3.60) and Ref. [26].

PMMs are not topologically protected, in the sense that detuning the system away from the ideal sweet spot makes the Majorana overlap and, eventually, split away from zero energy. For instance, if one QD is detuned away from 0 energy, the Majorana that was originally localized in the dot licks into the opposite QD. For example, if we take $\varepsilon_L \neq 0$, the two wavefunctions are modified as

$$\psi_1 = \frac{1}{A\sqrt{2}}(1, \delta, 1, \delta)^T \,, \qquad \psi_2 = \frac{i}{\sqrt{2}}(0, 1, 0, -1)^T \,. \tag{3.57}$$

(continued)

with $\delta = -\varepsilon_L/2t$ and $A = \sqrt{1+\delta^2}$. In this case, both wavefunctions are still Majorana. While ψ_2 is still localized at the right QD, ψ_1 has some weight in both dots, differently from the sweet spot result, Eq. (3.56). To the lowest order in the dots onsite energies, the splitting between the two ground states for $t = \Delta$ is given by

$$E_{1,2} = \pm\frac{\varepsilon_L\varepsilon_R}{2t} \tag{3.58}$$

up to second order corrections in $\varepsilon_{L,R}$. This expression makes it clear that shifting the energy of only of the QDs does not split the energy degeneracy, even though two Majorana states can overlap in the other dot, see Fig. 3.5a. Therefore, the energy degeneracy is linearly protected to local perturbations on the dots' energies, Fig. 3.5b. In contrast, deviations on the ECT and CAR amplitudes result in a linear splitting on the ground state energy, Fig. 3.5c

$$E_{1,2} = \pm(|\Delta| - |t|)\,. \tag{3.59}$$

Therefore, controlling the relative amplitude between CAR and ECT is a crucial condition to realize Majorana states in dot systems.

Many-Body Formulation

The many-body basis provides a complementary point of view to the problem, allowing also to introduce interactions between the QDs. This description

(continued)

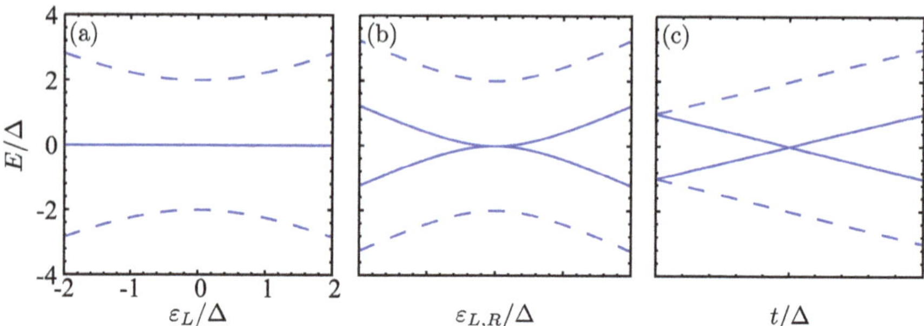

Fig. 3.5 Energy states of the PMM Hamiltonian (3.54), where the two states closer to 0 are represented with solid lines. (**a**) Energy levels for $t = \Delta$ and $\varepsilon_R = 0$ when detuning ε_L. Two states remain degenerate, regardless of the ε_L value. (**b**) Energy levels for $\varepsilon_L = \varepsilon_R$ and $t = \Delta$. (**c**) Energy levels for $\varepsilon_L = \varepsilon_R$ and different values of t. In all the panels, we used $\Delta = 1$

accounts for the fermion occupation of each of the two QDs. In the simple description of Eq. (3.54), where one of the spin spices is disregarded (limit of infinite exchange field in the dots) and the central superconductor is not described explicitly (it only mediates CAR and ECT processes), the four possible states are $|00\rangle$, $|11\rangle$ and $|01\rangle$, $|10\rangle$ for the even and odd subspaces. Here, 0 and 1 denote the fermion occupation of the left/right QDs. In this model, the Hamiltonian becomes block-diagonal, as there are no terms mixing the two sectors with different fermion parity

$$H = \begin{pmatrix} 0 & \Delta & 0 & 0 \\ \Delta & \varepsilon_R + \varepsilon_L & 0 & 0 \\ 0 & 0 & \varepsilon_R & t \\ 0 & 0 & t & -\varepsilon_L \end{pmatrix} . \tag{3.60}$$

Interestingly, the above equation written in terms of the two parity sectors is nothing but a non-local version of Eq. (3.22), where the spin in a single QD is replaced by a pseudospin living in the double QD system (the fermion occupation of the left/right QDs) and the hopping t acting as a pseudospin mixing term. This analogy is even more evident when explicitly including a Coulomb interaction between the QDs, see Eq. (3.64).

The Hamiltonian in Eq. (3.60) has two even and two odd fermion parity eigenvalues, given by

$$E_e^{\pm} = \frac{\varepsilon_L + \varepsilon_R \pm \sqrt{(\varepsilon_L + \varepsilon_R)^2 + 4\Delta^2}}{2} ,$$
$$E_o^{\pm} = \frac{\varepsilon_L + \varepsilon_R \pm \sqrt{(\varepsilon_L - \varepsilon_R)^2 + 4t^2}}{2} , \tag{3.61}$$

These expressions imply that the system ground state can be tuned from a total even to a total odd parity by either tuning the values of t and Δ, or by shifting the energy of the QDs, ε_L and ε_R. First experiments on minimal Kitaev chains have demonstrated that the ground state change is a good way to determine the position in parameter space of Majorana sweet spots [21], see discussion in Sect. 3.4.1.4. The ground state degeneracy occurs for

$$\varepsilon_L \, \varepsilon_R = t^2 - \Delta^2 . \tag{3.62}$$

For $t = \Delta$, this equation implies that the ground state is degenerate when either $\varepsilon_L = 0$ or $\varepsilon_R = 0$, see Fig. 3.6b. When both are zero, the system features a Majorana sweet spot with wavefunctions for the ground state (considering $t = \Delta > 0$) that are given by

(continued)

$$\psi_e^{\pm} = \frac{1}{\sqrt{2}} \left(|00\rangle - |11\rangle\right) ,$$

$$\psi_o^{\pm} = \frac{1}{\sqrt{2}} \left(|01\rangle - |10\rangle\right) . \tag{3.63}$$

It is straightforward to proof that the local Majorana operators $\gamma_L = (c_L + c_L^{\dagger})$ and $\gamma_R = i(c_R - c_R^{\dagger})$ induce transitions between the two opposite parity ground states, up to an irrelevant complex phase in the wavefunction. In the Majorana language, the two states in Eqs. (3.63) are eigenstates of the number operator $n = f^{\dagger} f$, corresponding to the non-local fermionic operator $f = (\gamma_L - i\gamma_R)/2$ that includes both Majorana states.

For $t \neq \Delta$, Eq. (3.62) describes two hyperbolas in the $(\varepsilon_L, \varepsilon_R)$ plane, as shown in Fig. 3.6a, c. Depending on the sign of $\Delta^2 - t^2$, either CAR or ECT processes dominate, changing the orientation of the avoided crossing between the two degeneracy lines. This provides an experimental way to determine which of the two processes dominate in the system and to tune it into the sweet spot configuration. In Sect. 3.4.1.3 we discuss the transport signatures in the three described regime.

We conclude the section by discussing the role of non-local Coulomb interaction between the PMM dots. This interaction can be described by adding an additional term $H_U = U_{LR}\, n_L\, n_R$ to the Hamiltonian in Eq. (3.54), with $n_\alpha = d_\alpha^{\dagger} d_\alpha$ being the number operator in the two dots. Therefore, the Hamiltonian matrix is given by

$$H = \begin{pmatrix} 0 & \Delta & 0 & 0 \\ \Delta & \varepsilon_R + \varepsilon_L + U & 0 & 0 \\ 0 & 0 & \varepsilon_R & t \\ 0 & 0 & t & -\varepsilon_L \end{pmatrix} , \tag{3.64}$$

from where it is clear that the Coulomb repulsion only affects to the states in the even parity sector, that now become

$$E_e^{\pm} = \frac{\varepsilon_L + \varepsilon_R + U_{LR} \pm \sqrt{(\varepsilon_L + \varepsilon_R + U_{LR})^2 + 4\Delta^2}}{2} , \tag{3.65}$$

while the odd-parity energies are described in Eq. (3.61). As shown in Eq. (3.65), the non-local charging energy only renormalizes the effective energy of the PMM QDs. Therefore, the energy splitting between the even and odd fermion parity sectors can be compensated by gating the QDs. The system shows even-odd degeneracy for

(continued)

$$U_{LR} = \frac{\left(\Delta^2 - t^2 + \varepsilon_L \varepsilon_R\right)\left(\varepsilon_L + \varepsilon_R + 2\sqrt{t^2 + (\varepsilon_L - \varepsilon_R)^2}\right)}{2\left(t^2 - \varepsilon_L \varepsilon_R\right)}. \quad (3.66)$$

Even if we can find sweet spots with ground state degeneracy, these points do not correspond to well-localized Majorana states in the two dots for $t = \Delta$, loosing the property of non-locality. This means that local measurements in one of the dots can determine the state of the system and the system becomes sensitive to local perturbations, in contrast to the quadratic protection described in the ideal scenario. Therefore, non-local interactions between the dots are detrimental for the realization of PMMs. This is in principle not an issue, as non-local interactions are screened by the central superconductor. In addition, new Majorana sweet spots can be found modifying the value of the dot levels, t, and Δ [176]. In Sect. 3.4.1.2 we comment on the role of local Coulomb interactions, an important in QDs that has been disregarded up to this point.

The local superconducting pairing is another important ingredient that has been ignored when considering a large Zeeman field, so one spin component can be disregarded. Adding local BCS pairing in the dots will hybridize the different spins of the dot, mixing them and ruining the Majorana properties. However, it is possible to exploit a local BCS pairing to induce PMM states [176, 177]. The simplest geometry to understand this problem are two QDs where one of them is strongly coupled to a grounded superconductor and features an ABS. In this case, the equivalent to ECT is a direct tunneling between the QDs, i.e. a first-order tunneling. In contrast, the equivalent to CAR is a third-order tunnel process, where the a Cooper pair splits from the superconductor into the two dots. In order to make these two processes equal, one can align opposite spins in the two QDs, leading to a potential sweet spot. The price to pay is a suppression of the CAR-like tunnel amplitude, that results in a lower superconducting gap at the sweet spot.

3.4.1.1 Microscopic Model for ECT and CAR

Tuning the energy of individual dots is relatively easy and can be done using external electrostatic gates that control the charge occupation of each dot, schematically shown in Fig. 3.4, and the tunnel amplitudes to the superconductor. However, tuning the relative amplitude of CAR and ECT is more challenging, as they scale in a similar way with external gates controlling the dots' properties. This issue was theoretically circumvented in Ref. [26] by considering non-colinear Zeeman fields in the dots, that may arise from an anisotropic Landé g-factor tensor in the dots. In this way, ECT and CAR amplitudes scale as $t = t_0 \cos\theta/2$ and $\Delta = \Delta_0 \sin\theta/2$,

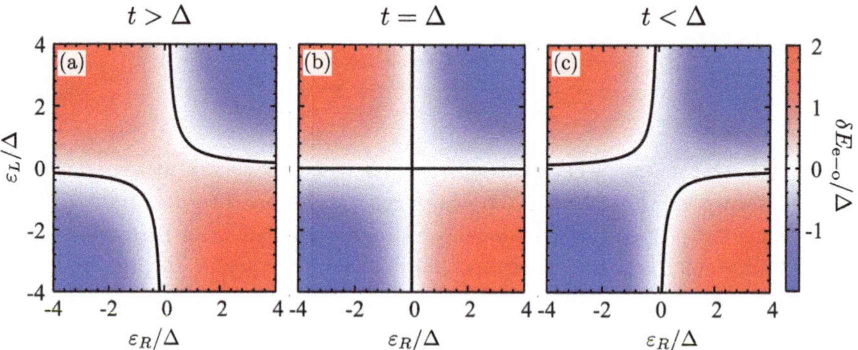

Fig. 3.6 Energy splitting between the even and odd ground states in a PMM system, $\delta E_{e-o} = E_o^- - E_e^-$, given in Eq. (3.61). We show results as a function of the energy of the PMM dots, ε_L and ε_R. The solid lines are the points with even–odd degeneracy, given by Eq. (3.62). (**a**) ECT-dominated regime ($t = 1.3\Delta$), (**b**) sweet spot regime ($t = \Delta$), and (**c**) CAR-dominated regime ($t = 0.7\Delta$). The direction of the avoided crossing can serve to distinguish between the different regimes and to find sweet spots

with θ being the angle between the spins and t_0 (Δ) is the maximum ECT (CAR) amplitudes for parallel (anti-parallel) spin configurations in the dot.

The spin-orbit coupling in semiconductor-superconductor and heavy compound superconductors provides a spin-mixing term for tunneling electrons. The spin-mixing term can lead to finite ECT and CAR amplitudes for spin-polarized QDs when the spin-orbit and magnetic fields are non-colinear. The tunability of the relative amplitudes between the two processes can be achieved by using a subgap state in the middle region [175]. This subgap state provides a low-excitation energy for ECT and CAR, therefore dominating the coupling between the two QDs. In the case where the central superconductor screens the external magnetic field and up to the lowest order in the tunnel amplitudes

$$t = f(\theta)\,\mathcal{E}_0 = f(\theta)\frac{t_L t_R}{\Delta_0}\left(\frac{2uv}{E/\Delta_0}\right)^2,$$

$$\Delta = g(\theta)\,C_0 = g(\theta)\frac{t_L t_R}{\Delta_0}\left(\frac{u^2 - v^2}{E/\Delta_0}\right)^2, \tag{3.67}$$

where

$$E = \sqrt{(\epsilon - \mu)^2 + \Delta_0^2} \tag{3.68}$$

is the energy of the central Andreev state, with BdG coefficients (see Sect. 3.1 for a discussion) given by $u^2 = 1 - v^2 = 1/2 + (\epsilon - \mu)/2E$. Here, Δ_0 is the gap of the central superconductor and μ controls the energy of the Andreev state, which has a minimum for $\mu = \epsilon$, Fig. 3.7a. For large values of μ, the Andreev state is mostly

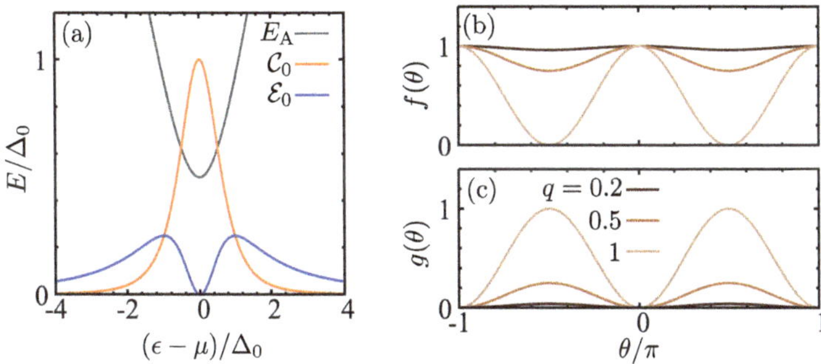

Fig. 3.7 CAR and ECT amplitudes, calculated using the model in Ref. [175]. (**a**) Gray line: energy of a subgap state, considering a parabolic model given in Eq. (3.68), with $\Delta_0 = 1$. The curve has been shifted down by $\Delta_0/2$. The orange and blue line show the amplitude for CAR and ECT amplitudes in Eq. (3.67). (**b**) and (**c**) show the angle dependence of the f and g functions in Eq. (3.67) for different values of the spin precession, parameterized with q

an electron ($u \approx 1$) or a hole ($v \approx 1$) like-state, dispersing linearly with μ. In this regime, the amplitude for ECT dominates, as CAR would depend on the product between u and v. On the other hand, the Andreev state has a minimum for $\mu = 0$, where $u = v = 1/\sqrt{2}$. In this regime, CAR reaches its maximum value, while ECT has a dip coming destructive interference between second-order processes, see discussion in Ref. [175]. Therefore, the ECT and CAR amplitudes intersect at two values as a function of the gate controlling the energy of the central superconductor, close to where Majorana sweet spots can appear. We note that the energy of the bound state in the middle region can be also tuned using the superconducting phase if it embedded in a Josepshon junction [178].

In this picture, we have ignore the spin texture, described by the functions f and g in Eq. (3.67), that account for the electron's spin-rotation due to the spin-orbit coupling. These functions depend on θ, the relative angle between the spin-orbit field and the local magnetization in the QDs. They are given by $f(\theta) = p^2 + q^2\cos^2\theta$ and $g(\theta) = q^2\sin^2\theta$ for equal spin polarization in the dots. The definitions are exchanged for dots with opposite magnetization. Here, $p = \cos(k_{\text{so}}L)$ and $q = \sin(k_{\text{so}}L)$ describe the spin precession probability for an electron traveling through the system, with $p^2 + q^2 = 1$, k_{so} being the spin-orbit wavevector and L the length of the system. When the exchange field aligns with the exchange field direction, $\theta = n\pi$, the electron spin remains invariant as it moves along the system. Therefore, $f = 1$ and $g = 0$, as illustrated in Fig. 3.7b, c. This situation is detrimental for spin-triplet superconductivity and, therefore, to generate Majorana states in the system, as CAR will be suppressed, see Eq. (3.67). On the other hand, $\theta = \pi/2$ boosts CAR. In this situation, ECT is only suppressed in the limit where $L = 1/k_{\text{so}}$. In a generic situation, there is a sweet spot where both amplitudes are equal and Majoranas are expected to appear in the system. A finite exchange field does not affect to the presence of crossings where CAR and ECT are equal, see Ref. [175] for a discussion.

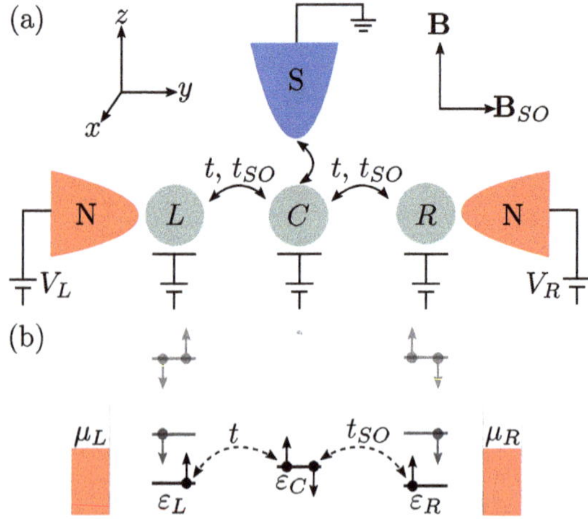

Fig. 3.8 (**a**) Sketch of the PMM system, containing three tunnel-coupled QDs (gray circles). The central QD is strongly coupled to a grounded superconductor (blue). We allow for spin preserving tunneling, with an amplitude t, and spin-flip tunneling due to spin-orbit coupling, t^{so}. The system couples weakly to two metallic electrodes to measure the local density of states and the non-local conductance. Panel (**b**) shows schematically the relevant energy scales and tunneling processes due to t and t^{so}. The left and right QDs are spin-polarized due to a external magnetic field, while the middle one is in the singlet state. Reprinted from Ref. [180] under CC-BY-4.0 license, ©2022, The Author(s)

3.4.1.2 Microscopic Models for the QDs

As shown before, the amplitudes for CAR and ECT are boosted if the two PMM QDs interact via a subgap Andreev state, which originates at the interface between a superconductor with other materials. To couple the two dots via the Andreev state, the length of the superconducting segment should be much smaller than the superconducting coherence length. In this regime, the state allows for an independent control of the ECT and CAR contributions via electrostatic gating of the superconducting segment between the QDs. In a realistic scenario, the situation is a bit more complex to the one described previously, as (i) the QDs exhibit interactions; (ii) the magnetic field that can be applied to the system, and therefore the Zeeman field in the QDs, is limited by the paramagnetic limit of the superconductor; and (iii) boosting the gap at the Majorana sweet spot requires to consider a non-perturbative coupling between the QDs and the central superconductor [179]. In this regime, the notions of CAR and ECT are not well-defied, as high-order tunneling processes between the dots dominate the physics. For this reason, in the following we will use a different nomenclature to refer to the coupling strength between the even and odd fermion parity subspaces: even and odd coupling.

The minimal microscopic model that describes all these effects is sketched in Fig. 3.8 and described by the Hamiltonian

$$H = H_D + H_S + H_T, \tag{3.69}$$

where dot Hamiltonian is described by

$$H_D = \sum_{\alpha=L,R} \left[\sum_{\sigma} (\varepsilon_\alpha + sE_{Z,\alpha}) d^\dagger_{\alpha\sigma} d_{\alpha\sigma} + U_\alpha n_{\alpha\uparrow} n_{\alpha\downarrow} \right], \tag{3.70}$$

with ε_α being the energy of the $\alpha = L, R$ QD, d_σ the annihilation operator for an electron with spin $\sigma =\uparrow, \downarrow$, E_Z the exchange field ($s = \pm 1$ for $\uparrow, \downarrow$ spin), U_α the charging energy, and $n_\sigma = d^\dagger_\sigma d_\sigma$. The physics of the central region is dominated by the onset of the bound state. The origin of this state is not important for the described physics below. For this reason, we consider a simple realization, based on a QD coupled to a bulk superconductor that screens charging energy and exchange field in the central region. We consider a regime where the gap in the central QD dominates over the charging energy. In this limit, the infinite gap limit provides a simple description for the Andreev state, see the large gap limit in Sect. 3.1.3.2 for details,

$$H_S = \sum_{\sigma} (\varepsilon_S + sE_{Z,S}) d^\dagger_{S\sigma} d_{S\sigma} + U_S n_{S\uparrow} n_{S\downarrow} + \Delta d^\dagger_{S\uparrow} d^\dagger_{S\downarrow} + \text{H.c.}, \tag{3.71}$$

where ε_S is the energy of the subgap state and Δ the induced pair amplitude in the central QD. The tunneling Hamiltonian is given by

$$H_T = \sum_{\alpha\sigma} t_\alpha d^\dagger_{\alpha\sigma} d_{S\sigma} + t^{\text{so}}_\alpha d^\dagger_{\alpha\sigma} d_{S\bar{\sigma}} + \text{H.c.}, \tag{3.72}$$

which describes the hopping of electrons between the QDs and the superconductor, with t and t^{so} being the amplitudes for normal and spin-flip tunneling due to the spin-orbit coupling ($\bar{\sigma}$ denotes the opposite spin to σ). This model recovers the one described the simple PMM models, described in Sect. 3.4.1 [26], in the limit $|t_\alpha|, |t^{\text{so}}_\alpha|, |U_\alpha|, |E_{Z,\alpha}| \ll |\Delta|, |E_{Z,\alpha}|$. In this limit, the outer QDs are occupied by a single electron for $|\varepsilon_\alpha| \ll E_{Z,\alpha}$ and the central is in a superposition between 0 and 2 electrons. A second-order perturbation theory leads to an effective coupling between L and R QDs that preserves spin ($\propto t_L t_R$), ECT, and flips spin ($\propto t_L t^{\text{so}}_R$, $\propto t_L t^{\text{so}}_R$), CAR.

In interacting Kitaev chains with finite magnetic field, there are no ideal Majorana sweet spots, as the presence of the other spin species introduces corrections to the simple picture described in Sect. 3.4.1. However, sweet spots with very good Majorana localization and (close to) degenerate ground states with even and odd parity can still appear in the system. The Majorana

(continued)

localization is quantified by the so-called Majorana polarization (MP), that describes how much the local wavefunction in the L/R QDs looks like a Majorana state [181]. For real Hamiltonians, i.e. with no complex phases, as the one described in Eq. (3.69), MP is defined as follows

$$M_\alpha = \frac{W_\alpha^2 - Z_\alpha^2}{W_\alpha^2 + Z_\alpha^2} \tag{3.73}$$

where,

$$\begin{aligned} W_\alpha &= \sum_\sigma \left\langle O \left| d_{\alpha\sigma} + d_{\alpha\sigma}^\dagger \right| E \right\rangle , \\ Z_\alpha &= \sum_\sigma \left\langle O \left| d_{\alpha\sigma} - d_{\alpha\sigma}^\dagger \right| E \right\rangle , \end{aligned} \tag{3.74}$$

with $|O(E)\rangle$ being the ground state wavevector with total odd (even) fermion parity. In an ideal sweet spot where two Majoranas are perfectly localized at the left/right QD, leading to $W_\sigma = 1$, $Z_\alpha = 0$ and $W_\alpha = 0$, $Z_\alpha = 1$ for the two sides. It means that $M_L = -M_R = \pm 1$. The presence of additional Majorana components on the dot, due either to a non-perfect localization or a finite Zeeman field results on an absolute MP value smaller than 1.

We note that MP does not contain information about the wavefunction localization, as it is a measurement that involves only local operators in one of the QDs. Therefore, having $|M_\alpha| \approx 1$ does not imply a perfect Majorana localization on the outer L/R QDs, as the Majorana wavefunctions can still overlap in the central region without splitting the ground state degeneracy. An ideal Majorana sweet spot would require having an almost unity value for W and Z in both dots. However, a state with a high MP still preserves the relevant Majorana properties, that implies that non-local experiments, including fusion and braiding, will also work in the setup, as we describe below in Sect. 3.6.4. Therefore, three conditions are required to define high-quality Majorana sweet spots: (i) degenerate ground states with even and odd fermion parity; (ii) localized MBSs of high quality, characterized by a high MP value in both QDs; and (iii) substantial gap to the excited states.

Figure 3.9a shows results for the energy spectrum as a function of the energy of the three QDs, that can be controlled with external gates. For sufficiently large exchange field in the external PMM dots, the system can exhibit a ground state with a total odd fermion parity (blue color in the figure). Around this region, the system has an even–odd degeneracy, needed to fulfill the energy degeneracy (i). The absolute value of MP in one of the QDs is shown in Fig. 3.9b. It peaks for some discrete values in gate configuration. For generic parameters, there are two spots at which conditions (i) and (ii) are simultaneously fulfilled, illustrated by the cuts shown in Fig. 3.9c and d, where also the energy spectrum is shown.

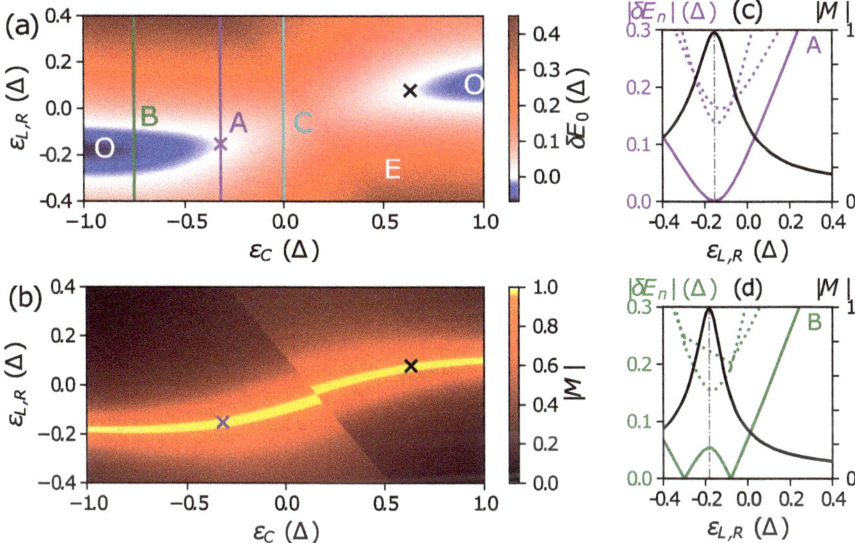

Fig. 3.9 (**a**) Energy spectrum as a function of the energy of the left and right QDs ($\varepsilon_{L,R}$) and the central QD (ε_C). Red (blue) colors indicate an even (odd) ground state, while the white stripes are regions with even–odd ground state degeneracy. (**b**) Absolute value of MP. The purple (black) cross indicates the spot where a ground state degeneracy coincides with a high MP values, appearing for $\varepsilon_{L,R} = -0.319\Delta$ (0.634Δ) and $\varepsilon_C = -0.151\Delta$ (0.0785Δ). (**c**) and (**d**) are cuts along lines A and B in panel (a). Color lines correspond to the excitation energies, where the solid line is the ground state. The black line represents the absolute value of MP and the vertical dashed-dotted line is a guide to the eye at the maximum MP value. The remaining parameters are $t = 0.5\Delta$, $t^{\mathrm{so}} = 0.2\Delta$, $U_{L,R} = 5\Delta$, $U_C = 0$, $E_{Z,L} = E_{Z,R} = 1.5\Delta$, $E_{ZC} = 0$. Reprinted from Ref. [180] under CC-BY-4.0 license, ©2022, The Author(s)

The system features low-MP sweet spot when either the QD energies shift from the sweet spot or the Zeeman field is reduced. Lowering the external magnetic field will re-introduce the second spin spices in the PMM dots, leading to overlapping Majorana states. Differently from the low-MP sweet spots found when varying the ratio between t and Δ from unity (or, equivalently, ε_C in Fig. 3.9), low-MP sweet spots can appear as (almost) zero excitation energy crossings in ε_L and ε_R. In this regime, the local charging energy in the outer PMM dots help at increasing the MP, as illustrated in Fig. 3.10. At the mean-field level, the effect of the local interactions is to increase the local exchange field.

3.4.1.3 Transport Characterization

Electron transport is one of the most broadly used methods to characterized the electronic properties of nanoscopic systems. In this section, we introduce three different experimental measurements that can be used to characterize the PMM system and to identify Majorana sweet spots with high MP. These measurements

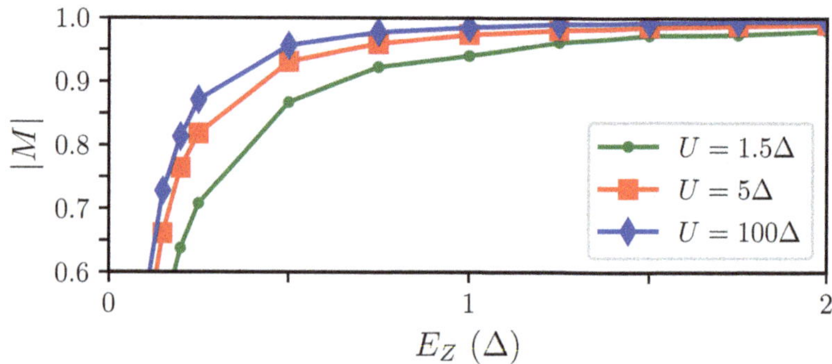

Fig. 3.10 Maximum MP values for sweet spots found with increasing exchange field and local charging energy in the left and right dots, U ($U_S = 0$). The sweet spots are energy crossings found by tuning ε_L, ε_C, and ε_R, and maximizing MP. Reprinted from Ref. [180] under CC-BY-4.0 license, ©2022, The Author(s)

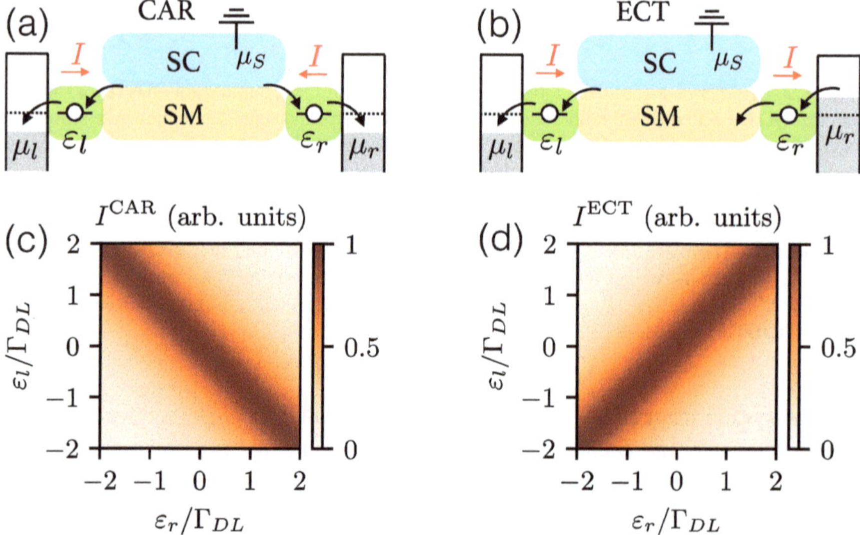

Fig. 3.11 Current measurement of CAR and ECT in the weak coupling regime. Top panels show the gate configurations used measure these two processes. The bottom panels show numerical calculations using rate equations on the model introduced in Sect. 3.4.1.1. Reprinted from Ref. [175] with permission, ©2022, American Physical Society. All rights reserved

involve coupling the PMM system to normal leads, that can be used to measure the local spectrum of the system.

The first experimental challenge consists on finding regimes where the coupling between the even and odd fermion parity subspaces (or CAR and ECT in the weak superconductor-dot tunneling regime) are equal. The experimental setup to measure these two amplitudes is shown schematically in the top panels of Fig. 3.11, taken from Ref. [175]. In the setup, two metallic leads couple to the outermost PMM dots, allowing for current to flow between the three different terminals (the two metallic electrodes and the grounded superconductor, blue rectangle in the figure).

For gate configurations where the dots have opposite energies, $\varepsilon_L = -\varepsilon_R$, processes involving the splitting of Cooper pairs are resonant in energy, Fig. 3.11a. The split electrons can flow out from the QDs to the metallic leads if their chemical potential is smaller than the QDs' energies, $\mu_L < \varepsilon_L$ and $\mu_R < \varepsilon_R$. Under these conditions, a net current will cross the system, flowing from the superconductor and dividing equally to the left and right metallic leads. There is a corresponding process where electrons recombine into Cooper pairs. This process is also resonant for $\varepsilon_L = -\varepsilon_R$ and dominates for a large enough bias voltage in the leads, $\mu_L > \varepsilon_L$ and $\mu_R > \varepsilon_R$, resulting in a net current flowing from the metallic electrodes into the central PMM superconductor.

In contrast, the elastic transference of electrons between the dots is resonant when the orbital energy of the two dots is the same, $\varepsilon_L = \varepsilon_R$, Fig. 3.11b. In this case, a finite current can flow between the leads if the bias is applied asymmetrically, so that electrons can tunnel on the right dot ($\mu_R > \varepsilon_R$) and leave to the left electrode ($\mu_L < \varepsilon_L$). This generates a current between the metallic leads, equal in magnitude but with opposite sign. The current direction is changed if the biasing conditions are reversed. In this picture, the central superconductor only mediates the coupling between the dots and does not add additional charges into the system. This is only true if the energy of the two dots is small enough (smaller than the lowest state in the superconductor) to avoid the tunneling of quasiparticles between the superconductor and the QDs.

Measuring ECT and CAR Amplitudes From Current

Simple expressions for the current in the two configurations described above can be found using the model described in Sect. 3.4.1.1 that considers very large exchange field (ignoring one of the spin-spices in the dots) and small tunneling coupling between the dots and the central superconductor. In the regime where also the coupling to the metallic leads is weak, perturbation theory provides an accurate way to obtain the current through the system. For the two chemical potential configurations described above, the current through the left lead is given by

$$I^{CAR} = \frac{e}{\hbar}\frac{\Gamma}{(\varepsilon_L + \varepsilon_R)^2 + \Gamma^2}|\Delta|^2\,, \qquad \text{for } \mu_S > \mu_{L,R}\,, \tag{3.75}$$

$$I^{ECT} = \frac{e}{\hbar}\frac{\Gamma}{(\varepsilon_L - \varepsilon_R)^2 + \Gamma^2}|t|^2\,, \qquad \text{for } \mu_L < \mu_S < \mu_L\,. \tag{3.76}$$

(continued)

Results for these current contributions as a function of the energy levels of both dots are shown in Fig. 3.11c, d. From Eqs. (3.76), it is easy to see that the relation between the maximum current in both configurations provides a way to measure the relative strength between CAR and ECT. In fact, $I^{CAR}/I^{ECT} = |\Delta/t|^2$ for $\varepsilon_L = \varepsilon_R = 0|$, providing a way to determine the effective CAR and ECT amplitudes.

Local and Non-local Conductance

The leads attached to the system can be also used to identify sweet spots with potentially high MP. In local spectroscopy, the low-bias transport and, therefore the zero-bias conductance, is enhanced when the full system has a degenerate ground state. This means that electrons at the Fermi level of the superconductor can resonantly tunnel through the system. This is illustrated in the top panels of Fig. 3.12, where the red lines correspond to the degeneracy between states with even and odd fermion parity (see also Fig. 3.6). Depending on whether the coupling between even or odd states dominates, we expect to see an avoided crossing between the zero-bias conductance lines that go through the anti-diagonal or the diagonal. In contrast, the zero-bias conductance lines cross at the sweet spot, located at the maximum of the local conductance value.

However, these features might be hard to resolve experimentally due to the broadening, mostly close to the sweet spot where the avoided crossing between the conductance lines might be small. The local conductance in Fig. 3.12 in the upper panels can serve as an illustration. For this reason, it is very convenient to study the non-local transport through the system, that provides additional information about the nature of the crossing. To measure the non-local current, a bias voltage is applied asymmetrically to the system, while current is measured in the other side. Therefore, the non-local conductance is defined as

$$G_{\mu\nu} = \frac{\partial I_\mu}{\partial V_\nu}, \tag{3.77}$$

with $\mu, \nu = L, R$. The lower panels of Fig. 3.12 show results for the non-local conductance, G_{LR}, through a PMM system around the Majorana sweet spot and at zero bias. The three cases correspond to the line cuts shown in Fig. 3.9.

In the regime where the even parity states couple more strongly than the odd ones, transport is dominated by CAR-like processes, changing simultaneously the occupation of both dots. For $\mu_R > \mu_S$ it means that electrons in both dots will recombine in Cooper pairs at the central superconductor,

(continued)

leading to a current flowing in from the left lead, Fig. 3.12d. In the opposite situation where the coupling between the odd parity states dominate, transport will be dominated by the injected electrons to the right dot will cotunnel to the left one. Therefore, current will mostly flow from right to left, Fig. 3.12f. At the sweet spot, both processes are equally important and the non-local conductance shows a symmetric pattern of positive and negative conductance, Fig. 3.12e. The Majorana sweet spot locates at the center of this pattern. In general, sweet spots with low-MP are characterized by a deformed non-local conductance pattern with respect to the one shown in Fig. 3.12e, see Ref. [180].

However, the differences between high and low-MP in the non-local conductance are sometimes faint, which makes it hard to distinguish between the two. For this reason, it is important to develop ways to identify high-MP sweet spots before proceeding to non-local experiments like braiding, Sect. 3.6.4.

The Majorana localization on one side of the PMM system can be proven by coupling the system to an additional dot, as theoretically proposed and analyzed in Refs. [150, 151] for Majorana wires. These proposals inspired experiments that showed similar patterns to the ones predicted by theory [148, 149]. In these references, they showed that a QD cannot split the ground state degeneracy when it only couples to one of the Majorana component. In contrast, the coupling between the dot and two Majoranas split the degeneracy. The energy splitting shows a "bowtie-like" shape as a function of the energy of the additional dot, becoming maximal when one of the additional dot's energy levels align with the chemical potential of the superconductor. The situation is reversed if the two Majoranas overlap but only one couple to the additional dot. In this case, the ground state degeneracy is split, except when one of the additional dots level align with the superconductor's chemical potential, leading to a "bowtie-like" pattern. This idea has been recently extended to PMM systems, showing that an additional QD can be used to identify high-MP sweet spots [182].

3.4.1.4 Experiments in Minimal Kitaev Chains

Early experiments studied CAR processes in superconducting heterostructures, with the aim of generating a source for entangled electrons, so-called Cooper pair splitters [171–174, 183–186]. The experimental realization of these splitters require electron channels separated by a distance smaller than the superconducting coherence length, that is usually of the order of tens of nanometers to a micron for conventional BCS superconductors. In this context, QDs offer a fundamental advantage as their charging energy suppresses local Andreev reflections for a sufficiently small coupling to the central superconductor [171, 185]: processes

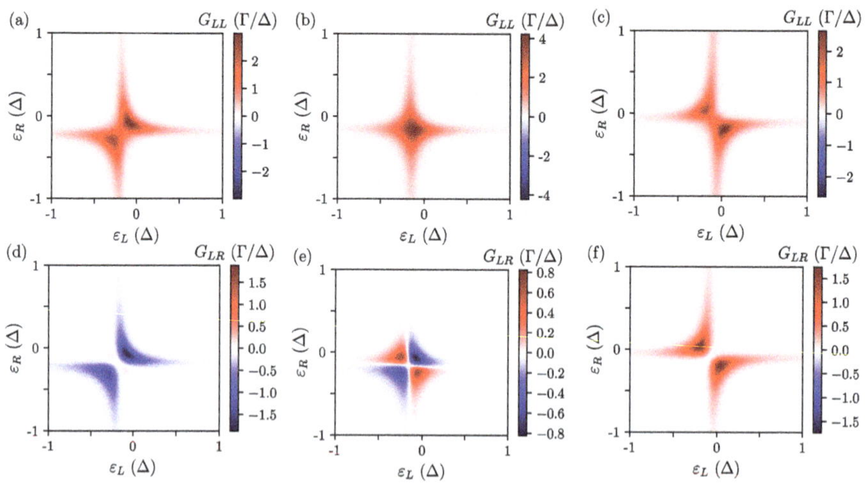

Fig. 3.12 Zero-bias conductance through a PMM system, as a function of the energy of the outer dots, which are attached to metallic electrodes, see Fig. 3.8. Top panels show local conductance at the left lead, where the bias is applied symmetrically in the two leads. Lower panels show the non-local conductance, where bias is applied to right lead and current measured in the left one. From left to right, we show results for even parity-dominated regime (the coupling between the even states is stronger), the sweet spot (equal coupling within the two fermion parity sectors), and the odd parity-dominated regime, corresponding to the ε_C values highlighted in Fig. 3.10. Reprinted from Ref. [180] under CC-BY-4.0 license, ©2022, The Author(s). Similar theoretical and experimental results were reported in Ref. [21], see also Sect. 3.4.1.4

where a Cooper pair tunnels to the same terminal. As mention in Sect. 3.4.1.1 (see also Ref. [175]), an Andreev bound state in the superconducting region can mediate the coupling between the electrons. These bound states, with energies below the superconducting gap, offer the dominant contribution for Cooper pairs tunneling to the dots due to their reduced energies. Therefore, they offer advantages, including the increase the Cooper pair splitting efficiency and the possibility of tuning the relative amplitudes between CAR and ECT by modulating the energy of the bound state, that can be achieved through electrostatic gates acting on the central superconductor. For this task, superconductor-semiconductor hybrid structures are advantageous, as they can exhibit subgap states that are gate tunable. Experiments have demonstrated the possibility of hybridizing a subgap state with a QD [107, 148, 149] and tuning the particle/hole component, i.e. the BCS charge of a subgap state, using electrostatic gates [108]. These are required ingredients to realize PMM systems and tune them to the sweet spots where localized Majoranas appear.

Independent measurements on semiconductor-superconductor hybrid devices demonstrated CAR processes of Cooper pairs with the same spin in nanowires [23, 24, 187] and two-dimensional electron gases [25]. This observation points towards the presence of equal-spin-triplet Cooper pairs due to the interplay between spin-orbit coupling and an external magnetic field. The existence of these Cooper pairs

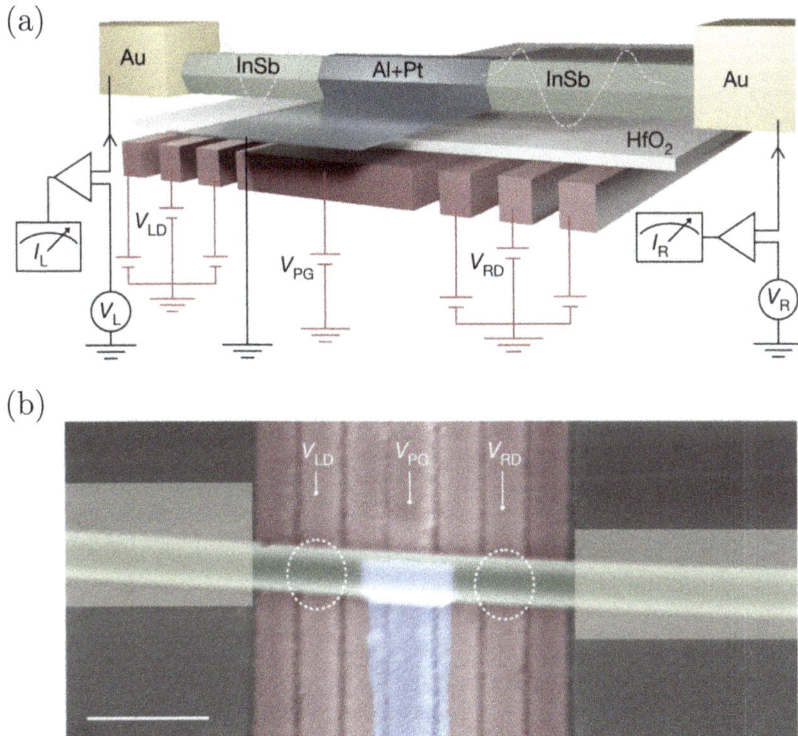

Fig. 3.13 Illustration of the PMM system used in Ref. [21]. (**a**) Two QDs are defined in an InSb nanowire and coupled via a central region that has proximity-induced superconductivity coming from a grounded superconductor. The outer Au leads are used for spectroscopy. The gates at the bottom (red) allow to tune the device into the desired configuration. (**b**) False-colored scanning electron microscopy image of the device, before the fabrication of the N leads (schematically shown in yellow). The scale bar is 300 nm. Reprinted from Ref. [21] with permission, ©2023, The Author(s), under exclusive licence to Springer Nature Limited. All rights reserved

is an essential requirement to realize topological superconductivity. In particular, Refs. [24, 25] determined the amplitudes for CAR and ECT using transport measurements across the device, as described in Sect. 3.4.1.3. They demonstrated a fine-tuned regime of parameters where the amplitude of both processes is the same and where PMMs can appear. Additionally, a Cooper pair splitting efficiency around 90% was demonstrated, that is over the threshold required for a Bell test experiment [185].

Additional local and non-local spectroscopic measurements in a PMM system were reported in Refs. [21,188,189]. The device in Ref. [21], sketched in Fig. 3.13a, is composed by a InSb nanowire fabricated using the shadow-wall lithography technique [190, 191]. The bottom gates (red) control the energy and the coupling of the QDs to the central superconductor (blue) and the metallic leads used for spectroscopy (yellow rectangles). The QDs are spin-polarized using an external

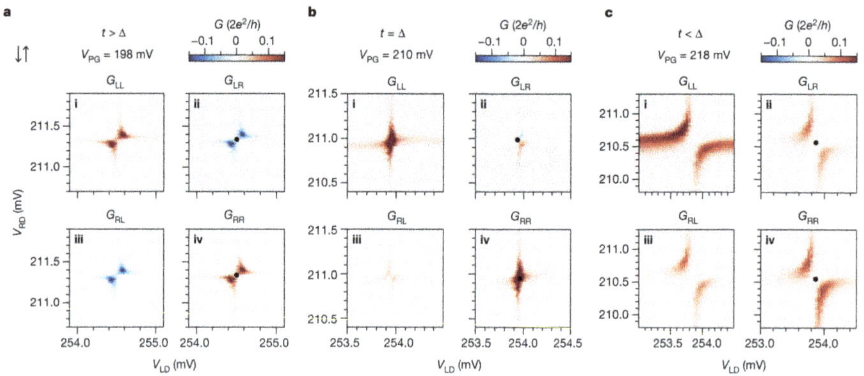

Fig. 3.14 Conductance matrices measured for different values of the gate controlling the semiconductor segment underneath the superconductor (V_{PG} =198, 210 and 218 mV, for panels **a**–**c**). The three panels are representative measurements of the three possible regimes (ECT-dominated regime, close to the sweet spot with ECT and CAR amplitudes similar, and CAR-dominated regime) and in qualitative agreement with theory, see Fig. 3.12. Reprinted from Ref. [21] with permission, ©2023, The Author(s), under exclusive licence to Springer Nature Limited. All rights reserved

magnetic field, boosted by the high Landé g-factor measured in InSb. The grounded superconductor at the center mediates CAR and ECT processes between the spin-polarized QDs. The central region hosts an Andreev bound state that mediates both processes and allows to tune their relative amplitude, as explained in Sect. 3.4.1.1.

Measurements for the local and non-local conductance are shown in Fig. 3.14 [21]. The figure shows representative results for the three important regimes: $t > \Delta$ (left panels), $t < \Delta$ (right panels), and $t \approx \Delta$ (middle panel). Local spectroscopy (subpanels panels i and iv in every panel) reveals an avoided crossing between two levels. The direction of the avoided crossing along the diagonal/anti-diagonal indicates the dominance of CAR/ECT processes. In contrast, for $t \approx \Delta$, the levels seem to cross, showing a high conductance peak at the sweet spot. We also note a change on the sign of the non-local conductance between the left and right panels (subpanels ii and iii), while at the sweet spot (middle panels) there seems to be an alternation between positive and negative features, although with a dominance of the positive ones.

In the simplest model proposed by Kitaev, the protection to deviations from the ideal sweet spot improves when adding additional dots to the chain. In particular, the truly topological phase is recovered in the limit of chain being much larger than the localization of the Majorana states. In this limit, deviations from the sweet spot does not affect to the topological properties of the system. This vision motivated recent experiments on Kitaev chains with three sites, i.e. three QDs that couple via two superconducting segments [192, 193]. Achieving longer Kitaev chains would require to tune many gate electrodes controlling the energies of the QDs, the intermediate superconducting segments, and the tunneling between these regions. With this motivation, approaches based on machine learning have been recently developed [194–196].

3.5 Hybrid Semiconductor-Based Superconducting Qubits

Superconductors constitute one of the most promising platforms for quantum technologies superconductors allow quantum effects to be seen and manipulated at scales much larger than individual atoms or particles, unlike typical quantum systems that only show coherent effects at the microscopic scale. At the heart of this phenomenon is the Cooper pair: two electrons that move through a superconductor without resistance. These pairs behave as a single quantum entity, displaying characteristic quantum properties like tunneling and entanglement over macroscopic distances. One of the most striking demonstrations of this is the Josephson effect, where an electric current flows between two superconductors separated by a thin barrier, without any voltage applied.

Transmon qubits are paradigmatic examples of the application of superconductivity for the implementation of quantum technologies. They are pivotal in the advancement of quantum computing, showcasing how macroscopic quantum phenomena can be harnessed for computational purposes. Originating from the Cooper pair box design, transmon qubits have been modified to increase their stability and coherence. Their key advantage lies in their diminished sensitivity to charge noise, which plagued earlier superconducting qubit designs. This is achieved through a combination of a Josephson junction and a large shunting capacitor, resulting in a system with distinct energy level separations [197]. Quantum information in transmon qubits is encoded in the energy states of the junction and manipulated using microwave pulses. This allows transmon qubits to maintain quantum states, like superpositions and entanglements, for extended durations, essential for effective quantum computation. Additionally, their relatively straightforward design and manufacturability facilitate integration into larger, more complex quantum circuits. Transmon qubits thus represent the realization of scalable and practical quantum computers, poised to tackle computational challenges far beyond the reach of classical computing [198].

The semiconductor-superconductor platform allows to tune the critical current of Josephson junctions using external gate electrons. This capability allowed to design the so-called gatemon: a transmon qubit with a tunable critical current [199–202]. This gate-tunability provides a key advantage: it allows for a more flexible and precise control over the qubit's frequency, enabling to tune the qubit into and out of resonance with other qubits or microwave cavities. It facilitates the qubit coupling and entanglement operations. Moreover, gatemon qubits retain the charge noise insensitivity characteristic of transmons, making them a promising architecture for scalable quantum computing.

Even though transmon qubits are protected against charge noise, they are still sensitive to other sources of noise that limit their performance. Different proposals based on Josephson junctions have been designed to improve the performance of qubits based on Josephson junctions, including the so-called "$0-\pi$" qubit [203–205], parity protected qubits [206–210], and the fluxonium qubit [211]. These proposals are compatible with macroscopic superconductors featuring Josephson junctions. On the other hand, microscopic superconductors offer the possibility

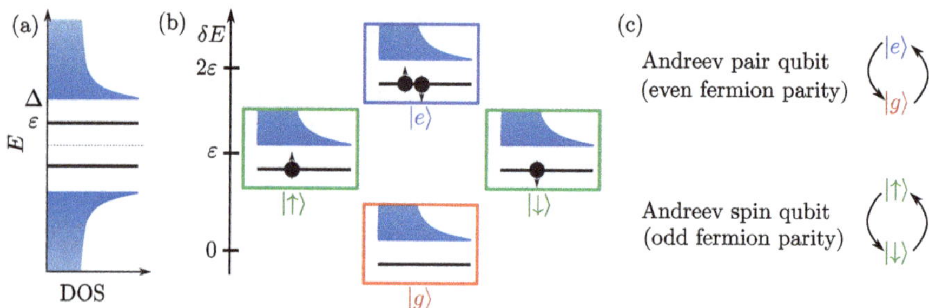

Fig. 3.15 Qubits based on single Andreev levels. (**a**) Energy spectrum of a superconducting system featuring one ABS at energy ε. (**b**) States of the spin-degenerate Andreev levels. Red (ground state) and blue (excited state with two opposite spin quasiparticles in the ABS) represent the even fermion parity states. The green squares show the odd fermion parity states. (**c**) Basis for the Andreev qubit and the Andreev spin qubit

of manipulating superconductivity at the nanoscale, as we have illustrated in the previous sections, providing a new realm for quantum technologies. In this context, superconductor-semiconductor platforms offer many advantages. In the following of this chapter, we present qubits based on nanoscopic superconductors that have been analyzed both, from the theory and the experimental side, in the recent years.

3.5.1 Qubits Based on a Single Andreev State

Andreev states, appearing inside the superconducting gap, are ideal candidates to storage and process quantum information. Spin-degenerate Andreev states can host up to two quasiparticle excitations. Therefore, the system has four states in total, two with even and two with odd fermion parity. Assuming that electrons in the superconductor form Cooper pairs, the fermion parity is a good quantum number and the even and odd fermion parity subspaces are disconnected. In the ground state, all the electrons are assumed to form Cooper pairs, occupying the states with energies below the superconductor's Fermi level ($|g\rangle$). The lowest-excited state with the same parity has two excited quasiparticles in the Andreev bound state

$$|e\rangle = \gamma_{\uparrow}^{\mathrm{ABS}}\gamma_{\downarrow}^{\mathrm{ABS}}|g\rangle\,, \tag{3.78}$$

with $\gamma_{\sigma}^{\mathrm{ABS}}$ being the BdG operator of the Andreev state. The state $|e\rangle$ has an energy 2ε higher than the ground state, as depicted in Fig. 3.15b. There are additional excited states with even fermion parity that involve the excitation of quasiparticles above the gap. However, quasiparticles in the continuum can diffuse fast and scape the system. For this reason, these states are avoided for quantum applications.

On the other hand, the odd fermion parity subspace has two states that correspond to excited quasiparticles with opposite spins

$$|\sigma\rangle = \gamma_{\sigma}^{\mathrm{ABS}}|g\rangle\,. \tag{3.79}$$

These states are degenerate in the absence of spin-breaking symmetry terms and split due to an external of magnetic field. Therefore a spinful subgap state defines two, in principle, disconnected subspaces where information can be encoded and processed.

The Andreev Pair Qubit

The first experimental demonstration of a qubit based on Andreev states focused on the even fermion parity state, therefore using the states $|g\rangle$ and $|e\rangle$ as qubit states, see Fig. 3.16. These experiments exploited the discrete Andreev states appearing in Josephson junctions between two superconductors. Andreev qubits have been realized in superconducting quantum point contacts [212] and semiconductor-superconductor nanowires [213]. To measure and manipulate the qubit, it is advantageous to embed the Josephson junction into a superconducting loop. The current through the loop depends on the state of the qubit, having different directions between the $|g\rangle$ and $|e\rangle$ states, due to their opposite curvature with phase, originated from particle-hole symmetry, see for example Fig. 3.1b. This means that the qubit can be measured by inductively coupling it to a microwave LC resonator, whose properties get renormalized depending on the qubit state.

Specifically, the full Hamiltonian for the LC resonator+qubit system is a Jaynes Cummings Hamiltonian of the form

$$H_{JC} = \hbar\omega_r a^\dagger a + H_A + \hbar g_c(a^\dagger\sigma_- + a\sigma_+), \tag{3.80}$$

with $a^\dagger(a)$ the creation (anhilation) operators for the LC resonator with frequency $\omega_r = 1/\sqrt{LC}$ and $H_A = \frac{\hbar\omega_A}{2}\sigma_z$ being the Hamiltonian defining the Andreev qubit, written in terms of the Pauli matrix $\sigma_z = |e\rangle\langle e| - |g\rangle\langle g|$ and $\omega_A = 2E_A(\phi)$ from Eq. (3.11). The Jaynes Cummings coupling term of strength g_c is written in terms of the corresponding Pauli matrices σ_+ and σ_-.

For this particular implementation, the Jaynes Cumming coupling is achieved through the inductive coupling. Specifically, one can write a direct coupling between the resonator and the supercurrent operator of the Andreev levels

$$\hat{I}_A(\phi) = \frac{2\pi}{\Phi_0}\frac{\partial H_A}{\partial\phi}, \tag{3.81}$$

with $\Phi_0 = h/2e$ being the superconducting flux quantum. The calculation of the supercurrent operator involves some subtleties since, apart from the obvious ϕ dependence of the qubit energy, one needs to consider the phase dependence of the Andreev level wavefunctions (since σ_z is written in the basis of the Andreev level). Putting everything together results in a

(continued)

supercurrent operator that involves off-diagonal elements (which in turn allow to drive in qubit transitions):

$$\hat{I}_A(\phi) = \frac{2\pi}{\Phi_0}\frac{\partial E_A(\phi)}{\partial \phi}[\sigma_z(\phi) + \sqrt{1-\tau}\tan(\phi/2)\sigma_x(\phi)]. \tag{3.82}$$

The direct coupling appears when Taylor-expanding the total flux-dependent hamiltonian $H = H_A + \Phi\frac{\partial H_A}{\partial \Phi} + \ldots$, which results in a Jaynes Cummings hamiltonian

$$H_c = \Phi\hat{I}_A(\phi) = \Phi_r(a^\dagger + a)\hat{I}_A(\phi) \approx g_c(\phi)(a^\dagger\sigma_- + a\sigma_+), \tag{3.83}$$

with the operator describing vacuum quantum phase fluctuations of the LC resonator (strength $\Phi_r = \sqrt{\frac{L\hbar\omega_r}{2}}$) defined as $\Phi = \Phi_r(a^\dagger + a)$. In the last step, the coupling is defined as $g_c(\phi) = 2\pi\frac{\Phi_r}{\Phi_0}\sqrt{1-\tau}\frac{\partial E(\phi)}{\partial \phi}\tan(\phi/2)$, where we have dropped the diagonal part of the current operator and performed a rotating wave approximation.

In circuit QED, experiments can be performed such that the frequency difference between the resonator and the qubit is much bigger than the coupling between them $g_c \ll |\omega_r - \omega_A|$, which is known as the dispersive regime. In such dispersive limit, the hamiltonian can be written as:

$$H_{dis} = \hbar(\omega_r + \chi\sigma_z)a^\dagger a + \frac{\hbar\omega_A}{2}\sigma_z, \tag{3.84}$$

with

$$\chi = \frac{g_c^2}{\omega_A - \omega_r}, \tag{3.85}$$

which is known as the dispersive shift. This equation has a profound meaning since it means that the Jaynes Cummings coupling in the dispersive regime results is a *qubit state dependent* shift of the resonator frequency by $\pm\chi$. By monitoring the resonator response to a microwave readout tone, the quantum state of the qubit can be resolved.

Interestingly, g_c is maximum at $\phi = \pi$ for the Andreev pair qubit. This is very convenient, since at half flux, the dispersive shift χ is maximized because the numerator in Eq. (3.85) becomes big while the denominator becomes

(continued)

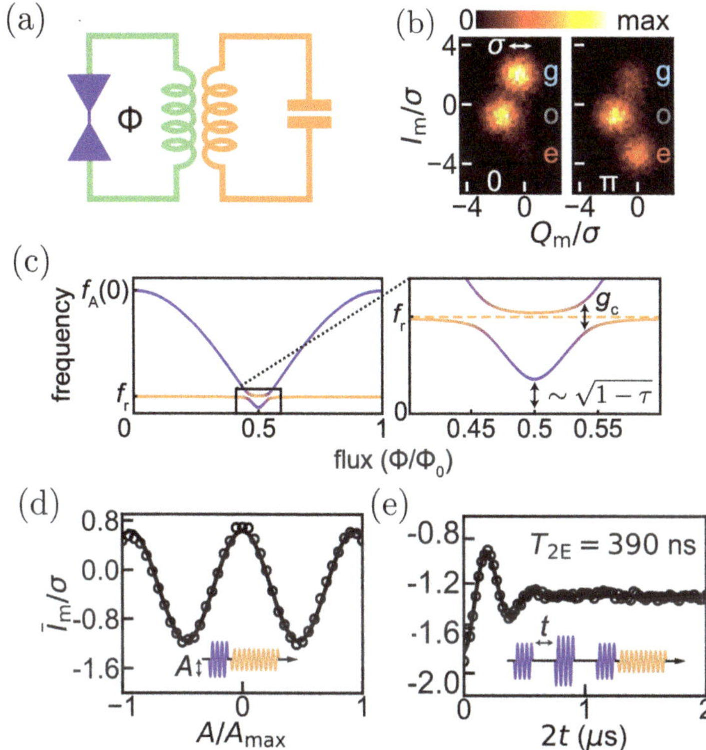

Fig. 3.16 (**a**) Sketch of the used setup to measure the coherent properties of the Andreev qubit, where a superconducting junction (purple) is embedded into a loop (green) and inductively coupled to a resonator (yellow). (**b**) Histogram of the $\bar{I}_m$ and Q_m quadratures of the readout tone with no driving (left) and after a π pulse (right). (**c**) Theory illustration of the spectrum of the system. (**d**) Rabi oscillations. (**e**). Coherence of the qubit measured using a Hahn-echo pulse sequence. Adapted from Ref. [213] with permission, ©2018, American Physical Society. All rights reserved

small. This results in a strong avoided crossing near $\phi = \pi$ as illustrated in Fig. 3.16c.

In the same way, resonant pulses, with frequency $\omega_A = 2\pi f_A$, induce Rabi oscillations, changing the relative population of $|g\rangle$ and $|e\rangle$ states, Fig. 3.16d. The decoherence times of such qubit can be extracted by pulse sequences such as Hahn-echoes, Fig. 3.16e. Finally, it is important to mention the recent achievement of a very long range coupling of two Andreev qubits mediated by a microwave photon in a superconducting cavity coupler [214]. Instead of using inductive coupling like in the previous cases, this experiment uses a coupling capacitor designed between two resonators at the respective voltage

(continued)

antinodes. This allows to perform fast readout of each qubit using the strongly coupled mode, while the weakly coupled mode is utilized to mediate a very long range coupling between the qubits (over a distance of six millimeters).

The Andreev Spin Qubit

The odd parity subspace is formed by two states characterized by a single BdG quasiparticle trapped in the ABSs. This quasiparticle has a well-defined spin, that can be decomposed into up/down components of a given basis, leading to an effective spin 1/2 degree of freedom where information can be encoded. This qubit [215] is the superconducting counterpart of conventional an spin qubit, realized in coupled QDs [216], and is hence dubbed Andreev spin qubit. Single qubit operations, like transitions between $|\uparrow\rangle$ and $|\downarrow\rangle$, require transitions between opposite spin states in the Andreev level. These transitions are allowed by terms that mix the two spins, like spin-orbit coupling and time-reversal breaking terms, like magnetic fields or a finite phase difference. Two implementations of Andreev spin qubits in semiconducting nanowire junctions have been studied. The first was experimentally realized using spin-split ABSs of a long Al/InAs semiconducting Josephson junction [217]. In these experimental implementation, the qubit was incorporated into the inductor of a lumped element resonator. By doing so, the intrinsic spin-supercurrent coupling of the superconducting spin qubit could be exploited to monitor the spin state of the qubit using circuit quantum electrodynamics techniques [15]. This first Andreev spin qubit implementation based on a long Josephson junction, however, presents a fundamental challenge: the ground state of the system is a singlet, making the two qubit states an excited manifold. Decay from the qubit states to the ground state, consequently, leads to leakage and constrains the qubit's coherence times. In this implementation, the singlet and doublet switching rates were found to be of the same order of magnitude, with spin lifetimes of tens of μs [15,217]. An alternative approach to implementing Andreev spin qubits in semiconducting nanowires, involves exploiting the finite charging energy within the Josephson junction when the semiconducting segment contains a QD. Importantly, this method enables the operation of the qubit's state in a regime where they are the lowest energy states of the system. As discussed in Sect. 3.1.3.2, such QD-based Josephson junctions can be tuned into a regime where the doublet states constitute the lowest energy manifold of the system, see Fig. 3.1d. The first step towards this QD-based Andreev spin qubit was demonstrated in Ref. [16] which presented microwave spectroscopy experiments in a hybrid superconductor-semiconductor transmon device in which the Josephson effect was controlled

(continued)

by a gate-defined quantum dot in an InAs-Al nanowire. In such setup, the microwave spectrum of the transmon depends on the ground-state parity of the QD. This allows to measure the parity phase diagram as a function of different experimental tuning knobs, such as gate voltages and external magnetic flux, in very good agreement with that predicted by a single-impurity Anderson model with superconducting leads. Furthermore, continuous-time monitoring of the circuit allows to resolve the quasiparticle dynamics which, owing to the QD in the Josephson junction, result in a notable increase in the doublet lifetime up to the order of 1–10 ms. Furthermore, the application of a magnetic field not only increases the spin splitting due to the Zeeman effect $f_s = E_\uparrow - E_\downarrow$, but, in combination with spin-orbit interaction, also allows for direct driving of the spin-flip transition [17], as opposed to Ref. [217].

Tunability of the doublet ground state [16] together with direct spin-flip transition [17] allowed to demonstrate an Andreev spin qubit based on QDs in Ref. [19]. Such configuration allows to to tune the qubit frequency by means of an external magnetic field over a frequency range of 10 GHz (Fig. 3.17 top right) and to investigate the qubit performance using direct spin manipulation. Specifically, the spin-split doublet phase (Fig. 3.17 bottom right) can be described by a minimal extension of the superconducting Anderson model of Sect. 3.1.3.2 in terms of a Josephson potential

$$U(\phi) = E_0 \cos(\phi) - E_{\mathrm{SO}}\, \boldsymbol{\sigma} \cdot \boldsymbol{n} \sin(\phi) + \frac{1}{2} \boldsymbol{E}_{\mathrm{Z}} \cdot \boldsymbol{\sigma} \,. \tag{3.86}$$

Here, $\boldsymbol{\sigma}$ is the spin operator, $\boldsymbol{n}$ is a unit vector along the spin-polarization direction set by the spin-orbit interaction, and E_{SO} and E_0 are spin-dependent and spin-independent Cooper pair tunneling rates. Note that the E_0 term proportional to $\cos(\phi)$ has a minimum at $\phi = \pi$, describing the doublet phase. Finally, $\boldsymbol{E}_{\mathrm{Z}}$ is the external Zeeman field.

Interestingly, using an all-electric microwave drive, Rabi frequencies exceeding 200 MHz can be achieved. Furthermore, the Andreev spin qubit can be embedded in a superconducting transmon qubit, which allows to demonstrate strong coherent qubit-qubit coupling [19]. Specifically, the transmon circuit consists of a capacitor, with charging energy E_C, shunting a superconducting quantum interference device (SQUID) formed by the parallel combination of a gate-tunable Josephson junction with Josephson energy E_J, and the quantum dot Josephson junction hosting the Andreev spin qubit, Fig. 3.17 top schematics. By operating in a regime where the ratio $E_J/\sqrt{E_0^2 + E_{SO}^2}$ is large, the phase difference ϕ across the quantum dot Josephson junction can be controlled through the magnetic flux through the SQUID loop $\Phi_{ext} = \phi_{ext}\Phi_0/(2\pi)$. Due to the presence of the $E_{SO}\sin(\phi)$ term, the transmon frequency becomes spin-dependent. This can be exploited

(continued)

to readout the Andreev spin qubit state by capacitively coupling the transmon circuit to a readout resonator. Finally, due to the transmon-resonator dispersive coupling, the resonator frequency in turn becomes spin-dependent. Therefore, spectroscopy of the spinful Andreev levels can be performed using standard two-tone circuit QED techniques. The spin-flipping qubit transition can be directly driven, while maintaining the transmon in its ground state, by applying a microwave tone on the central quantum dot gate. Since there is an intrinsic coupling between the spin degree of freedom and the supercurrent across the quantum dot Josephson junction, it is possible to tune the system such that a coherent coupling induces transitions that involve both qubits, in addition to the single-qubit transitions. Such transitions (Fig. 3.18 top right), $|g,\downarrow\rangle \leftrightarrow |e,\uparrow\rangle$ and $|g,\uparrow\rangle \leftrightarrow |e,\downarrow\rangle$, with $|g\rangle$ and $|e\rangle$ denoting the ground and first excited transmon states, respectively, are swap-like and could be used to construct entanglement and two qubit gates between the two different qubit platforms.

Finally, by performing a Taylor expansion of the phase-dependent SO term in the transmon+spin qubit hamiltonian:

$$H = -4E_C\partial_\phi^2 - E_J cos(\phi) + U(\phi - \phi_{ext}), \tag{3.87}$$

we can obtain an interesting coupling term of the form

$$H_c \approx E_{\mathrm{SO}}[\sin(\phi_{ext}) - \phi\cos(\phi_{ext})](\cos\theta\sigma_{\tilde{x}} + \sin\theta\sigma_{\tilde{z}}). \tag{3.88}$$

This term of the Hamiltonian couples the Andreev spin qubit to the transmon via the phase operator $\phi = \phi_{zpf}(a^\dagger + a)$, where $\phi_{zpf} = (2E_C/E_J)^{\frac{1}{4}}$ is the magnitude of zero-point fluctuation of the transmon phase, and is, thus, reminiscent of a dipole coupling. The coupling contains both longitudinal and transverse terms in the spin eigenbasis, with θ being the angle between the Zeeman field and the spin-orbit direction. By varying the external flux ϕ_{ext} such that the Andreev spin qubit frequency crosses the transmon frequency, one can obtain superpositions $1/\sqrt{2}(|e,\downarrow\rangle - |g,\uparrow\rangle)$ and $1/\sqrt{2}(|e,\downarrow\rangle + |g,\uparrow\rangle)$ separated by a frequency splitting $2J/2\pi \approx 100\,\mathrm{MHz}$ (Fig. 3.18 bottom right), greatly exceeding the decoherence rates of both qubits and hence demonstrating the first realization of a direct strong coupling between a spin qubit and a superconducting qubit [19]. The analytical estimation $J \approx E_{\mathrm{SO}}\phi_{zpf}\sin(\theta)$ suggests that by gating the system (or choosing a different material) to reach a larger E_{SO} and by aligning the magnetic field perpendicular to the spin-orbit direction, coupling rates of hundreds of MHz could be achieved, which would enable rapid two-qubit gates between the transmon and the Andreev spin qubit.

(continued)

We finish this part by mentioning another experimental breakthrough in the field: a supercurrent-mediated coupling between two distant Andreev spin qubits [20]. This qubit-qubit interaction is induced by a a shared Josephson inductance by using an early theoretical idea [12] that large spin-dependent supercurrents can lead to strong long-range and tunable spin-spin coupling. In the experiments, in particular, this qubit-qubit interaction is longitudinal type and tunable by both gate voltages and external fluxes.

Despite all this amazing progress, experiments with Andreev spin qubits still report rather small T_2 coherence times, of the order of tens of ns. The interaction to the surrounding spin bath of the nuclei is one of the dominant dephasing mechanisms. For this reason, the recent realization of hard-gap superconductivity in materials without a net nuclear spin, like Germanium [218,219], sets a promising route to improve the coherence times of the Andreev Spin qubits. First steps towards hybrid qubits using a two-dimensional hole gas based on Germanium quantum wells are already being demonstrated with the realization of a gate tunable transmon qubit in planar Germanium [220].

The spaces for the Andreev pair and Andreev spin qubits are disconnected as they have different fermion parity. However, random quasiparticle tunneling events can drive transitions between the two subspaces, bringing the system outside form the computational space. Therefore, the timescale of the quasi-particle poisoning events set an upper limit for operations of these systems. Fortunately, poisoning events (T_1 times in qubit language) of $\sim 100\,\mu\text{s}$ have been reported in superconductor-semiconductor systems, see for example Refs. [19,212,213,217].

3.5.2 Beyond Single Andreev States

The protection of superconductor qubits can be improved in systems of coupled Andreev states. These systems that can be engineered in, for example, sets of quantum dots coupled to superconductors. We have introduced in Sect. 3.4 a way to improve the coherent properties of qubits based on nanoscopic superconductors by making an artificial Kitaev chain. In this section, we focus on proposals that couple two Andreev states appearing, for example, in double QD systems coupled to superconductors [221–223]. These proposals exploit the superconducting properties and the tunability of QDs to find sweet spots where the qubit features some robustness to certain sources of noise. In this subsection, we discuss two different

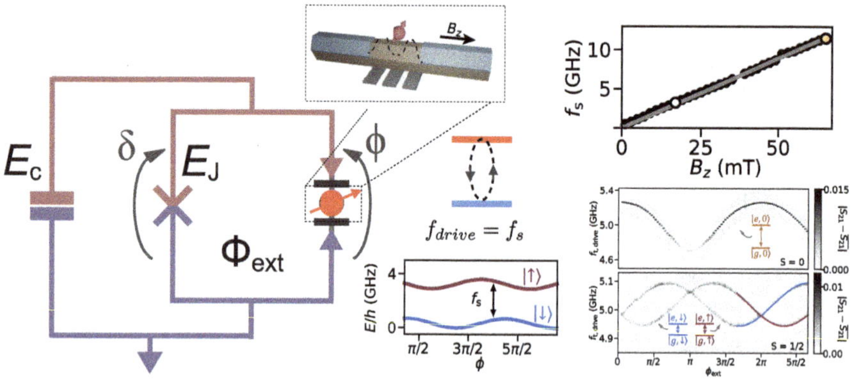

Fig. 3.17 Left: Circuit model of the Andreev spin qubit embedded in a transmon circuit. The transmon island with charging energy E_C is connected to ground by a SQUID formed by the parallel combination of the QD Josephson junction and a reference Josephson junction of energy E_J. ϕ and δ denote the superconducting phase difference across the quantum dot and reference junctions respectively, that can be tuned by means of an externally applied magnetic flux through the SQUID loop Φ_{ext}. A microwave drive of frequency f_{drive} applied to a central gate electrode (not shown) in the QD region (red circle with arrow) allows to manipulate the spin state. The dashed inset shows a sketch of the QD region with the Andreev spin qubit confined in a hybrid semiconductor-superconductor nanowire. Right Top: An applied magnetic field B_z along the nanowire axis allows to tune the frequency of the qubit f_s. Right bottom: Comparison of singlet and spin-split doublet ground states in transmon two-tone spectroscopy. Transmitted microwave signal versus external flux, $= \phi_{\text{ext}} = (2e/\hbar)\Phi_{\text{ext}}$, and transmon drive frequency, $f_{\text{t,drive}}$, for the quantum dot junction in the singlet state (top) and for the doublet state (bottom), revealing the spin-splitting of the doublet state. Adapted from Refs. [17] and [19] with permission, ©2023, American Physical Society. ©2023, The Author(s), under exclusive licence to Springer Nature Limited. All rights reserved

proposals of qubits based on coupled ABSs, engineered in a superconductor-QD-QD-superconductor device, as sketched in Fig. 3.19. Similarly to the previously discussed qubits based on single ABSs, one can define different qubits, depending on the total parity states.

The Parity Qubit

The first version we discuss is based on the total odd fermion parity state and introduced in Ref. [223]. In this qubit, information is encoded in a single quasiparticle occupying either the left or the right superconductor-QD subsystem, dubbed parity qubit, Fig. 3.19b. Therefore, the two half of the system have different fermion parity. The qubit can be seen as the superconducting version of the conventional charge qubit, studied in semiconductor QDs. In conventional charge qubits, charge noise coming from the coupling to the electrostatic environment provides the main source of decoherence. It has

(continued)

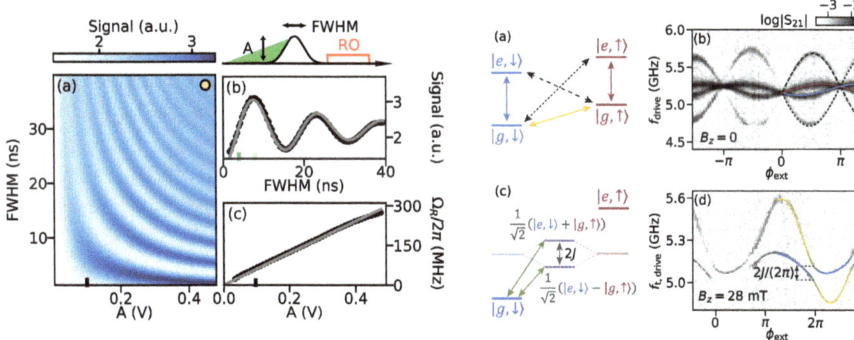

Fig. 3.18 Coherent manipulation of the Andreev spin qubit. Left: (**a**) Rabi oscillations for a range of Gaussian pulses characterized by their amplitude A at the waveform generator output and their full width at half maximum (FWHM). (**b**) Rabi oscillation corresponding to $A = 0.1V$. (**c**) Extracted Rabi frequencies versus pulse amplitude. Right: Coherent ASQ-transmon coupling. (**a**) Frequency diagram of the joint Andreev spin qubit-transmon system at large detuning between ASQ and transmon qubit energy levels. In addition to the two spin-conserving transmon transitions (solid red and blue) and the transmon-conserving spin qubit transition (solid yellow), two additional swap transitions $|g,\downarrow\rangle \leftrightarrow |e,\uparrow\rangle$ and $|g,\uparrow\rangle \leftrightarrow |e,\downarrow\rangle$ involving both qubits can take place in the presence of coherent coupling between them (dashed and dotted black). (**b**) Two tone spectroscopy of the joint two-qubit system at Bz = 0 showing the additional swap transitions. (**c**) In the presence of coherent coupling, the two qubits hybridize into states $1/\sqrt{2}(|e,\downarrow\rangle - |g,\uparrow\rangle)$ and $1/\sqrt{2}(|e,\downarrow\rangle + |g,\uparrow\rangle)$ with a frequency splitting of $2J$. Green arrows denote the transitions from ground to the two hybridized states. (**d**) Experimental demonstration of strong coupling between both qubits with Two-tone spectroscopy. Adapted from Ref. [19] with permission,

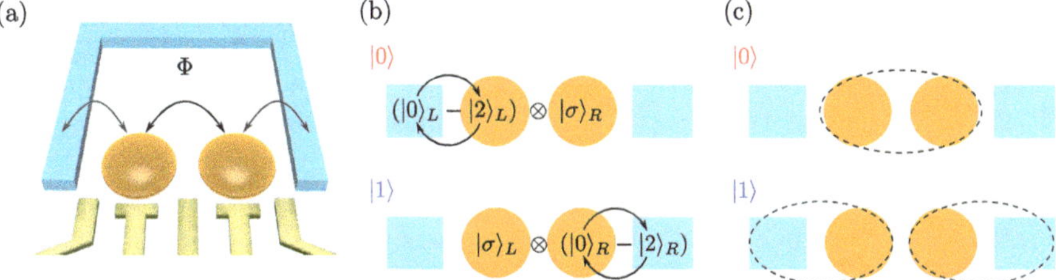

Fig. 3.19 Qubits based on coupled double dots. (**a**) Qubit sketch composed by two tunnel-coupled QDs. The two QDs are embedded into a superconducting loop that allows for phase control. Panel (**b**) and (**c**) show two possible qubits that can be defined in the total odd and even subspaces, respectively. (**b**) Two states with a single quasiparticle excited in either left/right QD, while the other QD is in the even parity state. (**c**) Singlet formed between the dots or between the dots and the superconductors

been recently shown that the use of charge neutral environments using, for example noble gases, can greatly enlarge the qubit's coherence times [224].

(continued)

The parity qubit uses a different strategy to enlarge the coherence times, exploiting the exotic properties of subgap states in superconductors. Andreev bound states, appearing for example in QD-superconductor systems, have sweet spots where the ground and the excited states have the same total charge and the system is insensitive to charge fluctuations. Mathematically, this means that the Bogoliubov-de Gennes coefficients describing the subgap state fulfill $|u| = |v|$. In superconductor-QD systems, this situation is achieved at discrete points in parameter space, so-called sweet spots. To achieve good protection, the parity qubit requires that the two halfs of the system are tuned to their respective sweet spots, where it is linearly insensitive to charge noise acting on the level energy of the QDs. The parity qubit is, however, still linearly sensitive to noise on the tunnel coupling between the QDs. This issue can be improved by increasing the tunnel coupling, at the cost of reducing the relaxation times. Quantum operations between the QDs can be performed by driving the tunnel coupling at the resonant frequency, that makes the excited quasiparticle jump across the junction. Finally, charge measurements in the QDs away from the sweet spot allows to determine the state of the system.

The YSR Bond Qubit
Similarly to the single Andreev qubits, one can also define a qubit in the total even subspace. One possibility is to define a qubit in the regime where both QDs have a single charge [225]. In this regime, there are two possibilities for screening the spin of the excess electron in the dots: by forming a singlet between the two QDs or by forming two YSR states with the two coupled superconductors. These two possibilities, illustrated in Fig. 3.19, have in general different energies, that are determined by the tunneling amplitudes, representing a two level system where one can encode quantum information. Importantly, the two states have the same charge and total spin. Therefore, the qubit is insensitive to small fluctuations of the level energies of the dots and magnetic fields. Quantum operations between the two qubit states require a quasiparticle being transfer through the full system, recombining/splitting a Cooper pair. These operations can be performed by driving the level energy of one of the QDs.

3.6 Coherent Experiments with Majorana States

In Sect. 3.2, we introduced the exotic properties of non-local Majorana states, which are predicted to emerge at the ends of one-dimensional topological superconductors. These superconductors can be engineered in semiconductor-superconductor heterostructures under strong magnetic fields. Section 3.3 reviewed experimental progress and identified signatures consistent with the presence of topological MBSs. However, the existence of low-energy trivial states complicates the picture as they can mimic most of the properties of MBSs. Therefore, demonstrating the topological origin of the measured subgap states would require probing the non-local properties of MBSs, related to their non-abelian statistics. Achieving this goal requires coherent control on the properties and the state of several MBSs.

In this section, we review the theoretical advancements and theoretical proposals for operating MBSs. We begin the section with a discussion on qubit designs based on topological superconductors. Following this, we examine experimental designs aimed at demonstrating the non-Abelian properties of MBSs and, consequently, their topological origin. The section includes proposals for fusion rules and braiding experiments. The section concludes with a discussion about operating on the state encoded in PMMs, realized in minimal Kitaev chains.

3.6.1 Topological Qubits

Majorana bound states, appearing at the ends of one-dimensional topological superconductors, are a promising route for the implementation of quantum technologies. The key feature that makes Majorana qubits apart from alternatives is their non-local nature: the information in a Majorana qubit is stored not in a local quasiparticle, but in the collective state of a pair of spatially separated Majorana states. This non-local encoding of information offers inherent protection against local perturbations, a significant advantage for quantum computing. In Majorana qubits, information is encoded in the fermion occupation of Majorana pairs, that can be in either the even or the odd fermion parity state. A single Majorana pair cannot encode all the possible states of a qubit. The reason is that transitions between even and odd states would require changing the overall fermion parity, a good quantum number in superconductors, except for undesired poisoning events. Therefore, four MBSs are needed to encode all the states of a qubit. The Majorana qubit states can be defined as $|0\rangle = |00\rangle$, $|1\rangle = |11\rangle$ for the total even parity subspace, where $0/1$ denotes the even/odd fermion occupation of a given Majorana pair. These two states have the same electron number and energy for decoupled MBSs, and differ from one Cooper pair that can split occupying the two non-local fermion modes defined by the Majoranas. We note there is another identical computational subspace with total odd fermion parity ($|01\rangle$, $|10\rangle$) that can be also used as a basis for computation.

The most intriguing property of Majorana qubits is their potential for topological quantum computation. In a topological quantum computer, information is stored and manipulated through the exchange of Majorana fermions. When these fermions are exchanged or "braided" around each other, the quantum state of the system changes in a way that depends only on which MBSs are exchanged, becoming independently from other details like the speed and the time of the exchange (if everything is done adiabatic) [226, 227]. The topological nature of Majorana qubits makes them inherently resistant to certain types of errors and decoherence, a major

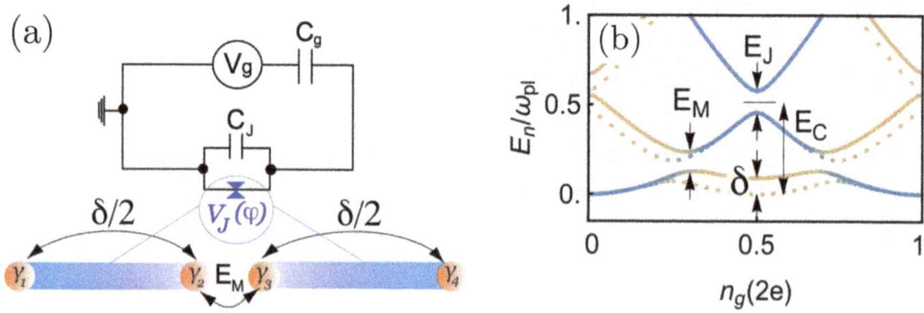

Fig. 3.20 (**a**) Sketch of a transmon Majorana qubit, where the connection between the two superconductors is made via two coupled Majorana wires, bottom panel. (**b**) Energy spectrum of the system in the charging-energy dominated regime, known as Cooper pair box regime ($E_J/E_C = 0.5$, $E_M/E_C \sim 0.12$), and close to the topological phase transition for the wires. Blue/orange colors denote fermionic even/odd parities of the inner MBSs. Adapted from Ref. [230] under CC-BY-4.0 license, ©2022, The Author(s)

challenge in conventional quantum computing systems. This resilience to local noise and perturbations is due to the fact that the quantum information is stored globally in a non-local state rather than in a local variable like an electron's spin or charge. As a result, disturbances or changes in the system do not easily disrupt the stored information, unless the gap is closed, offering a promising route to more stable and reliable quantum computation.

Topologically protected operations, like braiding, can be used to implement quantum gates, including the Hadamard and control-Z. However, the set of operations allowed by braiding do not form a universal gate set [228]. For this reason, braid operations need to be complemented with gates that are not topologically protected to achieve universal quantum computation. The $\pi/8$ rotation, also known as T gate or magic gate is an operation that can complement the previous two. Despite that this operation is not topologically protected, some schemes have been develop to cancel errors that may arise during unprotected gate operations [229].

Several qubit devices have been proposed, exploiting the Majorana degree of freedom to encode quantum information. In the following, we discuss some of these proposals.

The Majorana Transmon Qubit

The Majorana transmon is an example of a qubit that exploits the extra degree of freedom introduced by four Majoranas, see Fig. 3.20. This qubit is an extension of the celebrated transmon qubit, where the Josephson coupling between the superconductors includes a term that is originated from the coupling between two MBSs [230–239]. The Hamiltonian of the system can be written as

(continued)

$$H = H_C + H_J + H_M\,, \tag{3.89}$$

where $H_C = 4E_C(n - n_g)^2$, with n being the charge, n_g a charge offset, and E_C the charging energy of the island. The coupling between the Cooper pair box and the superconductor has two terms. The first one, $H_J = E_J \cos(\phi)$, describes the contribution from all the superconductors' states, except MBSs, with ϕ being the phase operator, conjugated variable to the number operator n. The sum of $H_C + H_J$ corresponds to the conventional transmon qubit. The Majorana physics appears in the last term of Eq. (3.89) that describes the coupling between two MBSs,

$$H_M = iE_M\gamma_2\gamma_3\,. \tag{3.90}$$

This coupling splits the ground state degeneracy, allowing to encode information in the degree of freedom of the coupled MBSs. The qubit can be manipulated and readout using microwave pulses, similar to what is currently done in transmon qubits.

The Majorana Box Qubit

The Majorana box qubit is another qubit geometry, where two topological superconductors couple via a trivial superconducting backbone, see Fig. 3.21. Therefore, the two wires and the superconductor form a floating island where charging energy dominates, fixing the total fermion parity [240]. For a given total fermion number, the system has two degenerate ground states, corresponding to the two possible occupations of the four MBSs. For instance, for a total even parity, the two relevant quantum states are $|0\rangle = |00\rangle \otimes |N_C\rangle$ and $|1\rangle = |11\rangle \otimes |N_C - 1\rangle$, where N_C denotes the number of Cooper pairs in the island. Therefore, the two states differ by one Cooper pair that splits, where the two electrons occupy the low-energy state defined by the MBSs. The measurement of the qubit can be done via charge sensing of two of the MBSs, coupling them to a nearby QD [241–244], or by measuring transport in an interferometry setup. The qubit initialization can be done using the current blockade through the system that fixes the parity of the system to a well-defined value [245]. The idea of fixing the parity of coupled Majorana wires inspired more complicated geometries that include several coupled wires [246] and networks of wires for Majorana color codes [247].

The ground state degeneracy in the blockaded superconducting island plays the role of an effective spin. Early theory works proposed to use this fact to demonstrate the so-called topological Kondo effect, that can be used as a demonstration of the non-locality of the Majorana degrees of freedom [248].

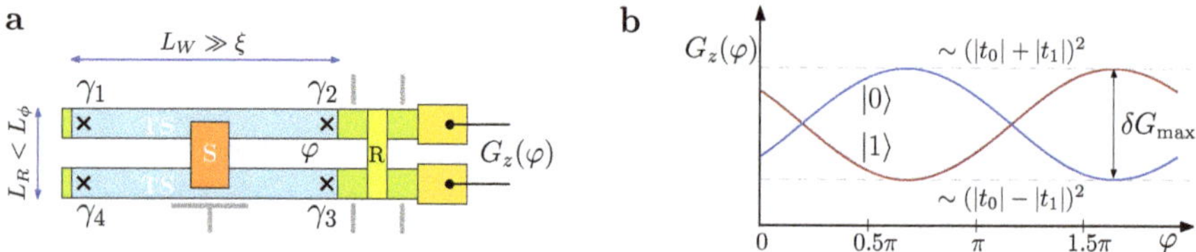

Fig. 3.21 (**a**) Sketch of the Majorana box qubit, where two topological superconductors (light blue) couple via a trivial superconductor (dark orange). The qubit readout can be done using conductance measurement via the leads attached to one end (yellow). (**b**) Conductance as a function of the flux. Reprinted from Ref. [240] under CC-BY-3.0 license, ©2022, The Author(s)

3.6.2 Majorana Fusion Rules

Majorana fusion rules are a key concept in understanding the non-Abelian nature of Majorana states. These rules describe how pairs of MBSs can combine or "fuse" to form different quantum states. The fusion of MBSs is not only a fundamental theoretical aspect but also has profound implications for demonstrating their non-Abelian properties, which are essential for topological quantum computation.

The basic idea of Majorana fusion rules is that, when two MBSs come together, they hybridize, resulting in a trivial (or vacuum) state, or they can fuse into a non-trivial fermionic state. This behavior is mathematically represented by the fusion rule

$$\sigma \times \sigma = I + \phi\,, \tag{3.91}$$

where σ represents a Majorana fermion, I the vacuum state, and ϕ a fermionic state. The outcome of this fusion process depends only on the collective topological state of the system and is not determined by local properties. This ambiguity in the fusion outcome is a manifestation of the Majorana fermions' non-abelian properties.

A less formal but equivalent way to understand Eq. (3.91) is that that the fusion (coupling and measurement) of two MBSs has two possible outcomes corresponding to the occupation of the fermion spanned by the measured MBS pair: empty (I) and full (ϕ). The probability for each outcome depends on the joint state of the two MBSs. The experimental demonstration of fusion rules requires the initialization of the system in a given basis and the measurement in a different one. It means that, pairs of Majoranas are initialized on a well-defined fermion parity state, while measurements are performed on different using a different MBS pair combination. The result of the fusion is topologically protected and insensitive to the details of the measurement. Different protocols have been proposed to demonstrate the MBS fusion rules [151, 249–258].

3.6.3 Majorana Braiding

The ground state of a system composed by N Majoranas has a degeneracy $2^{N/2}$, corresponding to every pair of MBSs being in the even/odd parity state. Majorana bound states are non-Abelian anyons: the exchange of a set of Majorana transforms the state of the system between the different ground states that are locally indistinguishable. This operation is topologically protected and does not depend on the details on how it is performed if MBSs remain well-separated and the exchange is adiabatic. Different set of Majoranas exchange correspond to distinct unitary operations that cannot be transformed into the other. This opens up the possibility of performing topologically-protected operations by just moving MBSs around. In addition, the demonstration of the non-trivial outcome after the Majorana exchange would constitute a direct proof of the topological origin of the measured zero-energy states, see Sect. 3.3.1. In this subsection, we describe different Majorana braiding protocols introduced in the literature, see also Ref. [259] for a review.

Real-space Braiding
The topological phase of a semiconductor-superconductor nanowire can be controlled using external gates. It means that the topological properties of the system can be controlled locally. When a segment of a nanowire changes phase, a MBS moves accordingly to the boundary between the trivial and the topological phases. It also implies that MBSs could be moved preserving coherence if a set of gate electrodes are controlled adiabatically between different values, but sufficiently fast to avoid poisoning events. Majorana braiding would require extending the wire to a second dimension, forming a T or a Y shape, using another ancillary wire. In this configuration, two Majoranas can be exchanged, using the ancillary wire to place one of the Majoranas to avoid overlap, see Ref. [260]. This braiding protocol would require full control over tens of gates [261] in a timescale much shorter than the typical quasiparticle poisoning time. For this reason, the proposal is considered to be challenging and another simplified schemes have been develop, where Majoranas are braid in parameter space instead.

Hybridization-Based Braiding
The approach that conceptually approximates the most to spatial braiding is the hybridization-induced braiding protocol, also referred to as the "three-point turn braid". For a visual representation, refer to Fig. 3.22, which schematically illustrates both the device structure and the braiding protocol. In

(continued)

this proposal, there is a central MBS capable of coupling with three additional MBSs. The key to this proposal is the controlled hybridization between these MBSs, as represented by solid lines in the figure. The hybridization, needs to be dynamically switched on and off to allow for effectively braid two MBSs. When two MBSs hybridize, their ground state degeneracy is lifted, leading to a low energy state with well-defined fermion parity. This is however not the case when 3 MBSs couple, where 2 of them can be grouped into a regular fermion and we remain with a single uncoupled MBS. This fact is used in the intermediate steps, where three MBSs hybridize, leading to a single MBS delocalized along the three places. Removing one of the connections adiabatically localizes the MBS again, leading to an effective move of the topological mode. For the demonstration of Majorana fermions' non-Abelian statistics, it is crucial to initialize MBSs in pairs. For this reason, we included the additional MBSs in subsystems A and B. Various versions of this protocol have been proposed and discussed in the literature [151, 251, 253, 262–267].

Charge-Transfer Based Braiding

Another way to test non-Abelian properties consist on shuttling single charges to a system composed by several MBSs [268, 269]. This charge can be provided by a QD whose energy is controllable using local gates. When the energy of the QD is adiabatically swept from below to above the system's Fermi level, one electron will be transferred between the QD and the system. In a successful protocol, the electron will be transferred to the zero-energy MBSs [270]. If the system couples to only one MBS (denoted by 1), the operation can be understood as acting with the operator $C_1 = \gamma_1$ to the state defined by the Majoranas. If the QD instead couples to 2 MBSs, the

(continued)

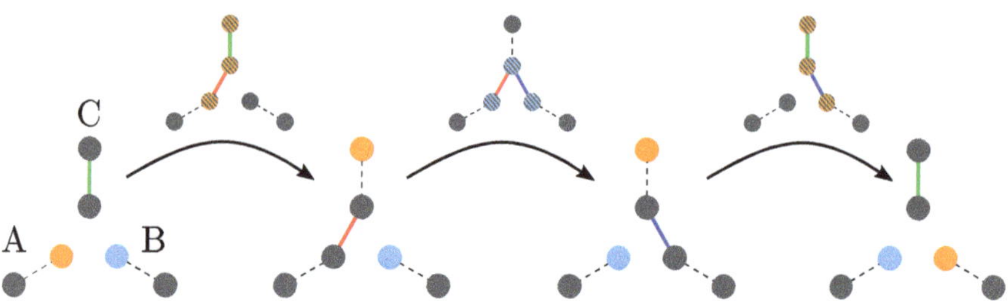

Fig. 3.22 Sketch of the hybridization-induced braiding. The system consists on 3 pairs of Majoranas (circles) that are initialize pairwise. By switching on and off the coupling between MBSs (thick lines) two of them are effectively braid, highlighted with orange and blue colors

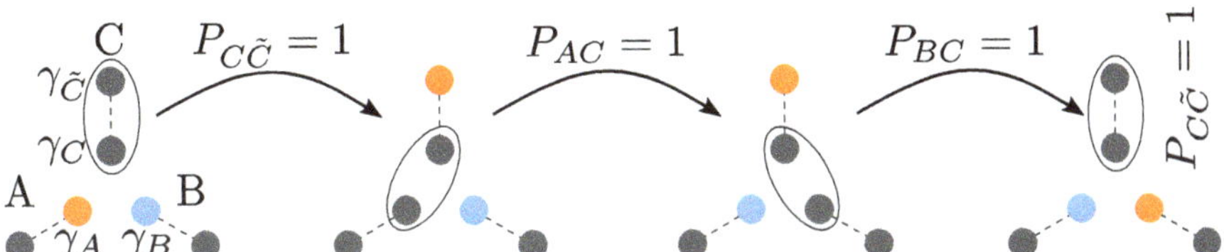

Fig. 3.23 Sketch of the measurement-based braiding. The fermion parity of two nearby MBSs is projected and read at every step, $P_{jk} = i\gamma_J\gamma_k$, schematically represented by the oval. The next step of the protocol is done only if the measured parity is 1. Otherwise, the protocol has to restart. After 4 successful measurements of the fermion parity, two MBSs are effectively braid

operator is written as $C_{1,2} = (\gamma_1+\gamma_2)/\sqrt{2}$. Therefore, the difference between applying these two operations in different orders, $C_1C_{1,2} = (1+\gamma_1\gamma_2)/\sqrt{2}$ and $C_{1,2}C_1 = (1+\gamma_2\gamma_1)/\sqrt{2}$, is due to the non-Abelian properties of MBSs: the final wavefunction is the same after the two sequences if Majorana operators are replaced by fermion operators. In this picture, we have ignored the dynamical phase that arise from the unprotected nature of these operations. A protocol has been recently developed to echo away this phase [269].

Measurement-Based Braiding
The measurement of the joint parity of two Majorana states provides a way to effectively braid MBSs [240, 246, 271]. The protocol is shown in Fig. 3.23. The setup is in essence similar to the one used for the hybridization-based braiding: it consists of 4 Majoranas, plus another two that only used for initialization and readout (lower two MBSs in systems A and B). The protocol consists on measuring alternatively the parity of two MBSs at every step, enclosed by an ellipse in Fig. 3.23. We post select for a given outcome, in this case even parity. If the outcome is odd at any of the steps the protocol has to be restarted. Mathematically, the measurement operation can be written as $M_{jk} = (1 + P_{jk})/2$, with $P_{jk} = i\gamma_j\gamma_k$. It can be shown mathematically that the braiding operation for γ_A (orange) and γ_B (blue) is equivalent to the set of measurements $B_{AB} \propto M_{C\tilde{C}}M_{AC}M_{BC}M_{C\tilde{C}}$.

In the proposals mentioned above, the operations to demonstrate Majorana braiding have to be adiabatic, ensuring that the system remains in the ground state during the protocol. On the other side, the full protocol has to be faster than the characteristic time for single quasiparticles to tunnel into the system, so-called quasiparticle poisoning time. In semiconductor-superconductor systems, poisoning times in the range of μs to ms [272–275] have been reported, also under relatively strong magnetic field [276].

We want to conclude this part by commenting on the initialization and readout of the state encoded in a set of MBSs. As mentioned above, these states are non-local and local perturbations cannot affect the state of the system. It means that local probes coupled to only one MBS cannot project the state of the system. Therefore, the detection schemes for the state encoded in a pair of Majoranas rely on measuring the properties of the system after the two (or more) MBSs couple, including the charge, the energy, or the transport properties through the system. On the other hand, two coupled MBSs have a single ground state with well-defined parity. This can be used to initialize the system into a given state.

3.6.4 Coherent Experiments with PMMs

In the previous subsection, we have introduced protocols to operate MBSs in topological superconducting wires, focusing specially in protocols to demonstrate Majorana non-Abelian statistics. However, the measurements reported so far on superconductor-semiconductor nanowires cannot determine unambiguously the topological origin of the measured low-energy states. For this reason, minimal Kitaev chains have emerged as interesting alternatives to demonstrate the topological properties of MBSs, so-called poor man's Majoranas (PMMs) due to their lack of topological protection, see Sect. 3.4 for a description of the system. In this subsection, we introduce a set of experiments that have been proposed in PMM systems to demonstrate the non-local properties of Majorana states. Before jumping into the proposals, we first discuss on the initialization and readout of the non-local fermionic state encoded in a pairs of PMM.

3.6.4.1 Initialization and Readout

The lack of topological protection of the Majorana states defined in minimal Kitaev chains turns into an advantage for initialization and readout. The ground state of the system is in general non-degenerate, except for some fine-tuned situations, including the sweet spot characterized by well-separated Majorana states. Therefore, the system can be initialize shifting the energy of the orbitals of the two QDs away from the sweet spot. The energy shift splits the ground state degeneracy, leading to a ground state with a well-defined fermion parity. If the splitting is much larger than the temperature, the system will relax to the ground state after a sufficiently long enough time, required to release the excess energy to the environment. A symmetric shift of the dots energy can be used to initialize the system in an even parity state, while shifting the orbitals of the dots in different directions can result on an initial odd fermion parity.

Regarding readout, the conceptually easiest way to readout the state of the system is through charge detection in one dot. This measurement can be achieved by using a capacitively-coupled system, for example a quantum point contact or an additional QD that senses changes on the local charge of one or both dots. At the Majorana sweet spot, a charge detector coupled to one of the QDs will not be able to measure, and therefore, project the PMM's state. This is a direct consequence of the inherent non-local character of well-separated PMMs: the local charge in each of the QDs is the same, independently from the state of the pair of PMMS. However, as the QDs are detuned from the sweet spot, there is a charge difference between the lowest even and odd parity states that is measurable by means of local charge detectors. This parity-to-charge conversion is similar to spin-to-charge conversion used for single-shot readout of spin qubits in double QDs [277].

An alternative way to measure the state of the system is based on measuring the quantum capacitance at the dots, see for example, Refs. [278–280]. At the sweet spot, the quantum capacitance measured for the even and odd ground states are the same, becoming indistinguishable for local measurements. Similar to the parity-to-charge conversion, a detectable difference between the even and the odd ground states signals appear when the system detunes form the sweet spot. This is also true at low-MP sweet spots. This fact can be also used to identify high-MP sweet spots using local charge measurements, see Ref. [281] for a discussion.

In both cases, it is only possible to measure the system state away from the Majorana sweet spot. Non-local detectors coupled to both dots simultaneously can measure the state of the system at the Majorana sweet spot. This would require adding an extra direct coupling between the dots, see for example proposals in Refs. [26, 282]. However, adding detectors capable of readout at the sweet spot not only complicates the device design and measurements (requiring either measurements sensitive to charge fluctuations or capacitive coupling to both QDs) but also introduces a decoherence mechanism at the sweet spot. These decoherence mechanisms might become relevant in experiments aiming at study the quantum coherent properties of PMMs.

The fermion parity degree of freedom is a good quantum number in superconducting systems, where electrons are paired together forming Cooper pairs, except from potentially unpaired electrons occupying low-energy states, like Majorana states. However, quasiparticle poisoning events can change the total femionic parity in the PMM system in an uncontrolled way, undermining coherence in the system, and being an obstacle toward quantum coherent experiments. Independent charge or capacitance measurements on both dots can help determining the poisoning timescale at and away from the Majorana sweet spot, providing the limiting timescale for future experiments.

3.6.4.2 Coherent Operations Between Two PMMs

Fermion parity is a good quantum number in PMM systems, meaning that the even and odd fermion parity states do not couple to each other. To couple both states and, therefore, perform coherent oscillations in the fermion parity degree of freedom, we need to couple the PMM system to another fermionic system, like a QD or another PMM system. This last situation is specially interesting, as this would be the corresponding poor man's version of the topological transmon qubit, introduced in Sect. 3.6.1. The system is sketched in Fig. 3.24, where the two inner dots can couple directly (direct hoping) or via an intermediate superconductor that mediates crossed

Andreev reflection and elastic cotunneling. Using the minimal model described in Sect. 3.4.1.2, the Hamiltonian can be written as

$$H = H^A_{\text{PMM}} + H^B_{\text{PMM}} + H_T + H_U \,, \tag{3.92}$$

where $H^{A,B}_{\text{PMM}}$ describe the left/right PMM system, given by Eq. (3.69). The tunneling is described by H_T, which depends on the details of the central region connecting the two PMM systems. For simplicity, we will focus on the case where the systems couple via normal tunneling, where the Hamiltonian can be written as

$$H_T = \sum_\sigma \left[t_{AB} d^\dagger_{AR\sigma} d_{BL\sigma} + t^{\text{so}}_{AB} d^\dagger_{AR\sigma} d_{BL\bar{\sigma}} + \text{H.c.} \right] , \tag{3.93}$$

where t and t^{so} describes the spin conserving and the spin-flip tunneling terms, respectively. Finally, we have included an extra term in Eq. (3.92) that accounts for non-local Coulomb interactions between the nearby QDs

$$H_U = U_{AB} n^B_L n^A_R \,, \tag{3.94}$$

with n^j_α being the charge of the $\alpha = L, R$ QD of the $j = A, B$ PMM. A finite U_{AB} will induce correlations between the charge on the innermost QDs and become an obstacle to find Majorana sweet spots with well-localized MBSs, i.e. with high MP, and zero energy. This contribution can be suppressed by placing a a grounded metallic gate between the dots, that can also tune tunneling amplitude. Another way to suppress the non-local Coulomb interaction is to couple the two PMM systems via a grounded superconductor, forming a 4-dot Kitaev chain. In the following, we are going to ignore the effect of U_{AB}. Numerical results for the low-energy states can be found in Ref. [281].

Low-Energy Model for Majorana Qubits

It is instructive to introduce a low-energy model describing the Majorana states and their couplings, that qualitatively captures the dominant physics in the PMM system. For the two PMM systems, the model is given by

$$H^{A,B}_{\text{Low-E}} = \frac{i}{2} \sum_{j=A,B} \xi_j \gamma_j \tilde{\gamma}_j + \frac{i}{4} \left[\lambda_{AB} \left(\gamma_A - i \zeta_A \tilde{\gamma}_A \right) \left(\gamma_B - i \zeta_B \tilde{\gamma}_B \right) - \text{H.c.} \right] , \tag{3.95}$$

where the first term describes the overlap between PMMs belonging to the same system, defined as $\gamma = d + d^\dagger$ and $\tilde{\gamma} = i(d - d^\dagger)$, causing an energy splitting of the ground states ξ. As we have shown previously in the chapter, two PMMs can overlap without splitting the ground state degeneracy ($\xi = 0$). They result in a zero-energy crossing characterized by a low MP.

(continued)

The model accounts for this scenario by allowing the tunneling to the second Majorana, given by ζ, with λ being the tunneling amplitude. The case $\zeta = 0$ describes the ideal sweet spot scenario with well-localized PMMs in each dot. In contrast, the limit $\zeta = \pm 1$ corresponds to a trivial fermionic state. The model interpolates between the two limiting situations.

The Hamiltonian in the low-energy model, Eq. (3.95), has 4 energy-degenerate ground states, given by either an even or an odd fermion parity, that we will denote as $|p_L, p_R\rangle$, with $p_j = 0, 1$ for even and odd parity. The total parity of the system is a good quantum number, so states with different total parity do not couple. Below, we restrict the discussion to the $|0, 0\rangle$, $|1, 1\rangle$ states, although similar results can be obtained for the odd fermion parity states. In the even subspace, the Hamiltonian in Eq. (3.95) can be rewritten as

$$H^{A,B}_{\text{Low-E}}\Big|_{\text{even}} = -\frac{\xi_+}{2}\sigma_z + \frac{\lambda_{AB}}{2}\left(1 - \zeta_A\zeta_B\right)\sigma_y\,, \tag{3.96}$$

where $\xi_+ = \xi_A + \xi_B$ and we have assumed that ζ_j are real. Here, σ are the Pauli matrices in the two-dimensional space given by the $|0, 0\rangle$, $|1, 1\rangle$ states, described by $\sigma_z = -i\gamma_A\tilde{\gamma}_A = -i\gamma_A\tilde{\gamma}_A$, $\sigma_y = i\gamma_A\gamma_B = -i\tilde{\gamma}_A\tilde{\gamma}_B$, and $\sigma_x = -i\gamma_A\tilde{\gamma}_B = -i\tilde{\gamma}_A\gamma_B$. Therefore, a finite coupling between the PMMs in the same system splits its energy degeneracy and can be used to initialize the system in a state with well-defined fermion parity on the left and the right, i.e. either the $|0, 0\rangle$ or $|1, 1\rangle$ state. The tunnel coupling between the sub-systems can be used to induce coherent transition between these 2 quantum states and demonstrate the coherent control on PMM systems.

Majorana Transmon Qubit with Minimal Kitaev Chains

The system in Fig. 3.24a has been analyzed in a transmon setup in Ref. [283], see Fig. 3.25, that is an extension of the Majorana transmon qubit introduced in Sect. 3.6.1. The minimal Hamiltonian describing the model is given by

$$H = 4E_C(n - n_g)^2 + E_J\cos(\phi) + V^J_{\text{PMM}}\,, \tag{3.97}$$

where the first term accounts for the charging energy, with $n = -i\frac{\partial}{\partial\phi}$, and the second for the tunneling of Cooper pairs via the trivial Josephson junction. The third term describes the Josephson coupling between the two PMM systems that allows for tunneling processes of a single electron across the junction, proportional to $\cos(\phi/2)$, that change the fermion parity of the left

(continued)

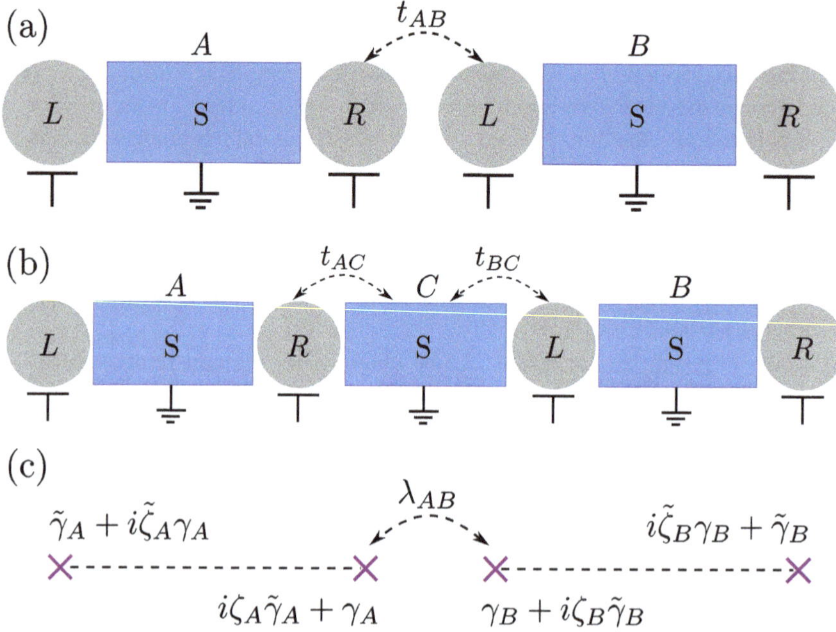

Fig. 3.24 Setup to demonstrate the coherent coupling between two PMM systems. Panels (**a**) and (**b**) show two different ways of coupling the two system, through direct tunneling and via a central superconductor. In both cases, the tunnel coupling results in a hybridization of the inner Majoranas. Panel (**c**) shows a minimal model, composed by four Majoranas, γ_j and $\tilde{\gamma}j$ belonging to sub-system $j = A, B$, to describe the coupling between the PMM systems

and right systems (the total fermion parity remains unchanged). Expressions for V^J_{PMM} are given in Ref. [283].

In this setup, the coupling to a microwave resonator that allows measuring the excitation spectrum of the system. The 4π Josephson effect manifests itself as some suppressed transitions close to phase difference $\phi = \pi$ that only appear in sweet spots with well-localized PMMs. In addition, the microwave response can be used to determine the MP, see Eq. (3.73) and the discussion around for a definition, being a way to quantitatively assess the localization of PMMs.

3.6.4.3 Testing Majorana Fusion Rules with PMMs

Majorana bound states are non-abelian quasiparticle that have non-trivial exchange statistics. Measuring non-abelian properties of MBSs is an open challenge in the field that would constitute the demonstration of a new non-local quasiparticle with a topological origin. Alongside with non-trivial exchange properties, MBSs have

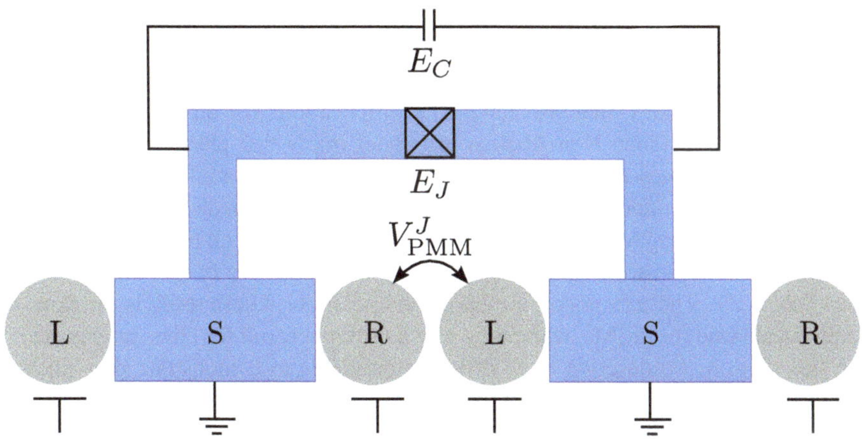

Fig. 3.25 Sketch of a PMM transmon qubit studied in Ref. [283], where two tunnel-coupled PMM systems are embedded into a SQUID with a trivial Josephson junction

non-trivial fusion rules, see Sect. 3.6.2 for a description. The outcome of fusing two MBSs, i.e. coupling two MBSs, is either an electron or no electron (a hole). The probability of having either of the two possible outcomes depend solely on the state of the system. Usually, fusion experiments aim at initializing and measuring MBSs in orthogonal basis, so the the measured outcome is either an electron or hole with equal probability.

Majorana fusion rules can be also tested in artificial Kitaev chains made out of QDs coupled via superconducting segment. In the minimal version, 4 QDs, similar to the device sketched in Fig. 3.24a or b. Shifting the energy of the QDs allow to initialize the system, while the tunnel between the left and right PMMs allows to couple the inner PMMs. Non-local charge measurements in a pair of QDs allows to infer the outcome of the fusion protocol [281, 282]. We note that some fusion protocols are not sensitive to important details, like the MP. Therefore, fusion experiments are believed to not be a conclusive demonstration of Majorana non-Abelian properties. However, in the case of PMMs, fusion experiments are crucial milestones for more sophisticated experiments probing non-abelian exchange properties, as the ones we discuss in the following section.

3.6.5 Non-abelian Experiments with PMMs

In the previous Sect. 3.4, we have analyzed different ways to identify Majorana sweet spots in minimal Kitaev chains, mostly based on transport measurements using metallic electrodes. However, there is a limit on the information that local probes can provide about non-local Majoranas. For this reason, there is always a risk that some trivial states can mimic the properties of PMMs. Additionally, metallic probes are a source of quasiparticles, undesired for quantum applications.

Non-Abelian properties are unique for MBSs, independently on whether they are protected or fine tuned in minimal Kitaev chains. A definitive demonstration of these properties remain as a main challenge in the field to confirm the existence of non-local Majorana states, that will have deep implications for quantum information storage and processing. Non-Abelian properties imply that physically exchanging two Majorana states rotates the state of the system in a non-trivial way within the degenerate ground state manyfold, see Sect. 3.2 for a discussion. However, spatially exchanging Majorana states is challenging and several protocols have been proposed to demonstrate Majorana's non-abelian properties in nanowire systems, see Sect. 3.6.3. These protocols are based on operations whose result is the same as a physical braid for PMMs, while they give a different result for other subgap states. In this section, we describe three different protocols to demonstrate non-Abelian properties of PMMs, based on the bulk proposals presented above.

Charge-Transfer Based Braiding with PMMs

Protocols based on transferring charges between a quantum dot and Majorana states provide a conceptually simple way of testing non-Abelian properties of MBSs [268–270]. The basic device for this proposal using minimal Kitaev chains is sketched in Fig. 3.26a, where two PMM systems couple to a single QD. For this protocol, we consider that the system is subject to a strong magnetic field, in such a way that the QDs' levels are spin-polarized. When the energy of the dentral QD (D) is ramped from negative to positive energies, a single electrons is transferred between the QD to the PMM systems. In this picture, we have fixed the zero energy reference to the chemical potential of the superconductors. If the QD couples to only one of the PMMs, a charge transfer operation will result on a parity change, mathematically represented as $C_j = \gamma_j$ in the Majorana basis. In contrast, if the QD couples to both PMM systems, A and B, the electron can tunnel with equal probability to any of the two, resulting on an operation $F_{AB} = (\gamma_A + \gamma_B)\sqrt{2}$. Now, the operation $B_{AB} = F_{AB}C_A$ leads the same result as a braiding the inner PMMs of the two subsystems, γ_A and γ_B. Therefore, the sequences $F_{AB}C_A$ and $C_A F_{AB}$ lead to different states, due to the non-Abelian character of PMMs. In contrast, the same protocol with trivial fermionic states will result in the same state for both sequences, independently from the order of the charge-transfer operations. In order to measure the state using the parity to charge conversion described in Sect. 3.6.4.1, it is convenient to add a third charge transfer operation, so the two protocols read as $F_{AB}C_A F_{AB}$ and $F_{AB}F_{AB}C_A$. So, if the system is initialized in $|0,0\rangle$, the outcome of protocols 1 and 2 is $|0,1\rangle$ and $|1,0\rangle$.

Figure 3.26b shows results of the two protocols as a function of the PMM localization, using the model introduced in Eq. (3.95) [281]. The figure shows the *visibility*, product of the probabilities of measuring the expected outcome after the two protocols [269], $\Lambda = P^{(1)}_{|1,0\rangle} P^{(2)}_{|0,1\rangle}$. As shown, the protocol

(continued)

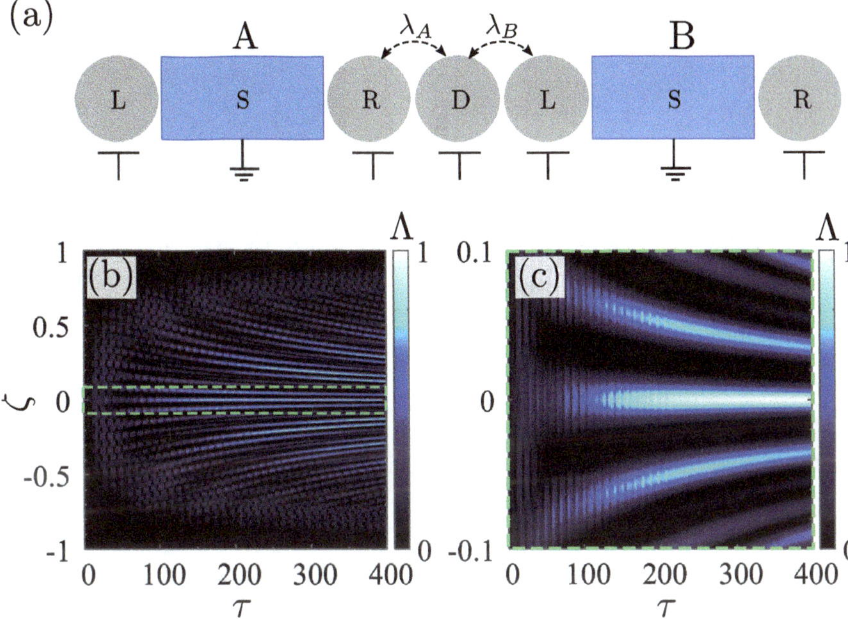

Fig. 3.26 (**a**) Device to demonstrate PMM non-Abelian properties based on charge-transfer to a central QD, D. Panels (**b**) and (**c**) show the visibility: probability of ending up in the target states after the braiding and the reference protocol (see text), as a function of the Majorana overlap in the same PMM system, ζ, and operation time, τ. Reprinted from Ref. [281] under CC-BY-4.0 license, ©2022, The Author(s)

only works independently from the protocol time, τ, for relatively small deviations from the ideal Majorana case with two-well separated PMMs, $\zeta = 0$. For some operation times, it is possible that the protocol gives false positives at finite ζ, originated from accumulated dynamical phases. These false positives are easy to rule out as they depend on τ. Therefore, charge-transfer based braiding can, in theory, distinguish PMMs from other trivial crossing. However, the short range of ζ values where it works can make it challenging to tune and keep the system to a very high MP sweet spot.

Measurement-Based Braiding with PMMs

Measurements provide a way to effectively braid MBSs in topological superconducting wires, as previously discussed in Sect. 3.6.3 for topological MBSs, see Fig. 3.23 for a sketch of the protocol. This proposal has been

(continued)

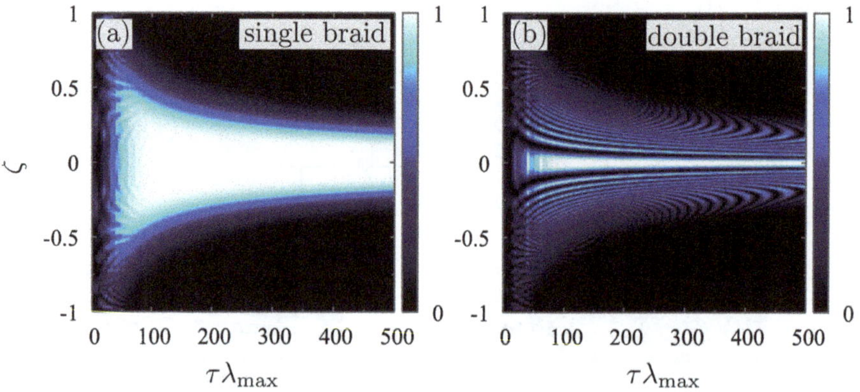

Fig. 3.27 Hybridization-induced braiding for a PMM system. Probability of measuring the target parity in both PMM systems after a single (**a**) and a double (**b**) braid protocol. Reprinted from Ref. [281] under CC-BY-4.0 license, ©2022, The Author(s)

adapted to PMM systems in Ref. [281], where its outcome has been studied as a function of different parameters. The measurement-based braiding could in principle be used to demonstrate PMM braiding. However, the interpretation might become difficult because the outcome depends, not only on the PMM localization parameter, ζ, but also on the device parameters.

Hybridization-Based Braiding with PMMs

We conclude the section discussing the hybridization-induced braiding for PMMs, already introduced in Sect. 3.6.3 for topological MBSs, see also Fig. 3.22. The introduced protocol, based on switching on and off the coupling between neighboring MBSs, provides the same result as spatially braiding two MBSs. In the ideal case, the system that is initialized $|0,0\rangle$ basis ends up in the state $(|0,0\rangle i \pm |1,1\rangle)/\sqrt{2}$, where the sign depends on the braid direction. For fast enough protocols, the main effect of a non-perfect Majorana localization is a dynamical phase that accumulates in the system. A projective measurement on the fermion parity state of the PMMs is not sensitive to the relative phase between the $|0,0\rangle$ and $|1,1\rangle$ states, explaining the relatively large plateau in Fig. 3.27a. For sufficiently large ζ or slow operations, also the relatives weights between $|0,0\rangle$ and $|1,1\rangle$ states become different. The accumulated phase can be revealed after a second braid, as a deviation from the expected $|1,1\rangle$ result, Fig. 3.27b.

Therefore, PMMs in minimal Kitaev chains are promising systems to demonstrate the non-Abelian properties of Majorana states, a main challenge in the field. This will pave the way toward topological quantum computation, allowing also to explore the combination of non-Abelian quasiparticles with other systems. However, the fine-tuned nature of PMMs lead to a sensitivity to some sources of noise. This is an obstacle toward the scaling up of the system toward realistic applications. Longer Kitaev can help at solving this issue.

Acknowledgments We would like to thank all our co-workers in the topic of superconducting devices. We acknowledge financial support from the Horizon Europe Framework Program of the European Commission through the European Innovation Council Pathfinder grant no. 101115315 (QuKiT), the Spanish Ministry of Science through Grants PID2022-140552NA-I00, PID2021-125343NB-I00 and TED2021-130292B-C43 funded by MCIN/AEI/10.13039/501100011033, "ERDF A way of making Europe", the Spanish CM "Talento Program" (project No. 2022-T1/IND-24070), and European Union NextGenerationEU/PRTR. Support by the CSIC Interdisciplinary Thematic Platform (PTI+) on Quantum Technologies (PTI-QTEP+) is also acknowledged.

References

1. R. Aguado, A. Cervera-Lierta, A. Correia, S. de Franceschi, R. Diez Muiño, J.J.G. Ripoll, A. Levi-Yeyati, G. Platero, S. Roche, D. Sanchez-Portal, When matter and information merge into "quantum". Commun. Phys. **6**(1), 266 (2023) 134
2. J. Preskill, Quantum computing in the NISQ era and beyond. Quantum **2**, 79 (2018) 134
3. M. Kjaergaard, M.E. Schwartz, J. Braumüller, P. Krantz, J.I.-J. Wang, S. Gustavsson, W.D. Oliver, Superconducting qubits: current state of play. Annu. Rev. Condens. Matter Phys. **11**(1), 369–395 (2020) 134
4. J. Koch, T.M. Yu, J. Gambetta, A.A. Houck, D.I. Schuster, J. Majer, A. Blais, M.H. Devoret, S.M. Girvin, R.J. Schoelkopf, Charge-insensitive qubit design derived from the cooper pair box. Phys. Rev. A **76**, 042319 (2007) 134
5. A. Blais, A.L. Grimsmo, S.M. Girvin, A. Wallraff, Circuit quantum electrodynamics. Rev. Mod. Phys. **93**, 025005 (2021) 134
6. A. Chatterjee, P. Stevenson, S. De Franceschi, A. Morello, N.P. de Leon, F. Kuemmeth, Semiconductor qubits in practice. Nat. Rev. Phys. **3**(3), 157–177 (2021) 134
7. P. Stano, D. Loss, Review of performance metrics of spin qubits in gated semiconducting nanostructures. Nat. Rev. Phys. **4**(10), 672–688 (2022)
8. G. Burkard, T.D. Ladd, A. Pan, J.M. Nichol, J.R. Petta, Semiconductor spin qubits. Rev. Mod. Phys. **95**, 025003 (2023) 134
9. T.W. Larsen, K.D. Petersson, F. Kuemmeth, T.S. Jespersen, P. Krogstrup, J. Nygård, C.M. Marcus, Semiconductor-nanowire-based superconducting qubit. Phys. Rev. Lett. **115**, 127001 (2015) 134
10. G. de Lange, B. van Heck, A. Bruno, D.J. van Woerkom, A. Geresdi, S.R. Plissard, E.P.A.M. Bakkers, A.R. Akhmerov, L. DiCarlo, Realization of microwave quantum circuits using hybrid superconducting-semiconducting nanowire Josephson elements. Phys. Rev. Lett. **115**, 127002 (2015)
11. L. Casparis, M.R. Connolly, M. Kjaergaard, N.J. Pearson, A. Kringhøj, T.W. Larsen, F. Kuemmeth, T. Wang, C. Thomas, S. Gronin, G.C. Gardner, M.J. Manfra, C.M. Marcus, K.D. Petersson, Superconducting gatemon qubit based on a proximitized two-dimensional electron gas. Nat. Nanotechnol. **13**(10), 915–919 (2018) 134
12. C. Padurariu, Yu.V. Nazarov, Theoretical proposal for superconducting spin qubits. Phys. Rev. B **81**, 144519 (2010) 134, 189

13. A. Zazunov, V.S. Shumeiko, E.N. Bratus', J. Lantz, G. Wendin, Andreev level qubit. Phys. Rev. Lett. **90**, 087003 (2003) 134
14. C. Janvier, L. Tosi, L. Bretheau, Ç.Ö. Girit, M. Stern, P. Bertet, P. Joyez, D. Vion, D. Esteve, M.F. Goffman, H. Pothier, C. Urbina, Coherent manipulation of Andreev states in superconducting atomic contacts. Science **349**(6253), 1199–1202 (2015) 134
15. M. Hays, V. Fatemi, K. Serniak, D. Bouman, S. Diamond, G. de Lange, P. Krogstrup, J. Nygård, A. Geresdi, M.H. Devoret, Continuous monitoring of a trapped superconducting spin. Nat. Phys. **16**(11), 1103–1107 (2020) 134, 186
16. A. Bargerbos, M. Pita-Vidal, R. Žitko, J. Ávila, L.J. Splitthoff, L. Grünhaupt, J.J. Wesdorp, C.K. Andersen, Y. Liu, L.P. Kouwenhoven, R. Aguado, A. Kou, B. van Heck, Singlet-doublet transitions of a quantum dot Josephson junction detected in a transmon circuit. PRX Quantum **3**, 030311 (2022) 186, 187
17. A. Bargerbos, M. Pita-Vidal, R. Žitko, L.J. Splitthoff, L. Grünhaupt, J.J. Wesdorp, Y. Liu, L.P. Kouwenhoven, R. Aguado, C.K. Andersen, A. Kou, B. van Heck, Spectroscopy of spin-split Andreev levels in a quantum dot with superconducting leads. Phys. Rev. Lett. **131**, 097001 (2023) 187, 190
18. V. Fatemi, P.D. Kurilovich, M. Hays, D. Bouman, T. Connolly, S. Diamond, N.E. Frattini, V.D. Kurilovich, P. Krogstrup, J. Nygård, A. Geresdi, L.I. Glazman, M.H. Devoret, Microwave susceptibility observation of interacting many-body Andreev states. Phys. Rev. Lett. **129**, 227701 (2022) 134
19. M. Pita-Vidal, A. Bargerbos, R. Žitko, L.J. Splitthoff, L. Grünhaupt, J.J. Wesdorp, Y. Liu, L.P. Kouwenhoven, R. Aguado, B. van Heck, A. Kou, C.K. Andersen, Direct manipulation of a superconducting spin qubit strongly coupled to a transmon qubit. Nat. Phys. **19**(8), 1110–1115 (2023) 135, 187, 188, 189, 190, 191
20. M. Pita-Vidal, J.J. Wesdorp, L.J. Splitthoff, A. Bargerbos, Y. Liu, L.P. Kouwenhoven, C.K. Andersen, Strong tunable coupling between two distant superconducting spin qubits. arXiv:2307.15654 (2023) 135, 189
21. T. Dvir, G. Wang, N. van Loo, C.-X. Liu, G.P. Mazur, A. Bordin, S.L.D. ten Haaf, J.-Y. Wang, D. van Driel, F. Zatelli, X. Li, F.K. Malinowski, S. Gazibegovic, G. Badawy, E.P.A.M. Bakkers, M. Wimmer, L.P. Kouwenhoven, Realization of a minimal Kitaev chain in coupled quantum dots. Nature **614**(7948), 445–450 (2023) 135, 165, 178, 179, 180
22. A.Y. Kitaev, Unpaired Majorana fermions in quantum wires. Physics-Uspekhi **44**(10S), 131–136 (2001) 135, 136, 149, 157, 161
23. G. Wang, T. Dvir, G.P. Mazur, C.-X. Liu, N. van Loo, S.L.D. ten Haaf, A. Bordin, S. Gazibegovic, G. Badawy, E.P.A.M. Bakkers, M. Wimmer, L.P. Kouwenhoven, Singlet and triplet cooper pair splitting in hybrid superconducting nanowires. Nature **612**(7940), 448–453 (2022) 135, 178
24. A. Bordoloi, V. Zannier, L. Sorba, C. Schönenberger, A. Baumgartner, Spin cross-correlation experiments in an electron entangler. Nature **612**(7940), 454–458 (2022) 135, 178, 179
25. Q. Wang, S.L.D. t. Haaf, I. Kulesh, D. Xiao, C. Thomas, M.J. Manfra, S. Goswami, Triplet correlations in cooper pair splitters realized in a two-dimensional electron gas. Nat. Commun. **14**(1), 4876 (2023) 135, 178, 179
26. M. Leijnse, K. Flensberg, Phys. Rev. B **86**, 134528 (2012) 135, 162, 163, 167, 171, 201
27. E. Prada, P. San-Jose, M.W.A. de Moor, A. Geresdi, E.J.H. Lee, J. Klinovaja, D. Loss, J. Nygård, R. Aguado, L.P. Kouwenhoven, From Andreev to Majorana bound states in hybrid superconductor–semiconductor nanowires. Nat. Rev. Phys. **2**(10), 575–594 (2020) 135
28. R. Aguado, L.P. Kouwenhoven, Majorana qubits for topological quantum computing. Phys. Today **73**(6), 44–50 (2020) 135
29. Z. Zhang, W. Song, Y. Gao, Y. Wang, Z. Yu, S. Yang, Y. Jiang, W. Miao, R. Li, F. Chen, Z. Geng, Q. Zhang, F. Meng, T. Lin, L. Gu, K. Zhu, Y. Zang, L. Li, R. Shang, X. Feng, Q.-K. Xue, K. He, H. Zhang, Proximity effect in PbTe-Pb hybrid nanowire Josephson junctions. Phys. Rev. Mater. **7**, 086201 (2023) 135
30. R. Li, W. Song, W. Miao, Z. Yu, Z. Wang, S. Yang, Y. Gao, Y. Wang, F. Chen, Z. Geng, L. Yang, J. Xu, X. Feng, T. Wang, Y. Zang, L. Li, R. Shang, Q. Xue, K. He, H. Zhang, Selective-

area-grown PbTe-Pb planar Josephson junctions for quantum devices Nanoletters **24**, 4658 (2024) 135
31. A. Maiellaro, J. Settino, C. Guarcello, F. Romeo, R. Citro, Hallmarks of orbital-flavored Majorana states in Josephson junctions based on oxide nanochannels Phys. Rev. B **107**, L201405 (2023) 135
32. A. Barthelemy, N. Bergeal, M. Bibes, A. Caviglia, R. Citro, M. Cuoco, A. Kalaboukhov, B. Kalisky, C.A. Perroni, J. Santamaria, D. Stornaiuolo, M. Salluzzo, Quasi-two-dimensional electron gas at the oxide interfaces for topological quantum physics. Europhys. Lett. **133**, 17001 (2021) 135
33. J.I.-J. Wang, D. Rodan-Legrain, L. Bretheau, D.L. Campbell, B. Kannan, D. Kim, M. Kjaergaard, P. Krantz, G.O. Samach, F. Yan, J.L. Yoder, K. Watanabe, T. Taniguchi, T.P. Orlando, S. Gustavsson, P. Jarillo-Herrero, W.D. Oliver, Coherent control of a hybrid superconducting circuit made with graphene-based van der waals heterostructures. Nat. Nanotechnol. **14**(2), 120–125 (2019)
34. A.V. Balatsky, I. Vekhter, J.-X. Zhu, Impurity-induced states in conventional and unconventional superconductors. Rev. Mod. Phys. **78**, 373–433 (2006) 135
35. L. Yu, Bound state in superconductors with paramagnetic impurities. Acta Phys. Sin. **21**(1), 75–9 (1965) 136
36. H. Shiba, Classical spins in superconductors. Prog. Theor. Phys. **40**(3), 435–451 (1968) 136, 146
37. A.I. Rusinov, Superconductivity near a paramagnetic impurity. JETP Lett. **9**, 85 (1969)
38. J.A. Sauls, Andreev bound states and their signatures. Philos. Trans. R. Soc. A **376**(2125) (2018) 136, 146
39. V. Meden, The Anderson–Josephson quantum dot—a theory perspective. J. Phys.: Condens. Matter **31**(16), 163001 (2019) 136
40. A. Martín-Rodero, A. Levy Yeyati, Josephson and Andreev transport through quantum dots. Adv. Phys. **60**(6), 899–958 (2011)
41. M. Tinkham, *Introduction to Superconductivity*. Dover Books on Physics Series (Dover Publications, Mineola, 2004) 136
42. P.D. Kurilovich, V.D. Kurilovich, V. Fatemi, M.H. Devoret, L.I. Glazman, Microwave response of an Andreev bound state. Phys. Rev. B **104**, 174517 (2021) 140
43. T. Vakhtel, B. van Heck, Quantum phase slips in a resonant Josephson junction. Phys. Rev. B **107**, 195405 (2023) 143
44. Y. Tanaka, A. Oguri, A.C. Hewson, Kondo effect in asymmetric Josephson couplings through a quantum dot. New J. Phys. **9**(5), 115 (2007) 143
45. J. Bauer, A. Oguri, A.C. Hewson, Spectral properties of locally correlated electrons in a Bardeen–Cooper–Schrieffer superconductor. J. Phys.: Condens. Matter **19**(48), 486211 (2007) 143, 144
46. T. Meng, S. Florens, P. Simon, Self-consistent description of Andreev bound states in Josephson quantum dot devices. Phys. Rev. B **79**, 224521 (2009)
47. E.J.H. Lee, X. Jiang, M. Houzet, R. Aguado, C.M. Lieber, S. De Franceschi, Spin-resolved Andreev levels and parity crossings in hybrid superconductor–semiconductor nanostructures. Nat. Nanotechnol. **9**(1), 79–84 (2014) 144
48. E.J.H. Lee, X. Jiang, R. Žitko, R. Aguado, C.M. Lieber, S. De Franceschi, Scaling of subgap excitations in a superconductor-semiconductor nanowire quantum dot. Phys. Rev. B **95**, 180502 (2017) 145, 161
49. M. Valentini, F. Peñaranda, A. Hofmann, M. Brauns, R. Hauschild, P. Krogstrup, P. San-Jose, E. Prada, R. Aguado, G. Katsaros, Nontopological zero-bias peaks in full-shell nanowires induced by flux-tunable Andreev states. Science **373**(6550), 82–88 (2021) 146, 161
50. R. Žitko, J.S. Lim, R. López, R. Aguado, Shiba states and zero-bias anomalies in the hybrid normal-superconductor Anderson model. Phys. Rev. B **91**, 045441 (2015) 145
51. R.M. Lutchyn, J.D. Sau, S. Das Sarma, Majorana fermions and a topological phase transition in semiconductor-superconductor heterostructures. Phys. Rev. Lett. **105**, 077001 (2010) 145, 146, 161

52. Y. Oreg, G. Refael, F. von Oppen, Helical liquids and Majorana bound states in quantum wires. Phys. Rev. Lett. **105**(17), 177002 (2010) 145, 156
53. R. Aguado, La Rivista del Nuovo Cimento **40**, 523 (2017)
54. R.M. Lutchyn, E.P.A.M. Bakkers, L.P. Kouwenhoven, P. Krogstrup, C.M. Marcus, Y. Oreg, Nat. Rev. Mater. **3**, 52 (2018)
55. M. Žonda, V. Pokorný, V. Janiš, T. Novotný, Perturbation theory of a superconducting 0 impurity quantum phase transition. Sci. Rep. **5**(1), 8821 (2015) 145, 156
56. M. Žonda, V. Pokorný, V. Janiš, T. Novotný, Perturbation theory for an anderson quantum dot asymmetrically attached to two superconducting leads. Phys. Rev. B **93**, 024523 (2016) 145, 148
57. M. Žonda, P. Zalom, T. Novotný, G. Loukeris, J. Bätge, V. Pokorný, Generalized atomic limit of a double quantum dot coupled to superconducting leads. Phys. Rev. B **107**, 115407 (2023) 148
58. M.R. Buitelaar, T. Nussbaumer, C. Schönenberger, Quantum dot in the kondo regime coupled to superconductors. Phys. Rev. Lett. **89**, 256801 (2002) 145
59. T. Sand-Jespersen, J. Paaske, B.M. Andersen, K. Grove-Rasmussen, H.I. Jørgensen, M. Aagesen, C.B. Sørensen, P.E. Lindelof, K. Flensberg, J. Nygård, Kondo-enhanced Andreev tunneling in INAS nanowire quantum dots. Phys. Rev. Lett. **99**, 126603 (2007) 146
60. J.O. Island, R. Gaudenzi, J. de Bruijckere, E. Burzurí, C. Franco, M. Mas-Torrent, C. Rovira, J. Veciana, T.M. Klapwijk, R. Aguado, H.S.J. van der Zant, Proximity-induced Shiba states in a molecular junction. Phys. Rev. Lett. **118**, 117001 (2017) 146
61. J.C. Estrada Saldaña, A. Vekris, V. Sosnovtseva, T. Kanne, P. Krogstrup, K. Grove-Rasmussen, J. Nygård, Temperature induced shifts of Yu–Shiba–Rusinov resonances in nanowire-based hybrid quantum dots. Commun. Phys. **3**(1), 125 (2020) 146
62. A. García Corral, D.M.T. van Zanten, K.J. Franke, H. Courtois, S. Florens, C.B. Winkelmann, Magnetic-field-induced transition in a quantum dot coupled to a superconductor. Phys. Rev. Res. **2**, 012065 (2020)
63. S. Trivini, J. Ortuzar, K. Vaxevani, J. Li, F. Sebastian Bergeret, M.A. Cazalilla, J.I. Pascual, Cooper pair excitation mediated by a molecular quantum spin on a superconducting proximitized gold film. Phys. Rev. Lett. **130**, 136004 (2023)
64. R. Bulla, T.A. Costi, T. Pruschke, Numerical renormalization group method for quantum impurity systems. Rev. Mod. Phys. **80**, 395–450 (2008) 146
65. R. Aguado, D.C. Langreth, Out-of-equilibrium kondo effect in double quantum dots. Phys. Rev. Lett. **85**, 1946–1949 (2000) 146
66. A. Levy Yeyati, A. Martín-Rodero, E. Vecino, Nonequilibrium dynamics of Andreev states in the Kondo regime. Phys. Rev. Lett. **91**, 266802 (2003) 146
67. R. López, M.-S. Choi, R. Aguado, Josephson current through a kondo molecule. Phys. Rev. B **75**, 045132 (2007) 146
68. J.S. Lim, R. López, R. Aguado, Josephson current in carbon nanotubes with spin-orbit interaction. Phys. Rev. Lett. **107**, 196801 (2011)
69. T. Yoshioka, Y. Ohashi, Numerical renormalization group studies on single impurity Anderson model in superconductivity: a unified treatment of magnetic, nonmagnetic impurities, and resonance scattering. J. Phys. Soc. Jpn. **69**(6), 1812–1823 (2000) 146
70. M.-S. Choi, M. Lee, K. Kang, W. Belzig, Kondo effect and Josephson current through a quantum dot between two superconductors. Phys. Rev. B **70**, 020502 (2004) 147
71. A. Oguri, Y. Tanaka, A.C. Hewson, Quantum phase transition in a minimal model for the Kondo effect in a Josephson junction. J. Phys. Soc. Jpn. **73**(9), 2494–2504 (2004)
72. Y. Tanaka, A. Oguri, A.C. Hewson, Kondo effect in asymmetric Josephson couplings through a quantum dot. New J. Phys. **9**(5), 115 (2007)
73. J.S. Lim, M.-S. Choi, Andreev bound states in the kondo quantum dots coupled to superconducting leads. J. Phys. Condens. Matter **20**(41), 415225 (2008)
74. J. Bauer, A. Oguri, A.C. Hewson, Spectral properties of locally correlated electrons in a Bardeen–Cooper–Schrieffer superconductor. J. Phys.: Condens. Matter **19**(48), 486211 (2007)

75. C. Karrasch, A. Oguri, V. Meden, Josephson current through a single anderson impurity coupled to bcs leads. Phys. Rev. B **77**, 024517 (2008)
76. R. žitko, O. Bodensiek, T. Pruschke, Effects of magnetic anisotropy on the subgap excitations induced by quantum impurities in a superconducting host. Phys. Rev. B **83**, 054512 (2011) 147
77. A. Martín-Rodero, A. Levy Yeyati, The Andreev states of a superconducting quantum dot: mean field versus exact numerical results. J. Phys.: Condens. Matter **24**(38), 385303 (2012)
78. J.-D. Pillet, P. Joyez, R. Žitko, M.F. Goffman, Tunneling spectroscopy of a single quantum dot coupled to a superconductor: from kondo ridge to Andreev bound states. Phys. Rev. B **88**, 045101 (2013) 148
79. R. Žitko, J.S. Lim, R. López, R. Aguado, Shiba states and zero-bias anomalies in the hybrid normal-superconductor Anderson model. Phys. Rev. B **91**, 045441 (2015)
80. F. Siano, R. Egger, Josephson current through a nanoscale magnetic quantum dot. Phys. Rev. Lett. **93**, 047002 (2004) 147
81. D.J. Luitz, F.F. Assaad, Weak-coupling continuous-time quantum monte carlo study of the single impurity and periodic anderson models with s-wave superconducting baths. Phys. Rev. B **81**, 024509 (2010) 147
82. D.J. Luitz, F.F. Assaad, T. Novotný, C. Karrasch, V. Meden, Understanding the Josephson current through a kondo-correlated quantum dot. Phys. Rev. Lett. **108**, 227001 (2012)
83. V. Pokorný, T. Novotný, Footprints of impurity quantum phase transitions in quantum monte carlo statistics. Phys. Rev. Res. **3**, 023013 (2021)
84. A. Kadlecová, M. Žonda, T. Novotný, Quantum dot attached to superconducting leads: relation between symmetric and asymmetric coupling. Phys. Rev. B **95**, 195114 (2017) 147
85. R. Seoane Souto, A.E. Feiguin, A. Martín-Rodero, A. Levy Yeyati, Transient dynamics of a magnetic impurity coupled to superconducting electrodes: exact numerics versus perturbation theory. Phys. Rev. B **104**, 214506 (2021) 147
86. E. Vecino, A. Martín-Rodero, A. Levy Yeyati, Josephson current through a correlated quantum level: Andreev states and π junction behavior. Phys. Rev. B **68**, 035105 (2003) 147, 148
87. F.S. Bergeret, A. Levy Yeyati, A. Martín-Rodero, Josephson effect through a quantum dot array. Phys. Rev. B **76**, 174510 (2007) 147
88. A. Zazunov, A. Levy Yeyati, R. Egger, Josephson effect for su(4) carbon-nanotube quantum dots. Phys. Rev. B **81**, 012502 (2010)
89. V.V. Baran, E.J.P. Frost, J. Paaske, Surrogate model solver for impurity-induced superconducting subgap states. Phys. Rev. B **108**, L220506 (2023) 147
90. J.-D. Pillet, C.H.L. Quay, P. Morfin, C. Bena, A. Levy Yeyati, P. Joyez, Andreev bound states in supercurrent-carrying carbon nanotubes revealed. Nat. Phys. **6**(12), 965–969 (2010) 148
91. C. Chamon, R. Jackiw, Y. Nishida, S.-Y. Pi, L. Santos, Quantizing Majorana fermions in a superconductor. Phys. Rev. B **81**, 224515 (2010) 148
92. D.A. Ivanov, Non-abelian statistics of half-quantum vortices in p-wave superconductors. Phys. Rev. Lett. **86**, 268–271 (2001) 148
93. L. Fu, C.L. Kane, Superconducting proximity effect and Majorana fermions at the surface of a topological insulator. Phys. Rev. Lett. **100**, 096407 (2008) 155
94. V. Mourik, K. Zuo, S.M. Frolov, S.R. Plissard, E.P.A.M. Bakkers, L.P. Kouwenhoven, Signatures of Majorana fermions in hybrid superconductor-semiconductor nanowire devices. Science **336**(6084), 1003–1007 (2012) 156
95. M.T. Deng, C.L. Yu, G.Y. Huang, M. Larsson, P. Caroff, H.Q. Xu, Anomalous zero-bias conductance peak in a Nb–InSb nanowire–Nb hybrid device. Nano Letters **12**(12), 6414–6419 (2012) 158
96. A. Das, Y. Ronen, Y. Most, Y. Oreg, M. Heiblum, H. Shtrikman, Zero-bias peaks and splitting in an Al–InAs nanowire topological superconductor as a signature of Majorana fermions. Nat. Phys. **8**(12), 887–895 (2012) 158
97. L.P. Rokhinson, X. Liu, J.K. Furdyna, The fractional a.c. Josephson effect in a semiconductor–superconductor nanowire as a signature of Majorana particles. Nat. Phys. **8**(11), 795–799 (2012)

98. H.O.H. Churchill, V. Fatemi, K. Grove-Rasmussen, M.T. Deng, P. Caroff, H.Q. Xu, C.M. Marcus, Superconductor-nanowire devices from tunneling to the multichannel regime: zero-bias oscillations and magnetoconductance crossover. Phys. Rev. B **87**, 241401 (2013)
99. P. Krogstrup, N.L.B. Ziino, W. Chang, S.M. Albrecht, M.H. Madsen, E. Johnson, J. Nygård, C.M. Marcus, T.S. Jespersen, Epitaxy of semiconductor–superconductor nanowires. Nat. Mater. **14**(4), 400–406 (2015) 158
100. W. Chang, S.M. Albrecht, T.S. Jespersen, F. Kuemmeth, P. Krogstrup, J. Nygård, C.M. Marcus, Hard gap in epitaxial semiconductor–superconductor nanowires. Nat. Nanotechnol. **10**(3), 232–236 (2015) 158
101. J.E. Sestoft, T. Kanne, A.N. Gejl, M. von Soosten, J.S. Yodh, D. Sherman, B. Tarasinski, M. Wimmer, E. Johnson, M. Deng, J. Nygård, T.S. Jespersen, C.M. Marcus, P. Krogstrup, Engineering hybrid epitaxial InAsSb/Al nanowires for stronger topological protection. Phys. Rev. Mater. **2**, 044202 (2018)
102. Ö. Gül, H. Zhang, J.D.S. Bommer, M.W.A. de Moor, D. Car, S.R. Plissard, E.P.A.M. Bakkers, A. Geresdi, K. Watanabe, T. Taniguchi, L.P. Kouwenhoven, Ballistic Majorana nanowire devices. Nat. Nanotechnol. **13**(3), 192–197 (2018)
103. K. Flensberg, F. von Oppen, A. Stern, Engineered platforms for topological superconductivity and Majorana zero modes. Nat. Rev. Mater. **6**(10), 944–958 (2021) 158
104. A. Yazdani, F. von Oppen, B.I. Halperin, A. Yacoby, Hunting for Majoranas. Science **380**(6651), eade0850 (2023) 158
105. J. Shabani, M. Kjaergaard, H.J. Suominen, Y. Kim, F. Nichele, K. Pakrouski, T. Stankevic, R.M. Lutchyn, P. Krogstrup, R. Feidenhans'l, S. Kraemer, C. Nayak, M. Troyer, C.M. Marcus, C.J. Palmstrøm, Two-dimensional epitaxial superconductor-semiconductor heterostructures: a platform for topological superconducting networks. Phys. Rev. B **93**, 155402 (2016) 158
106. M. Kjaergaard, H.J. Suominen, M.P. Nowak, A.R. Akhmerov, J. Shabani, C.J. Palmstrøm, F. Nichele, C.M. Marcus, Transparent semiconductor-superconductor interface and induced gap in an epitaxial heterostructure Josephson junction. Phys. Rev. Appl. **7**, 034029 (2017) 158
107. A. Pöschl, A. Danilenko, D. Sabonis, K. Kristjuhan, T. Lindemann, C. Thomas, M.J. Manfra, C.M. Marcus, Nonlocal signatures of hybridization between quantum dot and Andreev bound states. Phys. Rev. B **106**, L161301 (2022) 158
108. A. Pöschl, A. Danilenko, D. Sabonis, K. Kristjuhan, T. Lindemann, C. Thomas, M.J. Manfra, C.M. Marcus, Nonlocal conductance spectroscopy of Andreev bound states in gate-defined InAs/Al nanowires. Phys. Rev. B **106**, L241301 (2022) 158, 160, 178
109. A. Danilenko, A. Pöschl, D. Sabonis, V. Vlachodimitropoulos, C. Thomas, M.J. Manfra, C.M. Marcus, Spin spectroscopy of a hybrid superconducting nanowire using side-coupled quantum dots. Phys. Rev. B **108**, 054514 (2023) 160, 178
110. F. Nichele, A.C.C. Drachmann, A.M. Whiticar, E.C.T. O'Farrell, H.J. Suominen, A. Fornieri, T. Wang, G.C. Gardner, C. Thomas, A.T. Hatke, P. Krogstrup, M.J. Manfra, K. Flensberg, C.M. Marcus, Scaling of Majorana zero-bias conductance peaks. Phys. Rev. Lett. **119**, 136803 (2017) 158
111. E.C.T. O'Farrell, A.C.C. Drachmann, M. Hell, A. Fornieri, A.M. Whiticar, E.B. Hansen, S. Gronin, G.C. Gardner, C. Thomas, M.J. Manfra, K. Flensberg, C.M. Marcus, F. Nichele, Hybridization of subgap states in one-dimensional superconductor-semiconductor Coulomb islands. Phys. Rev. Lett. **121**, 256803 (2018) 158, 159
112. M. Aghaee, A. Akkala, Z. Alam, R. Ali, A. Alcaraz Ramirez, M. Andrzejczuk, A.E. Antipov, P. Aseev, M. Astafev, B. Bauer, J. Becker, S. Boddapati, F. Boekhout, J. Bommer, T. Bosma, L. Bourdet, S. Boutin, P. Caroff, L. Casparis, M. Cassidy, S. Chatoor, A.W. Christensen, N. Clay, W.S. Cole, F. Corsetti, A. Cui, P. Dalampiras, A. Dokania, G. de Lange, M. de Moor, J.C. Estrada Saldaña, S. Fallahi, Z.H. Fathabad, J. Gamble, G. Gardner, D. Govender, F. Griggio, R. Grigoryan, S. Gronin, J. Gukelberger, E.B. Hansen, S. Heedt, J. Herranz Zamorano, S. Ho, U.L. Holgaard, H. Ingerslev, L. Johansson, J. Jones, R. Kallaher, F. Karimi, T. Karzig, C. King, M.E. Kloster, C. Knapp, D. Kocon, J. Koski, P. Kostamo, P. Krogstrup, M. Kumar, T. Laeven, T. Larsen, K. Li, T. Lindemann, J. Love, R. Lutchyn, M.H. Madsen, M. Manfra, S. Markussen, E. Martinez, R. McNeil, E. Memisevic, T. Morgan, A. Mullally, C. Nayak, J.

Nielsen, W.H.P. Nielsen, B. Nijholt, A. Nurmohamed, E. O'Farrell, K. Otani, S. Pauka, K. Petersson, L. Petit, D.I. Pikulin, F. Preiss, M. Quintero-Perez, M. Rajpalke, K. Rasmussen, D. Razmadze, O. Reentila, D. Reilly, R. Rouse, I. Sadovskyy, L. Sainiemi, S. Schreppler, V. Sidorkin, A. Singh, S. Singh, S. Sinha, P. Sohr, T. Stankevič, L. Stek, H. Suominen, J. Suter, V. Svidenko, S. Teicher, M. Temuerhan, N. Thiyagarajah, R. Tholapi, M. Thomas, E. Toomey, S. Upadhyay, I. Urban, S. Vaitiekėnas, K. Van Hoogdalem, D. Van Woerkom, D.V. Viazmitinov, D. Vogel, S. Waddy, J. Watson, J. Weston, G.W. Winkler, C.K. Yang, S. Yau, D. Yi, E. Yucelen, A. Webster, R. Zeisel, R. Zhao, InAs-Al hybrid devices passing the topological gap protocol. Phys. Rev. B **107**, 245423 (2023)

113. Y. Liu, S. Vaitiekėnas, S. Martí'-Sánchez, C. Koch, S. Hart, Z. Cui, T. Kanne, S.A. Khan, R. Tanta, S. Upadhyay, M.E. Cachaza, C.M. Marcus, J. Arbiol, K.A. Moler, P. Krogstrup, Semiconductor–ferromagnetic insulator–superconductor nanowires: stray field and exchange field. Nano Letters **20**(1), 456–462 (2020) 158, 160
114. S. Vaitiekėnas, Y. Liu, P. Krogstrup, C.M. Marcus, Zero-bias peaks at zero magnetic field in ferromagnetic hybrid nanowires. Nat. Phys. **17**(1), 43–47 (2021) 159
115. S. Vaitiekėnas, R. Seoane Souto, Y. Liu, P. Krogstrup, K. Flensberg, M. Leijnse, C.M. Marcus, Evidence for spin-polarized bound states in semiconductor–superconductor–ferromagnetic-insulator islands. Phys. Rev. B **105**, L041304 (2022) 159
116. D. Razmadze, R. Seoane Souto, L. Galletti, A. Maiani, Y. Liu, P. Krogstrup, C. Schrade, A. Gyenis, C.M. Marcus, S. Vaitiekėnas, Supercurrent reversal in ferromagnetic hybrid nanowire Josephson junctions. Phys. Rev. B **107**, L081301 (2023) 159, 160
117. B.D. Woods, T.D. Stanescu, Electrostatic effects and topological superconductivity in semiconductor–superconductor–magnetic-insulator hybrid wires. Phys. Rev. B **104**, 195433 (2021) 159
118. S.D. Escribano, E. Prada, Y. Oreg, A.L. Yeyati, Tunable proximity effects and topological superconductivity in ferromagnetic hybrid nanowires. Phys. Rev. B **104**, L041404 (2021) 159
119. C.-X. Liu, S. Schuwalow, Y. Liu, K. Vilkelis, A.L.R. Manesco, P. Krogstrup, M. Wimmer, Electronic properties of InAs/EuS/Al hybrid nanowires. Phys. Rev. B **104**, 014516 (2021)
120. A. Khindanov, J. Alicea, P. Lee, W.S. Cole, A.E. Antipov, Topological superconductivity in nanowires proximate to a diffusive superconductor–magnetic-insulator bilayer. Phys. Rev. B **103**, 134506 (2021)
121. A. Maiani, R. Seoane Souto, M. Leijnse, K. Flensberg, Topological superconductivity in semiconductor–superconductor–magnetic-insulator heterostructures. Phys. Rev. B **103**, 104508 (2021)
122. K. Pöyhönen, D. Varjas, M. Wimmer, A.R. Akhmerov, Minimal Zeeman field requirement for a topological transition in superconductors. SciPost Phys. **10**, 108 (2021)
123. C.-X. Liu, M. Wimmer, Optimizing the topological properties of semiconductor-ferromagnet-superconductor heterostructures. Phys. Rev. B **105**, 224502 (2022) 159
124. S.D. Escribano, A. Maiani, M. Leijnse, K. Flensberg, Y. Oreg, A. Levy Yeyati, E. Prada, R. Seoane Souto, Semiconductor-ferromagnet-superconductor planar heterostructures for 1d topological superconductivity. npj Quantum Mater. **7**(1), 81 (2022) 159
125. W.A. Little, R.D. Parks, Observation of quantum periodicity in the transition temperature of a superconducting cylinder. Phys. Rev. Lett. **9**, 9–12 (1962) 159
126. S. Vaitiekėnas, G.W. Winkler, B. van Heck, T. Karzig, M.-T. Deng, K. Flensberg, L.I. Glazman, C. Nayak, P. Krogstrup, R.M. Lutchyn, C.M. Marcus, Flux-induced topological superconductivity in full-shell nanowires. Science **367**(6485) (2020) 159
127. A. Vekris, J.C. Estrada Saldaña, J. de Bruijckere, S. Lorić, T. Kanne, M. Marnauza, D. Olsteins, J. Nygård, K. Grove-Rasmussen, Asymmetric little–parks oscillations in full shell double nanowires. Sci. Rep. **11**(1), 19034 (2021) 159, 160
128. A. Ibabe, M. Gómez, G.O. Steffensen, T. Kanne, J. Nygård, A. Levy Yeyati, E.J.H. Lee, Joule spectroscopy of hybrid superconductor–semiconductor nanodevices. Nat. Commun. **14**(1), 2873 (2023)

129. D. Razmadze, R. Seoane Souto, E.C.T. O'Farrell, P. Krogstrup, M. Leijnse, C.M. Marcus, S. Vaitiekėnas, Supercurrent transport through 1e-periodic full-shell Coulomb Islands. Phys. Rev. B **109**, L041302 (2024)
130. F. Peñaranda, R. Aguado, P. San-Jose, E. Prada, Even-odd effect and Majorana states in full-shell nanowires. Phys. Rev. Res. **2**, 023171 (2020) 159
131. C. Payá, S.D. Escribano, A. Vezzosi, F. Peñaranda, R. Aguado, P. San-Jose, E. Prada, Phenomenology of Majorana zero modes in full-shell hybrid nanowires. Phys. Rev. B **109**, 115428 (2024) 159
132. M. Valentini, F. Peñaranda, A. Hofmann, M. Brauns, R. Hauschild, P. Krogstrup, P. San-Jose, E. Prada, R. Aguado, G. Katsaros, Nontopological zero-bias peaks in full-shell nanowires induced by flux-tunable Andreev states. Science **373**(6550), 82–88 (2021) 159
133. M. Valentini, M. Borovkov, E. Prada, S. Martí-Sánchez, M. Botifoll, A. Hofmann, J. Arbiol, R. Aguado, P. San-Jose, G. Katsaros, Majorana-like coulomb spectroscopy in the absence of zero-bias peaks. Nature **612**(7940), 442–447 (2022) 159, 161
134. P. San-Jose, C. Payá, C.M. Marcus, S. Vaitiekėnas, E. Prada, Theory of caroli–de gennes–matricon analogs in full-shell hybrid nanowires. Phys. Rev. B **107**, 155423 (2023) 159
135. O. Lesser, Y. Oreg, Majorana zero modes induced by superconducting phase bias. J. Phys. D: Appl. Phys. **55**(16), 164001 (2022) 159
136. F. Pientka, A. Keselman, E. Berg, A. Yacoby, A. Stern, B.I. Halperin, Topological superconductivity in a planar Josephson junction. Phys. Rev. X **7**, 021032 (2017) 159
137. M. Hell, M. Leijnse, K. Flensberg, Two-dimensional platform for networks of Majorana bound states. Phys. Rev. Lett. **118**, 107701 (2017) 159
138. T. Laeven, B. Nijholt, M. Wimmer, A.R. Akhmerov, Enhanced proximity effect in zigzag-shaped Majorana Josephson junctions. Phys. Rev. Lett. **125**, 086802 (2020)
139. S. Ikegaya, S. Tamura, D. Manske, Y. Tanaka, Anomalous proximity effect of planar topological Josephson junctions. Phys. Rev. B **102**, 140505 (2020)
140. P.P. Paudel, T. Cole, B.D. Woods, T.D. Stanescu, Enhanced topological superconductivity in spatially modulated planar Josephson junctions. Phys. Rev. B **104**, 155428 (2021)
141. I. Sardinero, R. Seoane Souto, P. Burset, Preprint. arXiv.2401.17670 (2024). https://doi.org/10.48550/arXiv.2401.17670
159
142. H. Ren, F. Pientka, S. Hart, A.T. Pierce, M. Kosowsky, L. Lunczer, R. Schlereth, B. Scharf, E.M. Hankiewicz, L.W. Molenkamp, B.I. Halperin, A. Yacoby, Topological superconductivity in a phase-controlled Josephson junction. Nature **569**(7754), 93–98 (2019) 159
143. A. Fornieri, A.M. Whiticar, F. Setiawan, E. Portolés, A.C.C. Drachmann, A. Keselman, S. Gronin, C. Thomas, T. Wang, R. Kallaher, G.C. Gardner, E. Berg, M.J. Manfra, A. Stern, C.M. Marcus, F. Nichele, Evidence of topological superconductivity in planar Josephson junctions. Nature **569**(7754), 89–92 (2019)
144. M.C. Dartiailh, W. Mayer, J. Yuan, K.S. Wickramasinghe, A. Matos-Abiague, I. Žutić, J. Shabani, Phase signature of topological transition in Josephson junctions. Phys. Rev. Lett. **126**, 036802 (2021)
145. A. Banerjee, M. Geier, M. Ahnaf Rahman, D.S. Sanchez, C. Thomas, T. Wang, M.J. Manfra, K. Flensberg, C.M. Marcus, Control of Andreev bound states using superconducting phase texture. Phys. Rev. Lett. **130**, 116203 (2023)
146. A. Banerjee, O. Lesser, M.A. Rahman, C. Thomas, T. Wang, M.J. Manfra, E. Berg, Y. Oreg, A. Stern, C.M. Marcus, Local and nonlocal transport spectroscopy in planar Josephson junctions. Phys. Rev. Lett. **130**, 096202 (2023) 159
147. C. Guarcello, R. Citro, Progresses on topological phenomena, time-driven phase transitions, and unconventional superconductivity. Europhys. Lett. **132**, 60003 (2021) 159
148. M.-T. Deng, S. Vaitiekenas, E.B. Hansen, J. Danon, M. Leijnse, K. Flensberg, J. Nygård, P. Krogstrup, C.M. Marcus, Majorana bound state in a coupled quantum-dot hybrid-nanowire system. Science **354**(6319), 1557–1562 (2016) 160, 177, 178
149. M.-T. Deng, S. Vaitiekėnas, E. Prada, P. San-Jose, J. Nygård, P. Krogstrup, R. Aguado, C.M. Marcus, Nonlocality of Majorana modes in hybrid nanowires. Phys. Rev. B **98**, 085125 (2018) 160, 177, 178

150. E. Prada, R. Aguado, P. San-Jose, Measuring Majorana nonlocality and spin structure with a quantum dot. Phys. Rev. B **96**, 085418 (2017) 160, 177
151. D.J. Clarke, Experimentally accessible topological quality factor for wires with zero energy modes. Phys. Rev. B **96**, 201109 (2017) 160, 177, 196, 198
152. J. Danon, A.B. Hellenes, E.B. Hansen, L. Casparis, A.P. Higginbotham, K. Flensberg, Nonlocal conductance spectroscopy of Andreev bound states: symmetry relations and bcs charges. Phys. Rev. Lett. **124**, 036801 (2020) 160
153. G.C. Ménard, G.L.R. Anselmetti, E.A. Martinez, D. Puglia, F.K. Malinowski, J.S. Lee, S. Choi, M. Pendharkar, C.J. Palmstrøm, K. Flensberg, C.M. Marcus, L. Casparis, A.P. Higginbotham, Conductance-matrix symmetries of a three-terminal hybrid device. Phys. Rev. Lett. **124**, 036802 (2020)
154. A. Maiani, M. Geier, K. Flensberg, Conductance matrix symmetries of multiterminal semiconductor-superconductor devices. Phys. Rev. B **106**, 104516 (2022) 160
155. D.I. Pikulin, B. van Heck, T. Karzig, E.A. Martinez, B. Nijholt, T. Laeven, G.W. Winkler, J.D. Watson, S. Heedt, M. Temurhan, V. Svidenko, R.M. Lutchyn, M. Thomas, G. de Lange, L. Casparis, C. Nayak, Protocol to identify a topological superconducting phase in a three-terminal device. arXiv:2103.12217 (2021) 160
156. L. Fu, Electron teleportation via Majorana bound states in a mesoscopic superconductor. Phys. Rev. Lett. **104**, 056402 (2010) 160
157. R. Hützen, A. Zazunov, B. Braunecker, A. Levy Yeyati, R. Egger, Majorana single-charge transistor. Phys. Rev. Lett. **109**, 166403 (2012) 160
158. A.M. Whiticar, A. Fornieri, E.C.T. O'Farrell, A.C.C. Drachmann, T. Wang, C. Thomas, S. Gronin, R. Kallaher, G.C. Gardner, M.J. Manfra, C.M. Marcus, F. Nichele, Coherent transport through a Majorana island in an Aharonov–Bohm interferometer. Nat. Commun. **11**(1), 3212 (2020) 160
159. J. Shen, S. Heedt, F. Borsoi, B. van Heck, R.L.M. Gazibegovic, S. Op het Veld, D. Car, J.A. Logan, M. Pendharkar, S.J.J. Ramakers, G. Wang, D. Xu, D. Bouman, A. Geresdi, C.J. Palmstrøm, E.P.A.M. Bakkers, L.P. Kouwenhoven, Parity transitions in the superconducting ground state of hybrid InSb–Al coulomb islands. Nat. Commun. **9**(1), 4801 (2018) 160
160. D.J. Carrad, M. Bjergfelt, T. Kanne, M. Aagesen, F. Krizek, E.M. Fiordaliso, E. Johnson, J. Nygård, T.S. Jespersen, Shadow epitaxy for in situ growth of generic semiconductor/superconductor hybrids. Adv. Mater. **32**(23), 1908411 (2020)
161. D.M.T. van Zanten, D. Sabonis, J. Suter, J.I. Väyrynen, T. Karzig, D.I. Pikulin, E.C.T. O'Farrell, D. Razmadze, K.D. Petersson, P. Krogstrup, C.M. Marcus, Photon-assisted tunnelling of zero modes in a Majorana wire. Nat. Phys. **16**(6), 663–668 (2020) 160
162. E. Prada, P. San-Jose, R. Aguado, Transport spectroscopy of *ns* nanowire junctions with Majorana fermions. Phys. Rev. B **86**, 180503 (2012) 160
163. G. Kells, D. Meidan, P.W. Brouwer, Near-zero-energy end states in topologically trivial spin-orbit coupled superconducting nanowires with a smooth confinement. Phys. Rev. B **86**, 100503 (2012)
164. C. Moore, C. Zeng, T.D. Stanescu, S. Tewari, Quantized zero-bias conductance plateau in semiconductor-superconductor heterostructures without topological Majorana zero modes. Phys. Rev. B **98**, 155314 (2018)
165. A. Vuik, B. Nijholt, A.R. Akhmerov, M. Wimmer, Reproducing topological properties with quasi-Majorana states. SciPost Phys. **7**, 61 (2019) 160
166. O.A. Awoga, J. Cayao, A.M. Black-Schaffer, Supercurrent detection of topologically trivial zero-energy states in nanowire junctions. Phys. Rev. Lett. **123**, 117001 (2019)
167. H. Pan, S. Das Sarma, Physical mechanisms for zero-bias conductance peaks in Majorana nanowires. Phys. Rev. Res. **2**, 013377 (2020)
168. J. Avila, F. Peñaranda, E. Prada, P. San-Jose, R. Aguado, Non-hermitian topology as a unifying framework for the Andreev versus Majorana states controversy. Commun. Phys. **2**(1), 133 (2019) 160
169. R. Hess, H.F. Legg, D. Loss, J. Klinovaja, Trivial Andreev band mimicking topological bulk gap reopening in the nonlocal conductance of long Rashba nanowires. Phys. Rev. Lett. **130**, 207001 (2023) 160

170. J.D. Sau, S.D. Sarma, Nat. Commun. **3**(1), 964 (2012) 162
171. P. Recher, E.V. Sukhorukov, D. Loss, Phys. Rev. B **63**, 165314 (2001) 162, 177
172. L. Hofstetter, S. Csonka, J. Nygård, C. Schönenberger, Nature **461**(7266), 960–963 (2009)
173. L.G. Herrmann, F. Portier, P. Roche, A. Levy Yeyati, T. Kontos, C. Strunk, Phys. Rev. Lett. **104**, 026801 (2010)
174. G. Fülöp, F. Domínguez, S. d'Hollosy, A. Baumgartner, P. Makk, M.H. Madsen, V.A. Guzenko, J. Nygård, C. Schönenberger, A. Levy Yeyati, S. Csonka, Phys. Rev. Lett. **115**, 227003 (2015) 162, 177
175. C.-X. Liu, G. Wang, T. Dvir, M. Wimmer, Phys. Rev. Lett. **129**, 267701 (2022) 162, 168, 169, 174, 175, 178
176. W. Samuelson, V. Svensson, M. Leijnse, Minimal quantum dot based Kitaev chain with only local superconducting proximity effect. Phys. Rev. B **109**, 035415 (2024) 167
177. S. Miles, D. van Driel, M. Wimmer, C.-X. Liu, Kitaev chain in an alternating quantum dot-Andreev bound state array. arXiv:2309.15777 (2023) 167
178. J.D.T. Luna, A.M. Bozkurt, M. Wimmer, C.-X. Liu, Flux-tunable Kitaev chain in a quantum dot array. arXiv:2402.07575 (2024) 169
179. C.-X. Liu, A.M. Bozkurt, F. Zatelli, S.L.D. ten Haaf, T. Dvir, M. Wimmer, Enhancing the excitation gap of a quantum-dot-based Kitaev chain. arXiv:2310.09106 (2023) 170
180. A. Tsintzis, R. Seoane Souto, M. Leijnse, Phys. Rev. B **106**, L201404 (2022) 170, 173, 174, 177, 178
181. S.V. Aksenov, A.O. Zlotnikov, M.S. Shustin, Phys. Rev. B **101**(12), 125431 (2020) 172
182. R.S. Souto, A. Tsintzis, M. Leijnse, J. Danon, Probing Majorana localization in minimal Kitaev chains through a quantum dot. Phys. Rev. Res. **5**, 043182 (2023) 177
183. D. Beckmann, H.B. Weber, H.v. Löhneysen, Evidence for crossed Andreev reflection in superconductor-ferromagnet hybrid structures. Phys. Rev. Lett. **93**, 197003 (2004) 177
184. S. Russo, M. Kroug, T.M. Klapwijk, A.F. Morpurgo, Experimental observation of bias-dependent nonlocal Andreev reflection. Phys. Rev. Lett. **95**, 027002 (2005)
185. J. Schindele, A. Baumgartner, C. Schönenberger, Near-unity cooper pair splitting efficiency. Phys. Rev. Lett. **109**, 157002 (2012) 177, 179
186. A. Das, Y. Ronen, M. Heiblum, D. Mahalu, A.V. Kretinin, H. Shtrikman, High-efficiency Cooper pair splitting demonstrated by two-particle conductance resonance and positive noise cross-correlation. Nat. Commun. **3**(1), 1165 (2012) 177
187. A. Bordin, G. Wang, C.-X. Liu, S.L.D. ten Haaf, N. van Loo, G.P. Mazur, D. Xu, D. van Driel, F. Zatelli, S. Gazibegovic, G. Badawy, E.P.A.M. Bakkers, M. Wimmer, L.P. Kouwenhoven, T. Dvir, Tunable crossed Andreev reflection and elastic cotunneling in hybrid nanowires. Phys. Rev. X **13**, 031031 (2023) 178
188. S.L.D. ten Haaf, Q. Wang, A.M. Bozkurt, C.-X. Liu, I. Kulesh, P. Kim, D. Xiao, C. Thomas, M.J. Manfra, T. Dvir, M. Wimmer, S. Goswami, Engineering Majorana bound states in coupled quantum dots in a two-dimensional electron gas. arXiv:2311.03208 (2023) 179
189. F. Zatelli, D. van Driel, D. Xu, G. Wang, C.-X. Liu, A. Bordin, B. Roovers, G.P. Mazur, N. van Loo, J.C. Wolff, A. Mert Bozkurt, G. Badawy, S. Gazibegovic, E.P.A.M. Bakkers, M. Wimmer, L.P. Kouwenhoven, T. Dvir, Robust poor man's Majorana zero modes using Yu-Shiba-Rusinov states. arXiv:2311.03193 (2023) 179
190. S. Heedt, M. Quintero-Pérez, F. Borsoi, A. Fursina, N. van Loo, G.P. Mazur, M.P. Nowak, M. Ammerlaan, K. Li, S. Korneychuk, J. Shen, M.A.Y. van de Poll, G. Badawy, S. Gazibegovic, N. de Jong, P. Aseev, K. van Hoogdalem, E.P.A.M. Bakkers, L.P. Kouwenhoven, Shadow-wall lithography of ballistic superconductor–semiconductor quantum devices. Nat. Commun. **12**(1), 4914 (2021) 179
191. F. Borsoi, G.P. Mazur, N. van Loo, M.P. Nowak, L. Bourdet, K. Li, S. Korneychuk, A. Fursina, J.-Y. Wang, V. Levajac, E. Memisevic, G. Badawy, S. Gazibegovic, K. van Hoogdalem, E.P.A.M. Bakkers, L.P. Kouwenhoven, S. Heedt, M. Quintero-Pérez, Single-shot fabrication of semiconducting–superconducting nanowire devices. Adv. Funct. Mater. **31**(34), 2102388 (2021) 179

192. A. Bordin, X. Li, D. van Driel, J.C. Wolff, Q. Wang, S.L.D. ten Haaf, G. Wang, N. van Loo, L.P. Kouwenhoven, T. Dvir, Crossed Andreev reflection and elastic cotunneling in three quantum dots coupled by superconductors. Phys. Rev. Lett. **132**, 056602 (2024) 180
193. A. Bordin, C.-X. Liu, T. Dvir, F. Zatelli, S.L.D. ten Haaf, D. van Driel, G. Wang, N. van Loo, T. van Caekenberghe, J.C. Wolff, Y. Zhang, G. Badawy, S. Gazibegovic, E.P.A.M. Bakkers, M. Wimmer, L.P. Kouwenhoven, G.P. Mazur, Signatures of Majorana protection in a three-site Kitaev chain. arXiv:2402.19382 (2024) 180
194. R. Koch, D. van Driel, A. Bordin, J.L. Lado, E. Greplova, Adversarial Hamiltonian learning of quantum dots in a minimal Kitaev chain. Phys. Rev. Appl. **20**, 044081 (2023) 180
195. J. Benestad, A. Tsintzis, R. Seoane Souto, M. Leijnse, E. van Nieuwenburg, J. Danon, Machine-learned tuning of artificial Kitaev chains from tunneling-spectroscopy measurements. arXiv:2405.01240 (2024)
196. D. van Driel, R. Koch, V.P.M. Sietses, S.L.D. ten Haaf, C.X. Liu, F. Zatelli, B. Roovers, A. Bordin, N. van Loo, G. Wang, J.C. Wolff, G.P. Mazur, T. Dvir, I. Kulesh, Q. Wang, A.M. Bozkurt, S. Gazibegovic, G. Badawy, E.P.A.M. Bakkers, M. Wimmer, S. Goswami, J.L. Lado, L.P. Kouwenhoven, E. Greplova, Cross-platform autonomous control of minimal Kitaev chains. arXiv:2405.04596 (2024) 180
197. J. Koch, T.M. Yu, J. Gambetta, A.A. Houck, D.I. Schuster, J. Majer, A. Blais, M.H. Devoret, S.M. Girvin, R.J. Schoelkopf, Charge-insensitive qubit design derived from the cooper pair box. Phys. Rev. A **76**, 042319 (2007) 181
198. F. Arute, K. Arya, R. Babbush, D. Bacon, J.C. Bardin, R. Barends, R. Biswas, S. Boixo, F.G.S.L. Brandao, D.A. Buell, B. Burkett, Y. Chen, Z. Chen, B. Chiaro, R. Collins, W. Courtney, A. Dunsworth, E. Farhi, B. Foxen, A. Fowler, C. Gidney, M. Giustina, R. Graff, K. Guerin, S. Habegger, M.P. Harrigan, M.J. Hartmann, A. Ho, M. Hoffmann, T. Huang, T.S. Humble, S.V. Isakov, E. Jeffrey, Z. Jiang, D. Kafri, K. Kechedzhi, J. Kelly, P.V. Klimov, S. Knysh, A. Korotkov, F. Kostritsa, D. Landhuis, M. Lindmark, E. Lucero, D. Lyakh, S. Mandrà, J.R. McClean, M. McEwen, A. Megrant, X. Mi, K. Michielsen, M. Mohseni, J. Mutus, O. Naaman, M. Neeley, C. Neill, M.Y. Niu, E. Ostby, A. Petukhov, J.C. Platt, C. Quintana, E.G. Rieffel, P. Roushan, N.C. Rubin, D. Sank, K.J. Satzinger, V. Smelyanskiy, K.J. Sung, M.D. Trevithick, A. Vainsencher, B. Villalonga, T. White, Z. Jamie Yao, P. Yeh, A. Zalcman, H. Neven, J.M. Martinis, Quantum supremacy using a programmable superconducting processor. Nature **574**(7779), 505–510 (2019) 181
199. T.W. Larsen, K.D. Petersson, F. Kuemmeth, T.S. Jespersen, P. Krogstrup, J. Nygård, C.M. Marcus, Semiconductor-nanowire-based superconducting qubit. Phys. Rev. Lett. **115**, 127001 (2015) 181
200. G. de Lange, B. van Heck, A. Bruno, D.J. van Woerkom, A. Geresdi, S.R. Plissard, E.P.A.M. Bakkers, A.R. Akhmerov, L. DiCarlo, Realization of microwave quantum circuits using hybrid superconducting-semiconducting nanowire Josephson elements. Phys. Rev. Lett. **115**, 127002 (2015)
201. L. Casparis, M.R. Connolly, M. Kjaergaard, N.J. Pearson, A. Kringhøj, T.W. Larsen, F. Kuemmeth, T. Wang, C. Thomas, S. Gronin, G.C. Gardner, M.J. Manfra, C.M. Marcus, K.D. Petersson, Superconducting gatemon qubit based on a proximitized two-dimensional electron gas. Nat. Nanotechnol. **13**(10), 915–919 (2018)
202. H. Zheng, L.Y. Cheung, N. Sangwan, A. Kononov, R. Haller, J. Ridderbos, C. Ciaccia, J.H. Ungerer, A. Li, E.P.A.M. Bakkers, A. Baumgartner, C. Schönenberger, Coherent control of a few-channel hole type gatemon qubit. arXiv:2312.06411 (2023) 181
203. P. Brooks, A. Kitaev, J. Preskill, Protected gates for superconducting qubits. Phys. Rev. A **87**, 052306 (2013) 181
204. P. Groszkowski, A. Di Paolo, A.L. Grimsmo, A. Blais, D.I. Schuster, A.A. Houck, J. Koch, Coherence properties of the 0-qubit. New J. Phys. **20**(4), 043053 (2018)
205. A. Gyenis, P.S. Mundada, A. Di Paolo, T.M. Hazard, X. You, D.I. Schuster, J. Koch, A. Blais, A.A. Houck, Experimental realization of a protected superconducting circuit derived from the 0-π qubit. PRX Quantum **2**, 010339 (2021) 181

206. W.C. Smith, A. Kou, X. Xiao, U. Vool, M.H. Devoret, Superconducting circuit protected by two-Cooper-pair tunneling. npj Quantum Inf. **6**(1), 8 (2020) 181
207. T.W. Larsen, M.E. Gershenson, L. Casparis, A. Kringhøj, N.J. Pearson, R.P.G. McNeil, F. Kuemmeth, P. Krogstrup, K.D. Petersson, C.M. Marcus, Parity-protected superconductor-semiconductor qubit. Phys. Rev. Lett. **125**, 056801 (2020)
208. C. Schrade, C.M. Marcus, A. Gyenis, Protected hybrid superconducting qubit in an array of gate-tunable Josephson interferometers. PRX Quantum **3**, 030303 (2022)
209. W.C. Smith, M. Villiers, A. Marquet, J. Palomo, M.R. Delbecq, T. Kontos, P. Campagne-Ibarcq, B. Douçot, Z. Leghtas, Magnifying quantum phase fluctuations with cooper-pair pairing. Phys. Rev. X **12**, 021002 (2022)
210. A. Maiani, M. Kjaergaard, C. Schrade, Entangling transmons with low-frequency protected superconducting qubits. PRX Quantum **3**, 030329 (2022) 181
211. F. Bao, H. Deng, D. Ding, R. Gao, X. Gao, C. Huang, X. Jiang, H.-S. Ku, Z. Li, X. Ma, X. Ni, J. Qin, Z. Song, H. Sun, C. Tang, T. Wang, F. Wu, T. Xia, W. Yu, F. Zhang, G. Zhang, X. Zhang, J. Zhou, X. Zhu, Y. Shi, J. Chen, H.-H. Zhao, C. Deng, Fluxonium: an alternative qubit platform for high-fidelity operations. Phys. Rev. Lett. **129**, 010502 (2022) 181
212. C. Janvier, L. Tosi, L. Bretheau, Ç. Ö. Girit, M. Stern, P. Bertet, P. Joyez, D. Vion, D. Esteve, M.F. Goffman, H. Pothier, C. Urbina, Coherent manipulation of Andreev states in superconducting atomic contacts. Science **349**(6253), 1199–1202 (2015) 183, 189
213. M. Hays, G. de Lange, K. Serniak, D.J. van Woerkom, D. Bouman, P. Krogstrup, J. Nygård, A. Geresdi, M.H. Devoret, Direct microwave measurement of Andreev-bound-state dynamics in a semiconductor-nanowire Josephson junction. Phys. Rev. Lett. **121**, 047001 (2018) 183, 185, 189
214. L.Y. Cheung, R. Haller, A. Kononov, C. Ciaccia, J.H. Ungerer, T. Kanne, J. Nygård, P. Winkel, T. Reisinger, I.M. Pop, A. Baumgartner, C. Schönenberger, Photon-mediated long range coupling of two Andreev level qubits. arXiv:2310.15995 (2023) 185
215. N.M. Chtchelkatchev, Y.V. Nazarov, Andreev quantum dots for spin manipulation. Phys. Rev. Lett. **90**, 226806 (2003) 186
216. G. Burkard, T.D. Ladd, A. Pan, J.M. Nichol, J.R. Petta, Semiconductor spin qubits. Rev. Mod. Phys. **95**, 025003 (2023) 186
217. M. Hays, V. Fatemi, D. Bouman, J. Cerrillo, S. Diamond, K. Serniak, T. Connolly, P. Krogstrup, J. Nygård, A. Levy Yeyati, A. Geresdi, M.H. Devoret, Coherent manipulation of an Andreev spin qubit. Science **373**(6553), 430–433 (2021) 186, 187, 189
218. A. Tosato, V. Levajac, J.-Y. Wang, C.J. Boor, F. Borsoi, M. Botifoll, C.N. Borja, S. Martí-Sánchez, J. Arbiol, A. Sammak, M. Veldhorst, G. Scappucci, Hard superconducting gap in germanium. Commun. Mater. **4**(1), 23 (2023) 189
219. M. Valentini, O. Sagi, L. Baghumyan, T. de Gijsel, J. Jung, S. Calcaterra, A. Ballabio, J.A. Servin, K. Aggarwal, M. Janik, T. Adletzberger, R. Seoane Souto, M. Leijnse, J. Danon, C. Schrade, E. Bakkers, D. Chrastina, G. Isella, G. Katsaros, Parity-conserving cooper-pair transport and ideal superconducting diode in planar germanium. Nat. Commun. **15**(1), 169 (2024) 189
220. G. Katsaros, Private communication (2024) 189
221. A. Mishra, P. Simon, T. Hyart, M. Trif, Yu-Shiba-Rusinov qubit. PRX Quantum **2**, 040347 (2021) 189
222. F.K. Malinowski, Preprint. arXiv.2303.14410 (2023). https://doi.org/10.48550/arXiv.2303.14410
223. M. Geier, R.S. Souto, J. Schulenborg, S. Asaad, M. Leijnse, K. Flensberg, A fermion-parity qubit in a proximitized double quantum dot. Phys. Rev. Research **6**, 023281 (2024). https://doi.org/10.1103/PhysRevResearch.6.023281 189, 190
224. X. Zhou, X. Li, Q. Chen, G. Koolstra, G. Yang, B. Dizdar, Y. Huang, C.S. Wang, X. Han, X. Zhang, D.I. Schuster, D. Jin, Electron charge qubit with 0.1 millisecond coherence time. Nat. Phys. **20**, 116–122 (2024) 191
225. G.O. Steffensen, A.L. Yeyati, YSR bond qubit in a double quantum dot with cQED operation. arXiv:2402.19261 (2024) 192

226. A.Yu. Kitaev, Fault-tolerant quantum computation by anyons. Ann. Phys. **303**(1), 2–30 (2003) 193
227. S.D. Sarma, M. Freedman, C. Nayak, Majorana zero modes and topological quantum computation. npj Quantum Inf. **1**(1), 15001 (2015) 193
228. C. Nayak, S.H. Simon, A. Stern, M. Freedman, S. Das Sarma, Non-Abelian anyons and topological quantum computation. Rev. Mod. Phys. **80**(3), 1083 (2008) 194
229. T. Karzig, Y. Oreg, G. Refael, M.H. Freedman, Robust Majorana magic gates via measurements. Phys. Rev. B **99**, 144521 (2019) 194
230. J. Ávila, E. Prada, P. San-Jose, R. Aguado, Majorana oscillations and parity crossings in semiconductor nanowire-based transmon qubits. Phys. Rev. Res. **2**, 033493 (2020) 194
231. E. Ginossar, E. Grosfeld, Microwave transitions as a signature of coherent parity mixing effects in the Majorana-transmon qubit. Nat. Commun. **5**(1), 4772 (2014)
232. K. Yavilberg, E. Ginossar, E. Grosfeld, Fermion parity measurement and control in Majorana circuit quantum electrodynamics. Phys. Rev. B **92**, 075143 (2015)
233. T. Li, W.A. Coish, M. Hell, K. Flensberg, M. Leijnse, Four-Majorana qubit with charge readout: dynamics and decoherence. Phys. Rev. B **98**, 205403 (2018)
234. F. Hassler, A.R. Akhmerov, C.W.J. Beenakker, The top-transmon: a hybrid superconducting qubit for parity-protected quantum computation. New J. Phys. **13**(9), 095004 (2011)
235. T. Hyart, B. van Heck, I.C. Fulga, M. Burrello, A.R. Akhmerov, C.W.J. Beenakker, Flux-controlled quantum computation with Majorana fermions. Phys. Rev. B **88**, 035121 (2013)
236. A. Keselman, C. Murthy, B. van Heck, B. Bauer, Spectral response of Josephson junctions with low-energy quasiparticles. SciPost Phys. **7**, 050 (2019)
237. J. Ávila, E. Prada, P. San-Jose, R. Aguado, Superconducting islands with topological Josephson junctions based on semiconductor nanowires. Phys. Rev. B **102**, 094518 (2020)
238. T.B. Smith, M.C. Cassidy, D.J. Reilly, S.D. Bartlett, A.L. Grimsmo, Dispersive readout of Majorana qubits. PRX Quantum **1**, 020313 (2020)
239. E. Lupo, E. Grosfeld, E. Ginossar, Implementation of single-qubit gates via parametric modulation in the Majorana transmon. PRX Quantum **3**, 020340 (2022) 194
240. S. Plugge, A. Rasmussen, R. Egger, K. Flensberg, Majorana box qubits. New J. Phys. **19**(1), 012001 (2017) 195, 196, 199
241. M.I.K. Munk, J. Schulenborg, R. Egger, K. Flensberg, Parity-to-charge conversion in Majorana qubit readout. Phys. Rev. Res. **2**, 033254 (2020) 195
242. J.F. Steiner, F. von Oppen, Readout of Majorana qubits. Phys. Rev. Res. **2**, 033255 (2020)
243. J. Schulenborg, M. Burrello, M. Leijnse, K. Flensberg, Multilevel effects in quantum dot based parity-to-charge conversion of Majorana box qubits. Phys. Rev. B **103**, 245407 (2021)
244. T.B. Smith, M.C. Cassidy, D.J. Reilly, S.D. Bartlett, A.L. Grimsmo, Dispersive readout of Majorana qubits. PRX Quantum **1**, 020313 (2020) 195
245. M. Nitsch, R. Seoane Souto, M. Leijnse, Interference and parity blockade in transport through a Majorana box. Phys. Rev. B **106**, L201305 (2022) 195
246. T. Karzig, C. Knapp, R.M. Lutchyn, P. Bonderson, M.B. Hastings, C. Nayak, J. Alicea, K. Flensberg, S. Plugge, Y. Oreg, C.M. Marcus, M.H. Freedman, Scalable designs for quasiparticle-poisoning-protected topological quantum computation with Majorana zero modes. Phys. Rev. B **95**, 235305 (2017) 195, 199
247. D. Litinski, F. von Oppen, Quantum computing with Majorana fermion codes. Phys. Rev. B **97**, 205404 (2018) 195
248. B. Béri, N.R. Cooper, Topological kondo effect with Majorana fermions. Phys. Rev. Lett. **109**, 156803 (2012) 195
249. W. Bishara, P. Bonderson, C. Nayak, K. Shtengel, J.K. Slingerland, Interferometric signature of non-abelian anyons. Phys. Rev. B **80**, 155303 (2009) 196
250. J. Ruhman, E. Berg, E. Altman, Topological states in a one-dimensional fermi gas with attractive interaction. Phys. Rev. Lett. **114**, 100401 (2015)
251. D. Aasen, M. Hell, R.V. Mishmash, A. Higginbotham, J. Danon, M. Leijnse, T.S. Jespersen, J.A. Folk, C.M. Marcus, K. Flensberg, J. Alicea, Milestones toward Majorana-based quantum computing. Phys. Rev. X **6**, 031016 (2016) 198

252. E.C. Rowell, Z. Wang, Degeneracy and non-abelian statistics. Phys. Rev. A **93**, 030102 (2016)
253. M. Hell, J. Danon, K. Flensberg, M. Leijnse, Time scales for Majorana manipulation using coulomb blockade in gate-controlled superconducting nanowires. Phys. Rev. B **94**, 035424 (2016) 198
254. M. Barkeshli, P. Bonderson, M. Cheng, Z. Wang, Symmetry fractionalization, defects, and gauging of topological phases. Phys. Rev. B **100**, 115147 (2019)
255. T. Zhou, M.C. Dartiailh, W. Mayer, J.E. Han, A. Matos-Abiague, J. Shabani, I. Žutić, Phase control of Majorana bound states in a topological X junction. Phys. Rev. Lett. **124**, 137001 (2020)
256. T. Zhou, M.C. Dartiailh, K. Sardashti, J.E. Han, A. Matos-Abiague, J. Shabani, I. Žutić, Fusion of Majorana bound states with mini-gate control in two-dimensional systems. Nat. Commun. **13**(1), 1738 (2022)
257. R.S. Souto, M. Leijnse, Fusion rules in a Majorana single-charge transistor. SciPost Phys. **12**, 161 (2022)
258. M. Nitsch, R. Seoane Souto, S. Matern, M. Leijnse, Transport-based fusion that distinguishes between Majorana and Andreev bound states. Phys. Rev. B **109**, 165404 (2024) 196
259. C.W.J. Beenakker, Search for non-Abelian Majorana braiding statistics in superconductors. SciPost Phys. Lect. Notes **15** (2020) 197
260. J. Alicea, Y. Oreg, G. Refael, F. von Oppen, M.P.A. Fisher, Non-abelian statistics and topological quantum information processing in 1d wire networks. Nat. Phys. **7**, 412 (2011) 197
261. F. Harper, A. Pushp, R. Roy, Majorana braiding in realistic nanowire y-junctions and tuning forks. Phys. Rev. Res. **1**, 033207 (2019) 197
262. D.J. Clarke, J.D. Sau, S. Tewari, Majorana fermion exchange in quasi-one-dimensional networks. Phys. Rev. B **84**, 035120 (2011) 198
263. B. van Heck, A.R. Akhmerov, F. Hassler, M. Burrello, C.W.J. Beenakker, Coulomb-assisted braiding of Majorana fermions in a Josephson junction array. New J. Phys. **14**(3), 035019 (2012)
264. M. Burrello, B. van Heck, A.R. Akhmerov, Braiding of non-abelian anyons using pairwise interactions. Phys. Rev. A **87**, 022343 (2013)
265. T. Karzig, F. Pientka, G. Refael, F. von Oppen, Shortcuts to non-Abelian braiding. Phys. Rev. B **91**, 201102(R) (2015)
266. M. Hell, M. Leijnse, K. Flensberg, Two-dimensional platform for networks of Majorana bound states. Phys. Rev. Lett. **118**(10), 107701 (2017)
267. J. Liu, W. Chen, M. Gong, Y. Wu, X. Xie, Minimal setup for non-abelian braiding of Majorana zero modes. Sci. China Phys. Mech. Astron. **64**(11), 117811 (2021) 198
268. K. Flensberg, Non-abelian operations on Majorana fermions via single-charge control. Phys. Rev. Lett. **106**, 090503 (2011) 198, 206
269. S. Krøjer, R. Seoane Souto, K. Flensberg, Demonstrating Majorana non-abelian properties using fast adiabatic charge transfer. Phys. Rev. B **105**, 045425 (2022) 198, 199, 206
270. R. Seoane Souto, K. Flensberg, M. Leijnse, Timescales for charge transfer based operations on Majorana systems. Phys. Rev. B **101**, 081407 (2020) 198, 206
271. P. Bonderson, M. Freedman, C. Nayak, Measurement-only topological quantum computation. Phys. Rev. Lett. **101**, 010501 (2008) 199
272. S.M. Albrecht, E.B. Hansen, A.P. Higginbotham, F. Kuemmeth, T.S. Jespersen, J. Nygård, P. Krogstrup, J. Danon, K. Flensberg, C.M. Marcus, Transport signatures of quasiparticle poisoning in a Majorana island. Phys. Rev. Lett. **118**, 137701 (2017) 200
273. G.C. Ménard, F.K. Malinowski, D. Puglia, D.I. Pikulin, T. Karzig, B. Bauer, P. Krogstrup, C.M. Marcus, Suppressing quasiparticle poisoning with a voltage-controlled filter. Phys. Rev. B **100**, 165307 (2019)
274. H.Q. Nguyen, D. Sabonis, D. Razmadze, E.T. Mannila, V.F. Maisi, D.M.T. van Zanten, E.C.T. O'Farrell, P. Krogstrup, F. Kuemmeth, J. P. Pekola, C.M. Marcus, Electrostatic control of quasiparticle poisoning in a hybrid semiconductor-superconductor island. Phys. Rev. B **108**, L041302 (2023)

275. M. Hinderling, S.C. ten Kate, D.Z. Haxell, M. Coraiola, S. Paredes, F. Krizek, E. Cheah, R. Schott, W. Wegscheider, D. Sabonis, F. Nichele, Flip-chip-based fast inductive parity readout of a planar superconducting island. arXiv:2307.06718 (2023) 200
276. M. Aghaee, A.A. Ramirez, Z. Alam, R. Ali, M. Andrzejczuk, A. Antipov, M. Astafev, A. Barzegar, B. Bauer, J. Becker, A. Bocharov, S. Boddapati, D. Bohn, J. Bommer, E.B. Hansen, L. Bourdet, A. Bousquet, S. Boutin, S.B. Markussen, J.C.E. Saldaña, L. Casparis, B.J. Chapman, S. Chatoor, C. Chua, P. Codd, W. Cole, P. Cooper, F. Corsetti, A. Cui, J.P. Dehollain, P. Dalpasso, M. de Moor, G. de Lange, A. Ekefjärd, T.E. Dandachi, S. Fallahi, L. Galletti, G. Gardner, D. Govender, F. Griggio, R. Grigoryan, S. Grijalva, S. Gronin, J. Gukelberger, M. Hamdast, F. Hamze, M.H. Madsen, J.H. Nielsen, J.M. Nielsen, S. Heedt, Z. Heidarnia, J.H. Zamorano, S. Ho, L. Holgaard, J. Hornibrook, W.H.P. Nielsen, J. Indrapiromkul, H. Ingerslev, L. Ivancevic, J. Jhoja, J. Jones, K.V. Kalashnikov, R. Kallaher, F. Karimi, T. Karzig, E. King, M.E. Kloster, C. Knapp, D. Kocon, J. Koski, P. Kostamo, U.K. Bhaskar, M. Kumar, T. Laeven, T. Larsen, K. Lee, J. Lee, G. Leum, K. Li, T. Lindemann, M. Looij, J. Love, M. Lucas, R. Lutchyn, N. Madulid, A. Malmros, M. Manfra, D. Mantri, E. Martinez, M. Mattila, R. McNeil, R.V. Mishmash, G. Mohandas, C. Mollgaard, T. Morgan, G. Moussa, C. Nayak, B. Nijholt, M. Nystrom, E. O'Farrell, T. Ohki, K. Otani, B.P. Wütz, S. Pauka, K. Petersson, L. Petit, D. Pikulin, G. Prawiroatmodjo, F. Preiss, M. Rajpalke, C. Ranta, K. Rasmussen, D. Razmadze, O. Reentila, D.J. Reilly, Y. Ren, K. Reneris, A.R. Mei, R. Rouse, I. Sadovskyy, L. Sainiemi, I. Sanlorenzo, E. Schmidgall, C. Sfiligoj, M.B. Shah, K. Simoes, S. Singh, S. Sinha, T. Soerensen, P. Sohr, T. Stankevic, L. Stek, E. Stuppard, H. Suominen, J. Suter, S. Teicher, N. Thiyagarajah, R. Tholapi, M. Thomas, E. Toomey, J. Tracy, M. Turley, S. Upadhyay, I. Urban, K. Van Hoogdalem, D.J. Van Woerkom, D.V. Viazmitinov, D. Vogel, G.W. Winkler, J. Watson, A. Webster, J. Weston, A.W. Christensen, C.K. Yang, E. Yucelen, R. Zeisel, G. Zheng, J. Zilke, Interferometric single-shot parity measurement in an InAs-Al hybrid device. arXiv:2401.09549 (2024) 200
277. C. Barthel, D.J. Reilly, C.M. Marcus, M.P. Hanson, A.C. Gossard, Rapid single-shot measurement of a singlet-triplet qubit. Phys. Rev. Lett. **103**, 160503 (2009) 201
278. K.D. Petersson, C.G. Smith, D. Anderson, P. Atkinson, G.A.C. Jones, D.A. Ritchie, Charge and spin state readout of a double quantum dot coupled to a resonator. Nano Lett. **10**(8), 2789–2793 (2010) 201
279. N.J. Lambert, A.A. Esmail, M. Edwards, F.A. Pollock, B.W. Lovett, A.J. Ferguson, Quantum capacitance and charge sensing of a superconducting double dot. Appl. Phys. Lett. **109**(11), 112603 (2016)
280. F. Vigneau, F. Fedele, A. Chatterjee, D. Reilly, F. Kuemmeth, M. Fernando Gonzalez-Zalba, E. Laird, N. Ares, Probing quantum devices with radio-frequency reflectometry. Appl. Phys. Rev. **10**(2), 021305 (2023) 201
281. A. Tsintzis, R. Seoane Souto, K. Flensberg, J. Danon, M. Leijnse, Majorana qubits and non-abelian physics in quantum dot-based minimal Kitaev chains. PRX Quantum **5**, 010323 (2024) 201, 202, 205, 206, 207, 208
282. C.-X. Liu, H. Pan, F. Setiawan, M. Wimmer, J.D. Sau, Fusion protocol for Majorana modes in coupled quantum dots. Phys. Rev. B **108**, 085437 (2023) 201, 205
283. D. Michel Pino, R.S. Souto, R. Aguado, Minimal Kitaev-transmon qubit based on double quantum dots. Phys. Rev. B **109**, 075101 (2024) 203, 204, 205

Introduction to Quantum Entanglement in Many-Body Systems*

4

Anubhav Kumar Srivastava, Guillem Müller-Rigat, Maciej Lewenstein, and Grzegorz Rajchel-Mieldzioć

Abstract

The quantum mechanics formalism introduced new revolutionary concepts challenging our everyday perceptions. Arguably, quantum entanglement, which explains correlations that cannot be reproduced classically, is the most notable of them. Besides its fundamental aspect, entanglement is also a resource, fueling emergent technologies such as quantum simulators and computers. The purpose of this *chapter* is to give a pedagogical introduction to the topic with a special emphasis on the multipartite scenario, i.e., entanglement distributed among many degrees of freedom. Due to the combinatorial complexity of this setting, particles can interact and become entangled in a plethora of ways, which we characterize here. We start by providing the necessary mathematical

*Anubhav Kumar Srivastava and Guillem Müller-Rigat have contributed equally.

A. K. Srivastava · G. Müller-Rigat
ICFO - Institut de Ciencies Fotoniques, The Barcelona Institute of Science and Technology, Castelldefels, Spain
e-mail: anubhav.srivastava@icfo.eu; guillem.muller@icfo.eu

M. Lewenstein (✉)
ICFO - Institut de Ciencies Fotoniques, The Barcelona Institute of Science and Technology, Castelldefels, Spain

ICREA, Barcelona, Spain
e-mail: maciej.lewenstein@icfo.eu

G. Rajchel-Mieldzioć
ICFO - Institut de Ciencies Fotoniques, The Barcelona Institute of Science and Technology, Castelldefels, Spain

NASK National Research Institute, Warszawa, Poland
e-mail: grzegorz.rajchel@icfo.eu

R. Aguado et al. (eds.), *New Trends and Platforms for Quantum Technologies*, Lecture Notes in Physics 1025, https://doi.org/10.1007/978-3-031-55657-9_4

tools and elementary concepts from entanglement theory. A part of this *chapter* will be devoted to classifying and ordering entangled states. Then, we focus on various entanglement structures useful in condensed-matter theory such as tensor-network states or symmetric states useful for quantum-enhanced sensing. Finally, we discuss state-of-the-art methods to detect and certify such correlations in experiments, with some relevant illustrative examples.

4.1 Introduction

While many entanglement reviews have already been written, the majority of them were written a decade ago (cf. [1–5]). To cover the recent discoveries, we provide a review with an easy and didactic introduction to the topic, and at the same time, we discuss some of the newest results in the field. The present text is based on a set of lectures by Prof. Lewenstein, and earlier reviews of the subject by some of us [6–8].

The main focus of this review is on multipartite entanglement, with experimentally viable setups, but we start with the discussion of the simplest and the most elementary bipartite case in Sect. 4.2. We introduce here the notion of bipartite entanglement and some conceptual topics and discuss its quantification and applications.

In Sect. 4.3, we focus on multipartite entanglement and the challenges associated with it. Here, after introducing the basics, we discuss the notion of partial separability with respect to a specific example of absolutely maximally entangled (AME) states. Furthermore, we consider the resource theory of entanglement to account for the permissible operations and the classification of quantum states in various entanglement classes based on these operations. We study the example of the simplest case of a multipartite system, i.e., a system of 3-qubits to study the entanglement classes and the invariants parameterizing such classes.

Section 4.4 concerns the use and detection of many-body quantum entanglement. Here, we define the entanglement depth and talk about the useful and useless entanglement. Tensor network states are presented here, and we discuss entanglement area laws. We also talk in this section about scalable entanglement certification, using entanglement witnesses. Finally, we introduce Bell correlations and discuss their implications.

We conclude in Sect. 4.5, by listing some open problems that are discussed within the scope of this chapter, particularly in the context of entanglement theory.

This chapter is based on the lectures Maciej Lewenstein delivered during the XXIII Training Courses in the Physics of Strongly Correlated Systems, conducted in October 2021, and organized by the International Institute for Advanced Scientific Studies "E.R. Caianiello" (IIASS), in collaboration with the Department of Physics, University of Salerno. Although the focus of the course was "Trends and Platforms for Quantum Technologies", Lewenstein's lectures focused explicitly on the introduction to entanglement in general, and in many-body systems in particular.

4.2 Mathematical Foundations and Bipartite Systems

The microscopic world is governed by linear laws, such as the Schrödinger equation, in opposition to what we can perceive with everyday experience. Thus, an appropriate description is given of modeling quantum objects as states from a linear space equipped with a scalar product for distinguishability—Hilbert space.

Definition 1 Hilbert space is a complex linear space $\mathcal{H}$, with the inner product[1] written as $\langle\cdot|\cdot\rangle$.

Quantum States Any state of the system must necessarily be physical, i.e., upon measurement in any basis the probabilities must sum up to one. For *pure* states, this condition can be stated as follows:

Definition 2 A pure state of the quantum system $|\psi\rangle \in \mathcal{H}$ from a Hilbert space $\mathcal{H}$ is a normalized vector $\langle\psi|\psi\rangle = 1$.

Description using vectors from a Hilbert space is useful because it carries a lot of physical meaning. For example, two states $|\psi\rangle$ and $|\phi\rangle$ can be perfectly distinguished if and only if they are orthogonal in the vectorial sense [9], $\langle\psi|\phi\rangle = 0$. Not all vectorial properties are physical though. Two orthogonal states will remain orthogonal after multiplication by a constant phase. As a more general rule, multiplying by a global phase $e^{i\varphi}$ does not change a state and is not detectable, $e^{i\varphi}|\psi\rangle \simeq |\psi\rangle$.

Measuring Quantum States If we want to characterize a state, it is necessary to measure it. There are two basic types of measurements in quantum information: projective and positive operator-valued measure (POVM). The first one can be treated as a projection into one of the orthonormal vectors and is generally more useful from an experimentalist's perspective. Since usually, projective measurements are sufficient in a laboratory setting (such as a quantum circuit), for the rest of this contribution we shall discuss only these measurements.[2]

Definition 3 A projective measurement Π is a set of projectors $\{\hat{\pi}_a\}$, pair-wise orthogonal $\hat{\pi}_a\hat{\pi}_b = \delta_{ab}\hat{\pi}_b$ and complete,[3] $\sum_a \hat{\pi}_a = \mathbb{I}$.

[1] For mathematical precision, we add that $\mathcal{H}$ is complete in the sense of metric induced by the inner product.

[2] However, from the foundational point of view, both projective measurements and POVMs are equivalent [10].

[3] We denote $\mathbb{I}$ the identity operator acting in $\mathcal{H}$.

Quantum phenomena do not occur in a Hilbert space, they occur in a laboratory [11]. Therefore, we need a law to map quantum states to the probabilities they predict.

Born's Rule
The probability of obtaining outcome a in measuring $\Pi = \{\hat{\pi}_a\}$ on the state $|\psi\rangle$ is given by [12]:

$$p_a(\psi) = \langle\psi|\,\hat{\pi}_a|\psi\rangle\,. \tag{4.1}$$

From it, one can deduce various statistics. The most important of them are:

- *Expectation* or *mean value* of its associated observable $\hat{O} = \sum_a a\hat{\pi}_a$,

$$\langle\hat{O}\rangle = \langle\psi|\,\hat{O}|\psi\rangle\,. \tag{4.2}$$

- *Variance* as a measure of the fluctuations with respect to the mean,

$$(\Delta\hat{O})^2 = \langle(\hat{O} - \langle\hat{O}\rangle)^2\rangle = \langle\hat{O}^2\rangle - \langle\hat{O}\rangle^2\,. \tag{4.3}$$

Contrary to our everyday intuition, not all observables are jointly measurable in the same quantum system. If two observables $\hat{A}$, $\hat{B}$ do not commute, $[\hat{A}, \hat{B}] := \hat{A}\hat{B} - \hat{B}\hat{A} \neq 0$, they are generally not well-defined simultaneously, i.e., they do not have a joint deterministic outcome. This unintuitive fact is enclosed in Heisenberg's uncertainty relation [13, 14]:

$$(\Delta\hat{A})^2(\Delta\hat{B})^2 \geq \frac{1}{4}\langle[\hat{A}, \hat{B}]\rangle^2\,. \tag{4.4}$$

Mixed States Pure states are important since they describe states that are, in some sense, elementary. However, for any pure state $|\psi\rangle$, there exists a projective measurement, $\{\hat{\pi}_\psi = |\psi\rangle\langle\psi|, \hat{\pi}_{\text{not}\psi} = \mathbb{I} - |\psi\rangle\langle\psi|\}$, with zero variance. This condition is often too stringent, which prompts us to relax it by incorporating into the theory also mixtures of pure states: that is, ensembles. Specifically, mixed states can be decomposed using outer product[4] of pure states as per

$$\hat{\rho} = \sum_i p_i|\psi_i\rangle\langle\psi_i|\,, \tag{4.5}$$

[4] From a physical perspective, outer product of two vectors $|\psi\rangle$ and $|\phi\rangle$, $|\psi\rangle\langle\phi|$, can be thought of as a matrix product of $|\psi\rangle$ and complex conjugated $\langle\phi|$.

where p_i is interpreted as the probability of finding the system in the state $|\psi_i\rangle$. Due to their statistical origin, the description of states using mixed states is alternatively called *density matrices*. To see that mixed states are more general than pure ones, observe that each pure state $|\psi\rangle$ can be written as a density matrix, $\hat{\rho} = |\psi\rangle\langle\psi|$; however, the converse is not true—not every state is pure.

One can also define mixed states mathematically, without any explicit reference to the ensemble:

Definition 4 Mixed state $\hat{\rho}$ is an operator, i.e., a matrix, on the Hilbert space $\mathcal{H}$ such that it is positive semidefinite (PSD)[5] $\hat{\rho} \succeq 0$ and with normalized trace $\text{Tr}(\hat{\rho}) = 1$.

Dynamics Evolution induced on a closed system is given by exponentiation of Hamiltonian (which is an Hermitian operator), $|\psi(t)\rangle = \exp\{-itH/\hbar\}|\psi\rangle$. Such an evolution is necessarily *unitary*, meaning that it preserves the inner product in the Hilbert space: $\langle\psi(t)|\psi(t)\rangle = \langle\psi|\psi\rangle$. Alternatively, if we think of states more concretely as vectors $|\psi\rangle$ of dimension d, then such a unitary evolution U can be treated as a square matrix of size d. Therefore, the unitarity condition is rewritten as

$$\text{for all states } |\psi\rangle : \langle\psi|U^\dagger U|\psi\rangle = \langle\psi|\psi\rangle \iff U^\dagger U = \mathbb{I}, \tag{4.6}$$

with an analogous $UU^\dagger = \mathbb{I}$. Consequently, the unitary transformations of quantum states can be represented as unitary matrices. This might seem trivial since the action of a unitary matrix of size d in a natural way should describe all valid transformations of a state, $|\psi\rangle \mapsto U|\psi\rangle$.

Quantum Channels Notwithstanding, the language of unitary operations is proper only for closed systems. In some cases, it is a useful approximation; however, no system is truly isolated from the environment. If the influence of the outside world is too large then the evolution is no longer unitary, it needs to be described using the language of quantum channels. Beyond dynamics, quantum channels represent the most general physical transformations.

Definition 5 Channel Φ is a completely positive trace-preserving[6] (CPTP) mapping from operators in $\mathcal{H}_X$ to operators in $\mathcal{H}_Y$.

---

[5] That is, all the eigenvalues of $\hat{\rho}$ are real and non-negative.

[6] This condition is added so that the final state is always physical, independently of the initial state. Non-completely positive maps can also be physical, provided some initial correlations with the environment [15].

There are many ways to describe quantum channels [16, 17], but the most common approach involves Kraus operators

$$\Phi(\hat{\rho}) = \sum_i K_i \hat{\rho} K_i^\dagger, \tag{4.7}$$

where $\sum_i K_i^\dagger K_i = \mathbb{I}$. By construction, Eq. (4.7), the map Φ is *positive*, that is for all PSD $\hat{\rho}$, $\Phi(\hat{\rho}) \succeq 0$. It is not so direct that Φ according to Eq. (4.7) is also *completely positive*, i.e., the extended map $\mathrm{Id} \otimes \Phi$, where Id is the identity map, is positive. Surprisingly, complete positivity is easier to verify than positivity alone. Indeed, thanks to Choi's theorem [18]:

Theorem 1 *A map Φ is completely positive if and only if the matrix,*

$$M_\Phi = \sum_{i,j} |i\rangle\langle j| \otimes \Phi(|i\rangle\langle j|)\,, \tag{4.8}$$

where $\{|i\rangle\}$ is an orthonormal basis of $\mathcal{H}_X$, is positive semidefinite.

Note that the Choi matrix, M_Φ, as defined earlier can be written as $M_\Phi = (\mathrm{Id} \otimes \Phi)(|\phi^+\rangle\langle\phi^+|)$, where $|\phi^+\rangle = \sum_{i=1}^d |i\rangle|i\rangle$ is a maximally entangled state.

4.2.1 Bipartite Systems

Sometimes it is useful to divide the Hilbert space of interest into two subsystems A and B, $\mathcal{H} = \mathcal{H}_A \otimes \mathcal{H}_B$. This is especially important when we have a clear division between two physical parts of the entire system.[7] With this picture in mind, any matrix M acting in $\mathcal{H}$ can be interpreted as $M = \sum_i A_i \otimes B_i$ where $\{A_i\}$ acts in $\mathcal{H}_A$ and $\{B_i\}$ acts in $\mathcal{H}_B$.

For future use, we introduce a couple of concepts that will become exceptionally useful for studying entanglement.

Definition 6 Partial trace over subsystem B of a bipartite matrix $M = \sum_i A_i \otimes B_i$ of order $d_A d_B$ is a matrix $\mathrm{Tr}_B M = \sum_i \mathrm{Tr}(B_i) A_i$ of order d_A (analogously for Tr_A).

The usefulness of the above definition in quantum information stems from the fact that if we are interested in only subsystem A, state $\hat{\rho}$ behaves exactly as would $\hat{\rho}_A = \mathrm{Tr}_B(\rho)$. Therefore, we conclude that partial trace encapsulates all information about the state that is accessible from its constituents.

[7] Nonetheless, in this chapter, we are not concerned with physical constraints that might govern a specific application of mathematical theory. Rather, we shall follow the path of introducing all the relevant notions for the general case.

Definition 7 Partial transposition over subsystem B of a bipartite matrix $M = \sum_i A_i \otimes B_i$ is $M^{T_B} = \sum_i A_i \otimes B_i^T$, where T is the usual transposition.

Such a concept will become useful for entanglement detection through the celebrated Peres-Horodecki criterion (PPT), which will be addressed in the next subsection.

Quantum Entanglement If the subsystems did not interact in the past in a meaningful way,[8] we can reasonably believe that they are not correlated. In such a case, the properties of each system are statistically independent. In particular, we have the factorization of joint observables, $\langle \hat{O}_A \hat{O}'_B \rangle = \langle \hat{O}_A \rangle \langle \hat{O}'_B \rangle$ where $\hat{O}_A$, $\hat{O}'_B$ are acting on subsystems A and B respectively.[9] If the underlying quantum state is pure, then it is of product form:

Definition 8 Pure state $|\psi\rangle \in \mathcal{H}_A \otimes \mathcal{H}_B$ is called separable when $|\psi\rangle = |\psi_A\rangle \otimes |\psi_B\rangle$.

Even if some correlation is exhibited $\langle \hat{O}_A \hat{O}'_B \rangle - \langle \hat{O}_A \rangle \langle \hat{O}'_B \rangle \neq 0$, it does not necessarily mean that the underlying state is entangled. The classical world is full of correlations which are described without invoking the Hilbert space formalism introduced in this chapter. Such cases are addressed with separable mixed states:

Definition 9 A mixed state $\hat{\rho}$ is separable if and only if it admits a decomposition into pure separable states

$$\hat{\rho} = \sum_i p_i \left|\psi_i^A\right\rangle\left\langle\psi_i^A\right| \otimes \left|\psi_i^B\right\rangle\left\langle\psi_i^B\right| , \tag{4.9}$$

with $\{p_i\}$ a probability distribution [19].

It is crucial that according to the previous definition, not all separable states are of product form $\hat{\rho} = \hat{\rho}_A \otimes \hat{\rho}_B$, as we allow for statistical mixtures.

The operational interpretation of such classical correlations is that parties A and B have access to a hidden shared "coin" sampling from probability distribution $\{p_i\}$.[10] When outcome i is produced A prepares $|\psi_i^A\rangle$, while B prepares $|\psi_i^B\rangle$ in their part. States that cannot be created using the above procedure, i.e., non-separable, are called entangled.[11]

---

[8] For example, one particle on Earth and the other on Mars.

[9] That is, $\hat{O}_A = \hat{O} \otimes \mathbb{I}_B$, $\hat{O}_B = \mathbb{I}_A \otimes \hat{O}$.

[10] Hidden in the sense that we do not have access to the sampled outcomes.

[11] Observe that here we distinguish between particles A and B. Yet another subset of entanglement study is concerned with *indistinguishable* particles [20,21] (see Sect. 4.4).

Definition 10 State $\hat{\rho}$ is entangled if it is not separable.

Entangled states represent rather strong correlations, in which no classical probability distribution $\{p_i\}$ exists. Along this line, the task of entanglement detection, which is precisely the topic of the next subsection, can be phrased as the task of telling apart quantum from classical correlations. It is not difficult to show that non-classical correlations, explained by entangled states necessarily involve correlation between incompatible (non-commuting) observables.

Understanding Quantum Entanglement with the Spin Singlet

We will settle the concepts defined above with an example. Imagine that you are given the following states in a two-qubit system:

$$|\psi^-\rangle = \frac{1}{\sqrt{2}}(|+1\rangle \otimes |-1\rangle - |-1\rangle \otimes |+1\rangle) \tag{4.10}$$

$$\hat{\rho} = \frac{1}{2}(|+1\rangle\langle+1| \otimes |-1\rangle\langle-1| + |-1\rangle\langle-1| \otimes |+1\rangle\langle+1|)\,, \tag{4.11}$$

and the spin observables with components in the basis $\{|+1\rangle, |-1\rangle\}$:

$$\hat{\sigma}_x = \begin{pmatrix} 0 & 1 \\ 1 & 0 \end{pmatrix}, \quad \hat{\sigma}_y = \begin{pmatrix} 0 & -i \\ i & 0 \end{pmatrix}, \quad \hat{\sigma}_z = \begin{pmatrix} 1 & 0 \\ 0 & -1 \end{pmatrix}. \tag{4.12}$$

Both states $|\psi^-\rangle$, $\hat{\rho}$ exhibit perfect anticorrelation with respect to z,

$$C_{zz} = \langle \hat{\sigma}_z \otimes \hat{\sigma}_z \rangle - \langle \hat{\sigma}_z \otimes \mathbb{I} \rangle \langle \mathbb{I} \otimes \hat{\sigma}_z \rangle = -1 \tag{4.13}$$

However, it turns out that $|\psi^-\rangle$ is entangled while $\hat{\rho}$ is separable. So, what is the difference between these two states?

The key observation is that $|\psi^-\rangle$ is also perfectly anticorrelated with respect to x, y, and, in general—because of the rotation invariance—for any linear combination, $n_x\hat{\sigma}_x + n_y\hat{\sigma}_y + n_z\hat{\sigma}_z$. A possible classical explanation would be that A and B have both access to a shared "coin" that determines A to prepare a state with well-defined spin in all three components "$|a_x b_y c_z\rangle$", $a, b, c \in \{\pm 1\}$ and B the opposite "$|-a_x - b_y - c_z\rangle$". However, the spin components x, y, z are incompatible (their corresponding operators do not commute), so the state "$|a_x b_y c_z\rangle$" does not exist as not all three components can be well-defined simultaneously. In such a case, quantum entanglement is necessary to explain these correlations.

4.2.2 Detection of Entanglement

One of the main topics of quantum information is the detection and quantification of entanglement. The first question that we raise is concerning the discrimination of entanglement:

> Given a bipartite $\hat{\rho}$ acting in $\mathcal{H}_A \otimes \mathcal{H}_B$, is it separable or entangled?

As simple as it is posed, deciding whether a state is entangled or separable is NP-hard [22]. The challenge concerns the mixed states. Conversely, for pure states, it is solved in a rather simpler way.

4.2.2.1 Pure States

In the case of bipartite pure states, this is usually done based on a simple observation concerning partial trace.

> **Purity Condition**
> Separable pure state $\hat{\rho} = |\psi_A\rangle\langle\psi_A| \otimes |\psi_B\rangle\langle\psi_B|$ is pure after partial trace, $\mathrm{Tr}_B\hat{\rho} = |\psi_A\rangle\langle\psi_A|$. Conversely, an entangled pure state is not pure after a partial trace.

Entropic Quantities To connect it with entanglement, we need to quantify how pure the state is.

Definition 11 The von Neumann entropy S of state $\hat{\rho}$ is defined as $S(\hat{\rho}) = -\mathrm{Tr}\,(\hat{\rho}\ln\hat{\rho})$, with ln being matrix logarithm.

To see that this is a valid quantifier, note that pure states have von Neumann entropy 0, since in the eigenbasis all elements are either 0, apart from the one that equals unity. The other extreme case is the maximally mixed state, defined as $\mathbb{I}/d$, where d is the dimension of the Hilbert space. It is maximally mixed in the sense that its von Neumann entropy is maximal and equal to $\ln d$.

Using the above observation concerning partial trace, we can connect the level of purity (or, conversely, its von Neumann entropy) of the partial state with the entanglement of the full state.

Definition 12 The entropy of entanglement E of a bipartite state $\hat{\rho}$ is defined as the von Neumann entropy of either of its reduced states, $E(\hat{\rho}) = S(\hat{\rho}_A) = S(\hat{\rho}_B)$.

It is not hard to prove that although reduced states $\hat{\rho}_A$ and $\hat{\rho}_B$ are *not* equal, their von Neumann entropies are, cf. Ref. [10]. In contrast to the least entangled, separable states (not entangled at all), we now can introduce a maximally entangled state called Bell state $|\phi^+\rangle = \frac{1}{\sqrt{d}} \sum_i^d |i\rangle|i\rangle$. Its partial trace is the maximally mixed state, thus it is truly a state of maximal possible entanglement. Clearly, acting with *local* unitary channel $U_A \otimes U_B$, we are unable to change the entanglement value of any state $\hat{\rho}$. Therefore, instead of one maximally entangled Bell state, we rather have a full family of them, defined as $\frac{1}{\sqrt{d}} \sum_i^d U_A|i\rangle \otimes U_B|i\rangle$. Furthermore, all of the bipartite states of maximal entropy of entanglement are local unitary equivalent to the Bell state $|\phi^+\rangle$, which encourages us to treat the Bell state as a gold standard for maximal entanglement [23].

Schmidt Decomposition Another useful notion for studying the entanglement of pure quantum states is the Schmidt decomposition, which is the singular value decomposition (SVD) of the density matrix.

Lemma 1 *Every bipartite pure state* $|\psi\rangle \in \mathcal{H}_A \otimes \mathcal{H}_B$ *can be written using its Schmidt decomposition, i.e.,* $|\psi\rangle = \sum_i \sqrt{\lambda_i}|\xi_i\rangle \otimes |\phi_i\rangle$, *with non-negative* λ_i *forming a probability distribution, while* $|\xi_i\rangle \in \mathcal{H}_A$ *and* $|\phi_i\rangle \in \mathcal{H}_B$.

The advantage of the Schmidt decomposition over an arbitrary basis of density matrices in the bipartite Hilbert space $\mathcal{H}_A \otimes \mathcal{H}_B$ is that it requires only one index, thus it is a form of diagonalization. Mathematically, the validity of the above lemma is proved using singular value decomposition [10].

The coefficients of the Schmidt decomposition $\boldsymbol{\lambda} = \{\lambda_i\}$ are a useful tool for entanglement classification since one can directly infer the entropy of entanglement, as well as other measures of entanglement such as the Schmidt rank. Importantly, in the pure bipartite case, essentially all entanglement measures are equivalent, thus it is enough to consider only one of them (typically the entropy of entanglement) [23].

4.2.2.2 Mixed States

Convex Roof Construction The entropic quantifiers defined in the previous paragraph are not useful for detecting entanglement in mixed states. Indeed, take for example the entanglement entropy E. Such quantity is able to capture correlations but fails to distinguish between those stemming from entanglement or from classical mixing. For instance, both states $|\phi^+\rangle = (|00\rangle + |11\rangle)/\sqrt{2}$ and $\hat{\rho} = (|00\rangle\langle 00| + |11\rangle\langle 11|)/2$ have maximal entropy of reduced states, but $\hat{\rho}$ is separable.

A rigorous way to extend entanglement measures of pure states into mixed states is via the convex roof construction:

Convex Roof Construction
Given an entanglement quantifier F valid for pure states $|\psi\rangle$, it can be extended to mixed states $\hat{\rho}$ by optimizing over all possible decompositions of $\hat{\rho}$ into pure states:

$$F(\hat{\rho}) = \min_{\sum_i p_i |\psi\rangle\langle\psi|_i = \hat{\rho}} \sum_i p_i F(|\psi_i\rangle) \tag{4.14}$$

In practice, the previous minimization is never carried out explicitly (which would be impossible), but implicitly via convex optimization techniques [24] which allow for some cases the derivation of closed formulas for $F(\hat{\rho})$.

PPT Criterion Entanglement discrimination in mixed states is more complicated than its pure counterpart—one needs to resort to other methods, out of which arguably the simplest is the positive partial transpose (PPT) criterion [25, 26].

Peres-Horodecki or Positive Partial Transpose (PPT) Criterion
Let $\hat{\rho}$ be a bipartite state acting on $\mathcal{H}_A \otimes \mathcal{H}_B$. If $\hat{\rho}$ is separable, then the partial transpose is PSD, $\hat{\rho}^{T_B} \succeq 0$. Conversely, if partially transposed $\hat{\rho}^{T_B}$ is not a PSD matrix, i.e., it has at least one negative eigenvalue, then $\hat{\rho}$ is entangled.

Since $\hat{\rho}^{T_A} = (\hat{\rho}^{T_B})^T$, the criterion does not depend on the choice of the subsystem. For small subsystems A and B—of dimensions $(d_A, d_B) = (2, 2)$ or $(2, 3)$—this condition is also sufficient for entanglement.

However, for higher-dimensional systems, PPT criterion is only one-way—there are entangled states with PSD partial transpose (so-called PPT entangled states) [27–29].

Detection Methods Here, we list well-established methods for entanglement detection, both bipartite and multipartite: entanglement witnesses (see subsequent paragraphs), Bell inequalities (see Sect. 4.2.3), matrix realignment [30, 31], nonlinear properties of more than one copy of a state [32], quantum switch-aided protocol [33], Gilbert's algorithm [34–36], Lewenstein-Sanpera decomposition [37], trace polynomial inequalities [38], moments of many-body systems [39–41], quantum Fisher information [42–45], randomized measurements [46, 47], and spin-squeezing inequalities [48–50]. The detection of entanglement is an active field also from the experimental perspective [51–54].

Lewenstein-Sanpera Decomposition
Any bipartite density matrix $\hat{\rho} \in \mathbb{C}^d \otimes \mathbb{C}^d$ can be decomposed as per

$$\hat{\rho} = \lambda \hat{\rho}_s + (1 - \lambda) \hat{P}_e \,, \tag{4.15}$$

where $\lambda \in [0, 1]$, $\hat{\rho}_s$ is a separable state and $\hat{P}_e = |\psi\rangle\langle\psi|$ is a rank-1 projector onto an entangled state $|\psi\rangle$.

The separable state $\hat{\rho}_s$ such that λ is maximal constitutes the *best separable approximation* of $\hat{\rho}$. In particular, if the optimal $\lambda = 1$, we deduce that $\hat{\rho} = \hat{\rho}_s$ is separable. Finally, such a method works for any convex subset of states, beyond the separable set.

Entanglement and Thermodynamics Usually entanglement is observed only in microscopic systems. Nonetheless, for specific choices of the Hamiltonian, entanglement can be related to several macroscopic thermodynamical properties, such as temperature [55], magnetic susceptibility [51,56], or heat capacity [57], that can act as an entanglement witnesses.

In many systems, the ground state is entangled and so are states with low energy. The entanglement gap for this case is defined as the energy difference between the ground state and the minimal energy attainable by separable states. Provided the existence of the entanglement gap, energy can act as an entanglement witness for thermal states of Hamiltonians with entangled ground state [58–60]. Similarly, internal energy outside of the separable bounds indicates the presence of entanglement even for states out of equilibrium [61].

For the purposes of this chapter, we shall not elaborate further on the topic of detection of entanglement as excellent positions already exist, such as the review by Gühne and Tóth [62]. Let us now focus on entanglement witness and linear maps methods.

4.2.2.3 Entanglement Witnesses

As such, the PPT is the most basic scheme for entanglement quantification; nevertheless, typically, one needs other criteria if this test is inconclusive. To this end, special operators have been introduced, called witnesses [25].

Definition 13 Entanglement witness $\hat{W}$ is an operator such that for all separable states $\mathrm{Tr}\,(\hat{W}\hat{\rho}) \geq 0$ and there exists at least one entangled state $\hat{\sigma}$ for which $\mathrm{Tr}\,(\hat{W}\hat{\sigma}) < 0$.

Similar to the PPT criterion, the entanglement witness method is not conclusive, for every $\hat{W}$ there are entangled states that are not certified to be entangled, as

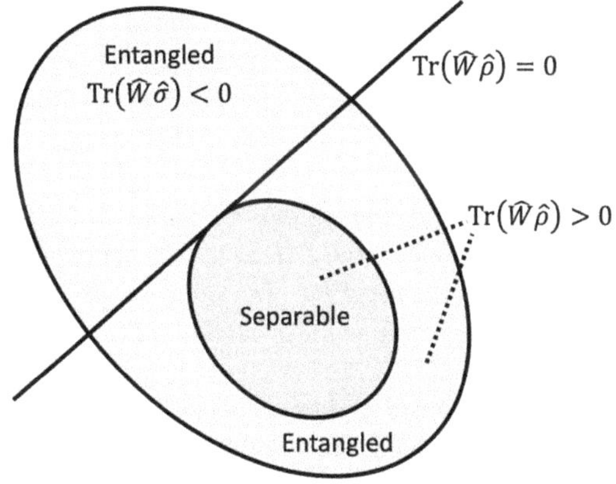

Fig. 4.1 Entanglement witness W divides the set of states into those which are entangled ($\mathrm{Tr}(\hat{W}\hat{\rho}) < 0$) and the rest, which might be entangled or not ($\mathrm{Tr}(\hat{W}\hat{\rho}) \geq 0$). We say $\hat{W}$ is optimal [28] if there exist separable state $\hat{\rho}_S$ on the boundary, i.e., $\mathrm{Tr}(\hat{W}\hat{\rho}_S) = 0$

illustrated by Fig. 4.1. Nevertheless, for every entangled state $\hat{\sigma}$ it is possible to find $\hat{W}_\sigma$ such that $\mathrm{Tr}\,(\hat{W}_\sigma\,\hat{\sigma}) < 0$, thus certifying its entanglement [63].

Entanglement witnesses can arise from experimentally feasible mutually unbiased bases [64], while connecting also to quantum state tomography [65–68]. Another way of constructing them is the semi-definite programming [69]. Furthermore, apart from a straight division of quantum states into two parts, there are also nonlinear entanglement witnesses [32, 70].

Now, we review an important tool to characterize entanglement witnesses via polynomials.

Entanglement Witnesses: A Polynomial Problem
The problem at hand is deciding whether an operator $\hat{W}$ is an entanglement witness. If it is so, for any separable state $\hat{\rho}_{\mathrm{SEP}}$, $\mathrm{Tr}(\hat{W}\hat{\rho}_{\mathrm{SEP}}) \geq 0$. By convexity, it is sufficient to prove such condition for pure states $|\Psi\rangle = |\psi_A\rangle \otimes |\psi_B\rangle$. Let $\{a_i = \langle i|\psi_A\rangle\rangle, b_i = \langle i|\psi_B\rangle\rangle\}$ be the components of the local vectors in an orthonormal basis. Then the positivity condition becomes:

$$\langle\Psi|\,\hat{W}|\Psi\rangle = \sum_{i,j,k,l} W_{ij,kl} a_i b_j a_k^* b_l^* \geq 0\,, \tag{4.16}$$

where $W_{ij,kl}$ are the coordinates of $\hat{W}$ with respect to the local basis.

We conclude that the expectation value is non-negative if and only if the quartic polynomial $p_4 = \sum_{i,j,k,l} W_{ij,kl} a_i b_j a_k^* b_l^*$ is non-negative in the complex domain.

Sum-of-Squares and Decomposable Entanglement Witnesses A certain class of polynomials that are positive are those that can be written as a sum of squares (SOS) of other polynomials of lower degree, $p_4 = \sum_i |p_2^{(i)}|^2$. Such class is feasible to characterize via convex hierarchies [71], and gives rise to entanglement witnesses of the form $\hat{W} = \hat{P} + \hat{Q}^{T_B}$ for $\hat{P}, \hat{Q} \succeq 0$, which are called *decomposable*. As proven in Refs. [28, 72],

Theorem 2 *Decomposable witnesses are equivalent to the PPT condition. That is, all entangled states detected by witnesses of the form* $\hat{W} = \hat{P} + \hat{Q}^{T_B}$ *for* $\hat{P}, \hat{Q} \succeq 0$ *are not PPT. The converse is also true: for any non-PPT state, there is a decomposable witness detecting it.*

Let us consider the converse statement: are all positive polynomials SOS? Hilbert proved that in general this is *not* the case [73]. Decades later, the first counterexample was found by Motzkin [74].[12] It is remarkable that such negative result has connections to the matter of this chapter. In our language, it implies that PPT is not sufficient, there exist PPT entangled states and, in general, that entanglement detection is hard (more precisely, NP-hard).

4.2.2.4 Linear Maps

Until now, we have developed the theory of entanglement detection and presented some of the most well-known strategies such as PPT and entanglement witnesses. Here, we will provide stronger criteria based on linear maps [75].

We divide this subsection into necessary and sufficient conditions for separability. The first one corresponds to witnesses, which rule out a separable description. Verifying that a state is *not* separable is much less challenging than proving that it actually *is*. In the second part, we address this latter problem via sufficient separability criteria.

Necessary Criteria for Separability Maps that are positive but not completely positive (PnCP) are generally not physical, but they are instrumental for entanglement detection.

Positive But Not Completely Positive (PnCP) Maps
Consider a bipartite state $\hat{\rho}$ acting on $\mathcal{H}_A \otimes \mathcal{H}_B$ and a positive map $\mathcal{E}$ transforming operators on $\mathcal{H}_B$. Then,

$$(\mathrm{Id}_A \otimes \mathcal{E})(\hat{\rho}) \not\succeq 0 \Longrightarrow \hat{\rho} \text{ is entangled} \tag{4.17}$$

Of course, if map $\mathcal{E}$ is CP, this cannot be used. The condition above generalizes the PPT criteria as presented earlier from the transposition to an arbitrary positive map.

[12] It reads $p(x, y) = x^4y^2 + x^2y^4 + 1 - 3x^2y^2$ over the real field.

There is a correspondence between maps and entanglement witnesses via the Choi matrix formalism that we introduced in Eq. (4.8). Specifically, the Choi matrix of the dual map $\mathcal{E}^\dagger$,[13] $\hat{M}_{\mathcal{E}^\dagger} = (\mathrm{Id}_A \otimes \mathcal{E}^\dagger)(|\phi^+\rangle\langle\phi^+|)$ is an entanglement witness:

$$\mathrm{Tr}(\hat{M}_{\mathcal{E}^\dagger}\hat{\rho}) < 0 \Longrightarrow \hat{\rho} \text{ is entangled .} \tag{4.18}$$

Note that, if $\mathcal{E}$ is CP, $\mathcal{E}^\dagger$ so is. In such case, the corresponding entanglement witness is trivial, as, according to Theorem 1, $\hat{M}_{\mathcal{E}^\dagger}$ would be PSD.

Evidently, the condition on maps Eq. (4.17) is stronger than a concrete entanglement witness realization Eq. (4.18). However, witnesses can be implemented in the experiments, while PnCP maps do not have a physical realization.

Sufficient Criteria for Separability Although lesser-known, linear maps can be employed for the complementary problem to derive *sufficient* conditions for separability beyond the PPT condition for low dimensionalities. In Ref. [76] new techniques in this regard were introduced, which were extended to the multipartite setting in Ref. [77]. Here, we offer an example of the proposed methodology.

Inverting Linear Maps to Detect Separable States
Let us consider a bipartite system $\mathbb{C}^2 \otimes \mathbb{C}^d$ and Λ_α be a family of reduction maps parametrized by $\alpha \in \mathbb{R}$, defined as:

$$\Lambda_\alpha(\hat{\varrho}) = \mathrm{Tr}(\hat{\varrho})\mathbb{I} + \alpha\hat{\varrho} \,. \tag{4.19}$$

One can show that:

- For the range $\alpha \in [-1, 2] := R$ the transformation maps any state to a separable state:

$$\forall \hat{\varrho} \succeq 0, \;\; \Lambda_{\alpha\in R}(\hat{\rho}) \text{ is separable .} \tag{4.20}$$

  The lower bound is already deduced by imposing positivity of the map. For the upper bound, see Ref. [76].
- The map can be readily inverted and its inverse reads: $\Lambda_\alpha^{-1}(\hat{\sigma}) = (\hat{\sigma} - \mathrm{Tr}(\hat{\sigma})\mathbb{I}/(2d + \alpha))/\alpha$.

These two observations allows us to assert $\Lambda_{\alpha\in R}^{-1}(\hat{\sigma}) \succeq 0 \Longrightarrow \hat{\sigma}$ is separable. Applying this finding for the extreme case $\alpha = 2$, yields the

(continued)

[13] The dual of the map $\mathcal{E}$ is defined as the map $\mathcal{E}^\dagger$ such that $\mathrm{Tr}[\hat{\rho}\mathcal{E}(\hat{\sigma})] = \mathrm{Tr}[\mathcal{E}^\dagger(\hat{\rho})\hat{\sigma}]$ for any pair of matrices $\hat{\rho}, \hat{\sigma}$.

following criterion:

$$\hat{\sigma} - \frac{\mathbb{I}}{2d+2} \geq 0 \Longrightarrow \hat{\sigma} \text{ is separable} \tag{4.21}$$

Such criterion detects, e.g., states in the separable ball around the completely mixed state, $\mathrm{Tr}(\hat{\rho}^2) \leq 1/(2d-1)$.

By definition, quantum entanglement is intrinsically linked to a Hilbert space. However, there exists a deeper notion of non-classicality, called Bell nonlocality, without explicit reference to such mathematical construction. In the next section, we give a pedagogical introduction to the field, which will be later developed in the last part of this chapter.

4.2.3 Bell Nonlocality

The inherent unpredictability of quantum mechanics is so different from the everyday world that even some of those who laid the groundwork for the formulation of quantum theory could not fully believe in it. Einstein, Podolsky, and Rosen in 1935 claimed that this unpredictability is a feature arising due to its incompleteness—which was later dubbed as EPR paradox [78]. In a nutshell, they argued that since entanglement can affect far objects in an instant, there must exist an underlying theory that is able to correctly attribute characteristics to a state of the system, such that the predictions of quantum mechanics are recoverable.

Those arguments were refuted by others [79]; however, for many years it was only a philosophical debate with physical implications nowhere to be found. Only in 1964, John Bell has proven that indeed, quantum mechanical unpredictability can be tested [80]. What was later named "Bell inequalities" are outcomes of certain measurements that must be confined to a specific region if we assume the incompleteness of quantum mechanics.[14]

Here, we shall discuss the most famous Bell inequality introduced by Clauser, Horne, Shimony, and Holt (CHSH) [81]. Two players, having access to a source of entanglement, are independently given one of the two possible inputs[15] $x \in \{1, 2\}$

[14] To be more precise, such requirement is called "local realism".

[15] Drawn uniformly and randomly.

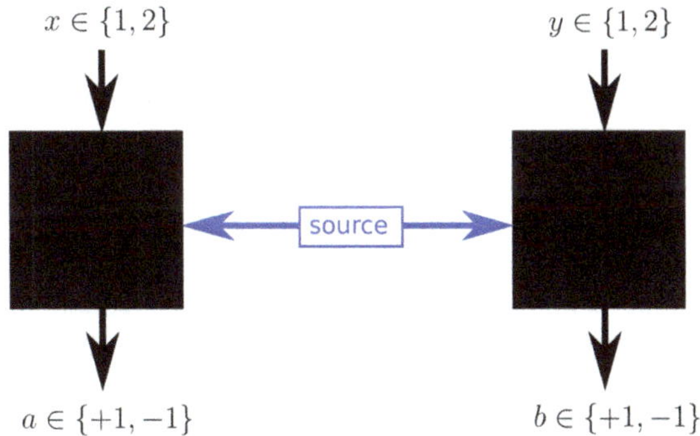

Fig. 4.2 Both parties to the CHSH game receive half of the entangled state from the source, which allows them to break the Bell inequality on the outputs a and b

to Alice and $y \in \{1, 2\}$ to Bob, as in Fig. 4.2. Then, they want to maximize the following sum of expectation values of four observables A_i and B_j

$$CHSH = \langle A_1 B_1 \rangle + \langle A_1 B_2 \rangle + \langle A_2 B_1 \rangle - \langle A_2 B_2 \rangle, \tag{4.22}$$

while the observables can take two values $\{+1, -1\}$.

Such value is bounded from above by 2 if the underlying theory is classical, i.e., the values of operators were chosen but are unknown to us. However, for maximally entangled Bell states of two qubits, it is possible to achieve theoretically $CHSH = 2\sqrt{2}$.

It took many years to achieve an experimental confirmation of this breaking of the Bell inequality; however, eventually, those efforts to disprove classical theory underlying the quantum one were awarded the Nobel prize in 2022 "for experiments with entangled photons, establishing the violation of Bell inequalities and pioneering quantum information science" [82].

The utility of Bell nonlocality and inequalities arising from it are abundant in quantum information; they can be used in generation as well as certification of true randomness, inaccessible by anyone else than the current user of a device [83]. Moreover, such inequalities can be used as entanglement witnesses [84]. In the following Sect. 4.4.2.2 we shall elaborate more on the applications of Bell inequalities in the many-body setting.

Without a doubt, entanglement is one of the cornerstones of quantum mechanics. We close the section by summarizing some of the applications in which quantum entanglement (beyond classical correlations) plays a crucial role.

4.2.4 Applications of Entanglement

Apart from its conceptual importance, entanglement is a necessary prerequisite for various quantum information protocols. A plethora of them are achievable only due to entanglement, e.g., quantum key distribution [85, 86], entanglement catalysis [87], quantum teleportation [88], quantum secret sharing [89], one-step quantum secure direct communication [90], quantum repeaters [91], state preparation [92], improvement of sensitivity of measurements [93–95], and enhanced quantum machine learning [96, 97]. These setups generally require maximal (or at least sufficiently high) entanglement in order to provide an improvement over their classical counterparts.

Entanglement Distillation Therefore, we also need a way to refine entanglement from many states of lower-quality entanglement via a process called entanglement distillation (purification) [98–100]. Distillation has been proposed [101–106] and performed [107–116] in laboratory settings, also in the multipartite scenario [117, 118].

Bound Entanglement Although entanglement distillation is a powerful technique, not all entangled states can be converted to a maximally entangled state, even with an arbitrary number of copies. We call such states *bound* entangled [27]. As an example of such states, all PPT states are undistillable.[16]

Even though bound entanglement is not useful at distilling maximally entangled states, it can still be used for entanglement activation [120], metrology [121], quantum steering [122], distillation of secure quantum keys [123, 124], and for teleportation [125]. Therefore, in analogy to "normal" entanglement witnesses, there exist some methods to witness bound entanglement [28, 120, 126, 127]. Another way of detecting bound entanglement involves using mutually unbiased bases [128], see the references therein for construction of bound entangled families and more about their detection.

Having explained the simplicity of the bipartite setting, let us now move on to exploring the richness of multipartite entanglement.

4.3 Entanglement in Multipartite Systems

There is more to this world than just two particles. In the previous sections, we have already discussed the concept of entanglement in quantum theory and its implications for the various physical concepts. As of now, we have only considered the simplest case of bipartite entanglement (entanglement between two subsystems), which automatically raises the question of whether we can extend this notion to entanglement between many subsystems. The answer is yes, and we refer to it as the multipartite entanglement. However, the concept of multipartite entanglement is not as simple and straightforward as bipartite entanglement, which has led to an extensive study of quantification and characterization of properties of multipartite entanglement over the past three decades [2, 16, 23, 129–133].

[16] In fact, one of the biggest open questions of entanglement theory concerns the reverse problem—are all bound entangled states also of positive partial transpose [119]?

The theory of multipartite entanglement is richer and much more extensive than the bipartite case as there are exponentially many number of ways in which an N-partite system can be partitioned; each partition giving rise to a different set of locality constraints. In the previous section, we have presented various ways for detection of bipartite entanglement and separability. Nonetheless, multipartite entanglement can be present even if the state is separable with respect to every bipartition [134]. Due to those reasons, it is difficult to characterize and come up with a single theory of multipartite entanglement, rather it depends on the resources and tasks in hand. In this section, we will start by defining the structure of states in the multipartite setting, and then discuss the elaborate resource theory of entanglement and various invariants and entanglement measures that allow us to classify the multipartite entanglement classes. We apply this discussion to the example of three qubits and study the various entanglement classes and structures that come out of it.

4.3.1 To Be *Separable* or Not to Be *Separable*

Starting with pure states, an unentangled state vector in a multipartite Hilbert space $\mathcal{H} = \mathcal{H}_1 \otimes \mathcal{H}_2 \otimes \ldots \otimes \mathcal{H}_N$ of N-distinguishable constituents, is given by the product state of the form

$$|\Psi\rangle = |\psi_1\rangle \otimes |\psi_2\rangle \otimes \ldots \otimes |\psi_N\rangle. \tag{4.23}$$

Any state vector which is *not* of this form is said to be entangled.

Unlike the case of bipartite systems, there are many ways in which one can create partitions in the multipartite system over the different parties, which gives rise to the notion of partial separability of a state based on these partitions.

Partial Separability in Pure States The Hilbert space of N-parties can be coarse-grained by creating partitions $\{I_1, I_2, \ldots, I_k\}$ such that $\cup_i^k I_i = I$ where $\{I_l\}$ are the disjoint subsets of indices, with $I = \{1, 2, 3, \ldots, N\}$. Due to this, a state that may have been previously entangled will now be separable under such partitions. This is called partial separability and a state which is partially separable under such partitions is expressed as

$$|\Psi\rangle = |\psi_{I_1}\rangle \otimes |\psi_{I_2}\rangle \otimes \ldots \otimes |\psi_{I_k}\rangle, \tag{4.24}$$

which looks similar to the state from Eq. (4.23), with the only difference that $|\psi_{I_i}\rangle$ is a collective pure state of the parties belonging to the subset I_i.

Based on this notion of partial separability, there exist states which are the multipartite generalization of the maximally entangled bipartite states introduced in the previous section, called the absolutely maximally entangled (AME) states.

Absolutely Maximally Entangled (AME) States

For an N-body d-dimensional quantum system, there exists a set of states called the absolutely maximally entangled (AME) states [135] which find applications in various quantum information tasks such as teleportation, secret sharing, error-correction [136], holography, and AdS/CFT correspondence [137]. The distinguishing property of these states is that these are multipartite generalizations of the bipartite maximally entangled states. In other words, an AME state is maximally entangled over all bipartitions of size $\lfloor \frac{N}{2} \rfloor$. For example, an AME state $|\Psi\rangle_{1234} \in \mathcal{H}_1^{d_1} \otimes \mathcal{H}_2^{d_2} \otimes \mathcal{H}_3^{d_3} \otimes \mathcal{H}_4^{d_4}$ is maximally entangled with respect to all partitions 12|34, 13|24, and 14|23, where $\{d_i\}$ are the local dimensions of the respective Hilbert spaces. An absolutely maximally entangled state $|\Psi\rangle \in \otimes_i^N \mathcal{H}_i^d$ of N-subsystems with local dimension-d is represented as AME(N, d).

The explicit construction of an AME state of four qutrits, AME(4,3), can be written as

$$|\Omega_{4,3}\rangle = \frac{1}{3}(|0000\rangle + |0112\rangle + |0221\rangle + |1011\rangle + |1120\rangle + |1202\rangle + |2022\rangle + |2101\rangle + |2210\rangle). \tag{4.25}$$

One can verify that all the reduced density matrices to two qutrits are $\hat{\rho} = \frac{1}{9}I_9$, with entropy $S = 2\log 3$ for every bipartition satisfying the condition of AME states. One of the *open problems* with respect to AME states is the classification of the states in equivalence classes of entanglement [138] (discussed later in Sect. 4.3.2).

Since the states are maximally entangled over the bipartitions, the reduced density matrices are maximally mixed, which is analogous to the condition of obtaining maximally mixed states when taking the partial trace of any of the subsystems in the maximally entangled bipartite states. It is known that AME states generally do not exist for every possible combination of (N, d). There are certain combinations where no AME states exist,[a] and one of the main focus of the current research is to find all such configurations [139]. One of the most recent breakthroughs in this direction was to show the existence of AME(4,6) by drawing a parallel with the age-old famous problem of 36 officers of Euler, by constructing quantum orthogonal Latin squares of this size [140].

[a] An elaborative list of discovered AME states is available at "http://www.tp.nt.uni-siegen.de/+fhuber/ame.html".

The next step is to relax the condition of purity and account for the mixing of states (using the density matrix formalism as described in the previous section), which further generalizes the definition of separability and partial separability to mixed states.

Separability of Mixed States The notion of full separability for pure states from Eq. (4.23) can be extended to mixed states, and a mixed separable state of a multipartite system is written as

$$\hat{\rho} = \sum_i p_i \hat{\rho}_i^{(1)} \otimes \hat{\rho}_i^{(2)} \otimes \ldots \otimes \hat{\rho}_i^{(N)}, \tag{4.26}$$

for the set of local density matrices $\hat{\rho}_i^{(j)}$ with probability distribution p_i [141, 142].

Similar to the bipartite case, all product states of the form $\hat{\rho}^{(1)} \otimes \ldots \otimes \hat{\rho}^{(N)}$ are separable, but the reverse is *not* true, as we allow for a mixture of states. Since there are many ways to create partitions in the mixed state multipartite scenario, this gives rise to the notion of partial separability where a state is not separable under any partition and yet not genuinely entangled. Any two states belong to the same *separability class* if they are separable under the same set of partitions.

Partial Separability in Mixed States The state $\hat{\rho}$ is separable under partitions $\{I_1, I_2, \ldots, I_k\}$ such that $\cup_i^k I_i = I$ where $\{I_l\}$ are the disjoint subsets of indices, with $I = \{1, 2, 3, \ldots, N\}$ [2]. The partially separable state has the form

$$\hat{\rho} = \sum_i p_i \hat{\rho}_i^{(I_1)} \otimes \hat{\rho}_i^{(I_2)} \otimes \ldots \hat{\rho}_i^{(I_k)}, \tag{4.27}$$

where any state $\hat{\rho}_i^{(I_j)}$ is defined on the tensor product of all elementary Hilbert spaces corresponding to indices belonging to set I_j. Surprisingly, the conditions for full separability are stronger than separability with respect to all bipartitions—there exist mixed states which are biseparable but still entangled [134].

The concept of partial separability automatically gives rise to the notion of *k-body* entanglement and entanglement depth which will be discussed in detail in the following Sect. 4.4.

Having defined the notion of entanglement in the multipartite setup and the complexity attached to it, the next step is to study the classification of states into various entanglement classes and class invariants. To do this, we consider the resource-theoretic framework of entanglement where entanglement is considered as a resource.

4.3.2 Multipartite Entanglement as a Resource

Entanglement as a resource has been studied extensively over the past few years [143–145], which has led to the development of a very elaborate and detailed resource theory of entanglement. Like any other resource theory, there are resource-free operations which are physical (or permissible) operations, and resource-free states which can be obtained from the action of resource-free operations. When considering entanglement as a resource:

- *Resource free operations*: entanglement non-increasing operations or physically realizable operations (specifically, local operations with classical communication-LOCC). The resource content of a state can be quantified using entanglement monotones (convex hulls of local entropy), i.e., non-increasing functions under resource- free operations. Usually resource free operations form convex sets, although extensions to non-convex sets have also been studied recently [146, 147].
- *Resource free states*: separable states which can be generated from the resource-free operations. An entangled state is automatically a resourceful state.

The resource theory of entanglement provides a framework for better characterization and quantification of entanglement, and studies the manipulation and transformation of quantum states, and their utilization for specific quantum information tasks. Since entanglement is a purely non-local quantum phenomenon, any local operation *cannot* increase the entanglement of the system. We start with the simplest class of resource-free operations which are the local unitary operations acting on each of the subsystems [130].

Local Unitary (LU) Operations Any two state vectors $|\Psi\rangle$, $|\Phi\rangle$ are considered to be equivalently LU-entangled if they differ by a local unitary basis change:

$$|\Psi\rangle \sim_{LU} |\Phi\rangle \equiv |\Psi\rangle = (U_1 \otimes U_2 \otimes \ldots \otimes U_N)|\Phi\rangle, \tag{4.28}$$

where U_i are the local unitary operations acting on the respective parties. Having defined the LU-equivalence relation, the next step is to classify the set of entangled states into LU-inequivalent classes.

For the bipartite case, this classification into entangled and non-entangled states is much simpler due to the notion of Schmidt coefficients as introduced in the previous section. Here, we will revisit the topic from the perspective of LU-equivalence: A bipartite pure state can be expressed in the form

$$|\Psi\rangle = \sum_{i=1} \sqrt{p_i}|\psi_i\rangle \otimes |\phi_i\rangle, \tag{4.29}$$

and the action of local unitaries U_1 and U_2 on the respective subsystems results in the basis change of the subsystems shown as

$$(U_1 \otimes U_2)|\Psi\rangle = \sum_{i=1} \sqrt{p_i}(U_1|\psi_i\rangle) \otimes (U_2|\phi_i\rangle). \tag{4.30}$$

The Schmidt coefficients[17] $\{p_i\}$ are invariant under the action of local unitaries. The LU-inequivalent classes are completely described by these Schmidt coefficients in the bipartite case.

However, for the multipartite case, the characterization becomes more challenging.[18] This is evident from counting the number of parameters necessary to describe a vector of N-qubits in the quotient space with respect to the given inequivalent relation.

LU-Parameter Counting in N-Qubits
It takes $2^{N+1} - 2$ real parameters to specify a normalized quantum state in $\mathcal{H} = (C^2)^{\otimes N}$, whereas the group of local unitary transformations $SU(2) \times \ldots \times SU(2)$ has $3N$ real parameters. This means that even in the case of $N-$qubits, one needs at least $2^{N+1} - 3N - 2$ real numbers to parameterize the sets of LU-inequivalent pure quantum states [5, 148, 149]. Unlike the Schmidt coefficients from the bipartite case, most of these parameters do not have a physical interpretation.

Another key difference in the multipartite scenario is that only very rarely pure multipartite states admit the generalized Schmidt decomposition [150, 151] unlike the bipartite case.

In contrast to local unitary operations, *global* unitary ones are able to create entanglement. Such features were quantified as *entangling power*, which measures the (average or maximal) entanglement created on separable pure states, studied for both bipartite systems [152, 153], as well as in the multipartite case [154]. As it turns out, its counterpart—the disentangling power, which measures how the entanglement degrades upon the action of the operation is closely related in certain cases [155], but not in general [156].

Now we shall extend the realm of unitary operations. Apart from LU, there exists a wider class of operations that cannot generate entanglement from separable states. It will give rise to a coarser notion of "equivalent entanglement".

Local Operations and Classical Communication (**LOCC**) A more general form of the LU operations which can still be completely defined classically are the local operations with classical communication (LOCC). LOCC allows for an exchange of classical information of the local measurement outcomes of the respective parties, hence, the name. Thus, the choice of the local operations of individual parties can be affected by the information of measurement outcomes by any other party.

[17] There is a physical interpretation of Schmidt coefficients in the bipartite case as they are the set of eigenvalues of each of the reduced density matrices of the bipartite system.

[18] There is no singular value decomposition for tensors of higher degree than two.

The best way to imagine this is with the help of the "distant laboratories model" [157]. There are N-particles each in their laboratory. The particles may or may not be entangled in the first place. Each laboratory is capable of performing local measurements on its own particle, and conveying the information regarding the outcome without exchanging the quantum systems among themselves. This whole process is called one round of LOCC, where all parties perform measurements based on the previously circulated information, and further pass on the information regarding the current measurement outcomes to the rest of the parties. The LOCC is difficult to characterize, but there exists a set of general operations called separable operations SEP, such that LOCC $\subset$ SEP, which can be represented as a channel Λ acting as

$$\Lambda(\hat{\rho}) = \sum_{a} \left(\otimes_{i=1}^{N} K_{a,i}\right) \hat{\rho} \left(\otimes_{i=1}^{N} K_{a,i}\right)^{\dagger}, \tag{4.31}$$

where $K_{a,i}$ are the local Kraus operators acting on the i-th subsystem respectively [158, 159]. This is a direct generalization of LU operations, where we considered the unitary matrices of Eq. (4.28), which are composed of only one Kraus operator, acting on the subsystem i, given by $U_i = K_{a=1,i}$.

LOCC are considered the resource-free operations in the resource theory of entanglement [99, 160, 161] since any separable state $|\Psi\rangle \in$ SEP can be obtained from any other state in the Hilbert space by the action of LOCC. On the other hand, it is impossible to generate any entangled state from a separable state through LOCC alone [158].

A typical LOCC protocol has many rounds,[19] where each laboratory performs a (POVM) measurement on their particle. The post-measurement state is stored and the classical outcome is broadcasted to the other laboratories. In the next round, another party performs measurement on their particle based on the results of the previous measurement outcomes, and so on. Any two states are LOCC-*equivalent* if they can be interconverted by this kind of protocol. For mathematical simplicity, there are several variants of this definition based on the allowed number of rounds, for example, two states are LOCC$_r$-equivalent if they can be interconverted using LOCC protocol with no more than r rounds. They are $\overline{\text{LOCC}}$-equivalent if starting from any of them, one can approximate to the other one with arbitrary precision if the number of rounds $r \to \infty$.

Although having a physical interpretation, there is no simple mathematical description of LOCC-equivalence in the multipartite scenario so far. There is no

[19] Some of the maps require infinite rounds of LOCC for implementation [157, 162].

known algorithm that decides whether two vectors are $\overline{\text{LOCC}}$-equivalent, even after an exponential runtime in the total dimension [5, 157]. However, in the case of bipartite entanglement, there exists a well-defined mathematical framework for studying the LOCC-equivalence classes, using the majorization condition of Schmidt coefficients (Nielsen's theorem) [163]:

Nielsen's Theorem
Let $|\Psi\rangle, |\Phi\rangle$ be two states in $\mathbb{C}^d \otimes \mathbb{C}^d$ with respective Schmidt coefficients $\boldsymbol{\lambda} = \{\lambda_i\}, \boldsymbol{\mu} = \{\mu_i\}$ in increasing order. Then, there exists a LOCC protocol converting $|\Psi\rangle$ exactly to $|\Phi\rangle$ if and only if $\boldsymbol{\lambda}$ is majorized by $\boldsymbol{\mu}$, i.e., $\forall k \in \{1, 2, .., d\}$,

$$\sum_{i=1}^{k} \lambda_i \leq \sum_{i=1}^{k} \mu_i. \tag{4.32}$$

There exist some generalizations of Nielsen's theorem to the multipartite scenario as shown in Ref. [164].

We can further have a stricter criterion by allowing for a level of randomness in local transformations. This classification scheme goes beyond the deterministic nature of LOCC, offering a more detailed understanding of the entanglement classification in multipartite systems.

***Stochastic*-LOCC (SLOCC)** Any two states are SLOCC equivalent if they can be converted into each other by LOCC with some finite probability. Similar to LOCC, the SLOCC protocol consists of several rounds with each party performing operations based on measurement outcomes by other parties. The whole process can be imagined as a tree, and every measurement results in a new branch of the tree. If at least any one of the branches leads to the target state $|\Phi\rangle$ starting from $|\Psi\rangle$, the two states are SLOCC-equivalent.

- A conversion from $|\Psi\rangle \xrightarrow{\text{SLOCC}} |\Phi\rangle$ is possible under SLOCC if there exists a set of operators $\{A_i\}$ (compared to Eq. (4.28)) such that

$$|\Psi\rangle \sim_{\text{SLOCC}} |\Phi\rangle \equiv |\Psi\rangle = (A_1 \otimes A_2 \otimes \ldots \otimes A_N)|\Phi\rangle, \tag{4.33}$$

- In particular, the two states $|\Psi\rangle \xleftrightarrow{\text{SLOCC}} |\Phi\rangle$ are SLOCC-equivalent iff the matrices $\{A_i\}$ are invertible (or determinant $\det A_i \neq 0$) [165, 166]. The matrices $\{A_i\}$ for which $\det A_i = 1$ which satisfy this property form the *special linear* group SL.

Similar to the case of $N-$qubits with LU-equivalence, one obtains a lower bound on the number of parameters required to define the SLOCC-equivalence classes given by $2^{N+1} - 6N - 2$ by substituting the SU(2) group with SL($\mathbb{C}^2$) group.

Now that we have defined the operations and transformations that are allowed within the resource theory framework of entanglement, we consider the example of three qubits to study the classification of states and the corresponding invariants for each of the classes.

Entanglement Classification of 3 Qubits
The simplest model to extend our discussion on multipartite entanglement is a tripartite system of three qubits. Based on the equivalence relations shown above with respect to different types of operations, quantum states of 3-qubits can be classified into six SLOCC-inequivalent entanglement classes [165].

Before advancing the discussion further toward the classification of 3-qubit entangled states, let us define the notion of class invariants.

Definition 14 SLOCC-invariants are referred to as the functions of the states that do not change under the action of SLOCC. These invariants play a crucial role in characterizing and distinguishing between different types of entangled states within the SLOCC framework. Tensor rank and hyperdeterminant are the two well-known and studied SLOCC-invariants that we will later use to classify the genuinely tripartite entangled states.

We study the classification of states based on these invariants. First, we restrict our discussion to 3-qubit pure states and later generalize it to mixed states to obtain a hierarchical structure of these sets informally known as the *onion ring structure* [165, 167, 168].

- The first class of states that exist are the set of separable states (or free states) written as product states of the form $|\Psi\rangle = |\psi_1\rangle \otimes |\psi_2\rangle \otimes |\psi_3\rangle$. In this particular case of separable states, $r(\hat{\rho}_1) = r(\hat{\rho}_2) = r(\hat{\rho}_3) = 1$, where $r(\rho_i)$ is the local rank of the reduced density matrix of i-th subsystem. The rank is invariant under the action of invertible SLOCC [5, 142, 165, 167].
- Next, there are three classes of bipartite states based on the bipartitions $1|23$, $2|13$, and $3|12$. For each bipartition $i|jk$, the set of pure product states forms an SLOCC class,[20] and are of the form $|\psi_i\rangle \otimes |\Phi_{j,k}\rangle$, where $|\Phi_{j,k}\rangle$ is a non-

[20] This follows from the fact that any two entangled pure states of two parties are equivalent under SLOCC.

product state of subsystems-j, k, where $r(\rho_i) = 1$, and $r(\rho_j) = r(\rho_k) = 2$. Thus, giving rise to three separate SLOCC-inequivalent classes of entanglement [165].
- Finally, the remaining states are genuinely 3-qubit entangled—of maximal local rank with respect to any single-qubit $r(\rho_1) = r(\rho_2) = r(\rho_3) = 2$. Nonetheless, among these states, one can distinguish two inequivalent classes [165, 167]. These two classes of states are the W and the GHZ states, which are discussed below in detail, and shown to be SLOCC-inequivalent using invariants such as hyperdeterminant and tensor rank.

Using Rank as an Invariant in Bipartite Systems
In the bipartite scenario, two state vectors $|\Psi\rangle$ and $|\Phi\rangle$ can be written in the product basis using coefficient matrices $T_{i,j}$ and $T'_{i,j}$ as follows:

$$|\Psi\rangle = \sum_{i,j}^{d} T_{i,j}|i\rangle \otimes |j\rangle, \qquad |\Phi\rangle = \sum_{i,j=1}^{d} T'_{i,j}|i\rangle \otimes |j\rangle. \tag{4.34}$$

Let us assume $|\Psi\rangle$ and $|\Phi\rangle$ are SLOCC-equivalent. Then, there exist invertible operators A_1 and A_2 such that $A_1 \otimes A_2|\Psi\rangle = |\Phi\rangle$ as shown in Eq. (4.33), and ranks of the coefficients are equal $r(T') = r(T)$ [5]. Hence, the rank of the coefficient matrix is invariant under invertible SLOCC operations.

A pure state of two qudits is entangled with maximal Schmidt rank iff $\det T \neq 0$.

Hyperdeterminant Hyperdeterminant is the generalization of the determinant defined for the matrices, to higher order tensors [169]. In 2003, Miyake [170] showed that the entanglement measures such as concurrence and tangle are special forms of hyperdeterminant, and, as such, are invariant under LU operations. Concurrence in the bipartite case is the determinant of the coefficient matrix of a state $C(\Psi) = 2|\det T|$, whereas tangle is the hyperdeterminant of second order, which is given by the $\tau_3(|\Psi\rangle) = |\det \Psi|$ [171, 172]. The strict positivity of the hyperdeterminant is an invariant under the local general linear group $\mathrm{GL}^{\times N}$ [170]. The hyperdeterminant is an example of LU-invariant which can be expressed as a polynomial function of the state's coefficients [173].

It has been shown that the det(W) $= 0 \neq$ det(GHZ), thus giving rise to two different SLOCC-equivalent classes within 3-qubit genuinely tripartite entangled states. Therefore, from the perspective of *3-tangle* (sum of tangles with respect to all bipartitions), the GHZ-states are more entangled than the W-states, thus detecting so-called *monogamy* of entanglement [171].

Tensor Rank We can also show the distinction between the GHZ and W SLOCC-inequivalent classes of pure states using tensor rank as an invariant. Let us decompose any pure state vector using the minimal decomposition

$$|\Psi\rangle = \sum_{i=1}^{R_{\min}} c_i |\psi_i^{(1)}\rangle \otimes \ldots \otimes |\psi_i^{(N)}\rangle, \tag{4.35}$$

where $\left\{|\psi_i^{(j)}\rangle\right\}_{i=1}^{R_{\min}}$ may or may not be orthogonal unlike the Schmidt decomposition (where all the vectors need to be pair-wise orthogonal to each other). The number of terms $R_{\min}$ defines the tensor rank[21] of the state $|\Psi\rangle$. Unlike the Schmidt rank for bipartite systems, for the multipartite case, the tensor rank is not easy to compute. It is invariant under SLOCC operations, and it distinguishes the two genuine tripartite entangled SLOCC-inequivalent classes of W and GHZ states which can be shown below:

The state vectors representing GHZ [174] and W [165] states respectively are

$$|\text{GHZ}\rangle = \frac{1}{\sqrt{2}}\left(|000\rangle + |111\rangle\right), \quad |\text{W}\rangle = \frac{1}{\sqrt{3}}\left(|001\rangle + |010\rangle + |100\rangle\right). \tag{4.36}$$

Based on the definition of tensor rank, we can see that it is not possible to express $|\text{W}\rangle$ state using only two product terms unlike $|\text{GHZ}\rangle$ state, i.e., $R_{\min}(|\text{GHZ}\rangle) = 2$ and $R_{\min}(|\text{W}\rangle) = 3$, hence they cannot be interconverted using SLOCC alone. However, pure states from the W class can be approximated to an arbitrary precision by states in the GHZ class [167, 175], while the converse is not true.

Based on the discussion above, we can represent the different SLOCC-inequivalent classes in Table 4.1 and the corresponding invariants used to distinguish the respective classes. Here, we show the local entropy for the reduced subsystems as described in Definition 12, local rank, and the 3-tangle τ as the class invariants for each of the six SLOCC classes. This set of invariants is complete—it allows us to fully distinguish between different LOCC classes in the case of pure states.

Mixed States The situation is more involved when we allow for a mixture of pure states—now not all states from a given class will be SLOCC interconvertible, but they can be created using a mixing of states from an appropriate class of pure state

[21] The logarithm of tensor rank is also called the *Schmidt measure*.

Table 4.1 Values of local entropies $\{S_1, S_2, S_3\}$ and the local ranks $\{r_1, r_2, r_3\}$ of the reduced subsystems, and the 3-tangle τ for different SLOCC-inequivalent entanglement classes

Class	S_1	S_2	S_3	r_1	r_2	r_3	τ
SEP	0	0	0	1	1	1	0
1|23	0	>0	>0	1	2	2	0
2|13	>0	0	>0	2	1	2	0
3|12	>0	>0	0	2	2	1	0
W	>0	>0	>0	2	2	2	0
GHZ	>0	>0	>0	2	2	2	>0

entanglement, up to an arbitrary precision. The hierarchical structure of entanglement classes in the mixed case scenario for 3 qubits [134,142,165,167,168,176,177] has been shown to be

$$S \subset B \subset W \subset GHZ, \tag{4.37}$$

where in this context:

- S is the class of all convex sum of projectors onto pure separable vectors,
- B is the closure of the set defined as the convex sum of projectors onto pure biseparable entangled vectors (from all bipartitions: 1|23, 2|13, 3|12),
- W is the closure of the set defined as the convex sum of projectors onto pure W states (in contrast to the pure case, W class is not of measure zero.),
- GHZ is the set of all possible physical states, i.e., the closure of the set containing convex sums of projectors onto pure GHZ states.

The reason for the introduction of the closure is that all states in S have a biseparable state[22] in the neighborhood, all states in B have a W state in the neighborhood, and all states in W have a GHZ state in the neighborhood [165]. However, there are states that belong to only one of those classes, e.g., W states that cannot be written a convex combination of projectors onto GHZ states—the simplest example are all pure states. We emphasize this distinction in Fig. 4.3, highlighting the internal structure of the biseparable class to provide more geometrical intuition.

Characterization of those classes can be performed using entanglement witnesses [168] via best approximation [37], as well as utilizing the Schmidt number [177, 178], or via Bell correlations [179]. Alternatively, multipartite entanglement classification for pure states involves studying the local purities of the reduced states. Such methods produce similar pictures to Fig. 4.3, yielding an object known as the entanglement polytope or the Kirwan polytope [180–185]. However, for a higher number of parties, it is a rudimentary division, not allowing to highlight the intricacies of different entanglement classes.

[22] More precisely, a state formed as a convex sum of projectors onto biseparable states. The same remark applies to other neighborhoods as well.

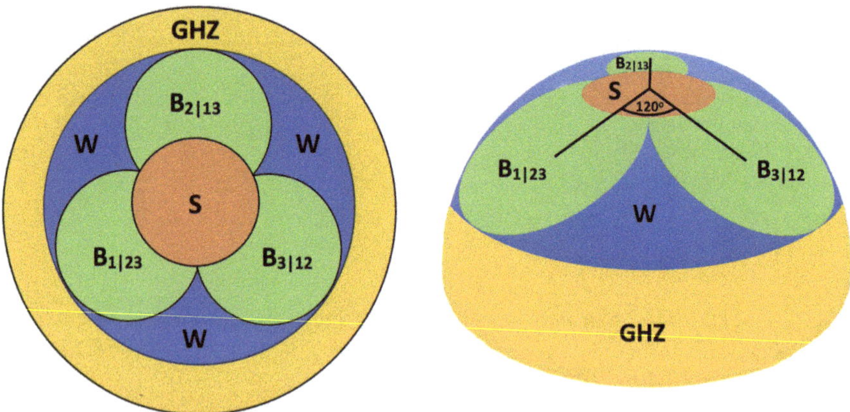

Fig. 4.3 The four possible entanglement classes based on the SLOCC operations in the case of tripartite systems. Such sketches were done many times before [62, 168], but we believe that a 3-dimensional structure is needed to fully understand the intricacies of the set—W class is smaller than GHZ, but there are W states that cannot be written as a convex combination of GHZ states alone [168]. Therefore, the inclusion (4.37) requires the closure of the sets—any W state can be *approximated* by GHZ states up to arbitrary precision, same as biseparable states can be approximated via W states and separable ones via biseparable. Note that the separable set S is a proper subset of the intersection of $B_{1|23} \cap B_{2|13} \cap B_{3|12}$; in other words, there exist states that are separable with respect to every bipartition yet entangled [134]. We express our gratitude to Karol Życzkowski for his valuable remarks regarding the figure

Increasing the number of parties, the natural question is the entanglement classification for four qubits (four parties). The difficulty to characterize entanglement increases as the number of parties grows—already for four qubits there are infinitely many SLOCC classes [166].

Since there does not exist a generalized theory of multipartite entanglement, it is more feasible to focus on concrete families of states of physical relevance. An example of them are matrix product states, which efficiently approximate ground states of one-dimensional quantum many-body systems and will be studied in the next section. Another approach is to characterize many-body entanglement from a given set of accessible few-body expectation values. This so-called *data-driven* approach will also be touched upon in the following section.

4.4 Use and Detection of Many-Body Quantum Entanglement

Certainly, there is more to this world than just a few particles: systems composed of many interacting constituents or *bodies* are ubiquitous in Nature. They can be found in the whole range of the energy scale, from elementary particles forming atoms, atoms bindings to form molecules, many atoms interacting in solid state materials or gases, planetary systems, and beyond. Such systems are intrinsically challenging to describe. Moreover, they feature collective phenomena in the form of emergent dynamics which are hard to predict from the properties of the individual constituents.

One may think that interactions will eventually decohere the system, washing out any possible quantum contribution to thermal noise. This is the norm when the classical description is sufficient as the microscopic degrees of freedom can be integrated out. However, some emergent phenomena require a quantum description, with entanglement playing a central role. Such systems include Bose-Einstein condensation at finite temperature [186] and exotic phases of matter at zero temperature driven by quantum fluctuations [187,188]. Remarkably, these experiments show that quantum physics can be manifested macroscopically and witnessed "with the naked eye", beyond its traditional scale of applicability.

The Many-Body Regime We consider a closed system formed by N elementary constituents (particles). In the many-body regime, the collective dynamics dominate, masking any detailed local degrees of freedom from being *relevant* in the observers' macroscopic scale. A necessary condition for the collective behaviors to take over is that N must be sufficiently large[23] such that the observed quantities (with a given scaling) become independent of N. For the purpose of this *section*, though, the main message is that for most practical applications, such N is too high for a generic (entangled) many-body state to be stored and processed in an ordinary computer:

Curse of Dimensionality
The state of N interacting particles with d internal degrees of freedom (e.g. spin) is written in the computational basis as,

$$|\Psi\rangle = \sum_{\{s\in[d]\}} \Psi_{s_1 s_2 .. s_N} |s_1, s_2, \ldots, s_N\rangle \in [\mathbb{C}^d]^{\otimes N} . \tag{4.38}$$

Notice how $O\left(d^N\right)$ parameters are needed to just specify the state. This number quickly becomes unfeasible to store in a conventional computer.

[23] Such number will generally depend on the nature of the interactions and can be as high as the Avogadro constant, $N_A \sim 10^{24}$.

However, not everything is lost, as:

The hopes for scalable descriptions:

1. While generic states in the Hilbert space are highly entangled, those physically relevant are typically weakly entangled.
2. Even if the global state is inaccessible, partial tomography can be performed and it might be sufficient to witness its resource content (e.g. in the form of many-body entanglement or stronger quantum correlations).

In the remainder of the *section*, we will formalize and elaborate on these two highly non-trivial aspects. We will show as well how despite the complexity of the scenario, entanglement can still be characterized and certified to be exploited in current applications.

Entanglement Depth In the previous *section*, we have studied entanglement with respect to different partitions or *parties*. Here, as a complementary view, we focus on the number of subsystems or *bodies*. There is a distinction between *bodies* and *parties* as defined here. In particular, many-body states can be studied with respect to different partitions. As a reminder, we define $\mathcal{P}_K$ as a partition of $[N] := \{1, 2, .., N\}$ in non-empty pairwise disjoint subsets of at most K elements (see Fig. 4.4). A global vector $|\Psi\rangle \in \mathcal{H}^{\otimes N}$ may factorize in a given partition $\mathcal{P}_K$ as

$$|\Psi_K\rangle = \bigotimes_{p \in \mathcal{P}_K} |\psi_p\rangle. \tag{4.39}$$

If such decomposition exists, we say that $|\Psi_K\rangle$ is *K-producible with respect to the partition $\mathcal{P}_K$*. Such a definition is extended to mixed states by the usual convex hull construction. We call a state $\hat{\rho} \in \mathcal{B}(\mathcal{H}^{\otimes N})$ *K-producible* if it can be expressed as a convex combination K-producible states (possibly with respect to different partitions), i.e.,

$$\hat{\rho}_K = \sum_{\Psi} p_\Psi |\Psi_K\rangle\langle\Psi_K|, \tag{4.40}$$

with $\{p_\Psi\}_\Psi$ a probability distribution.[24]

[24] If $\mathcal{H}$ is finite-dimensional, a sum is sufficient to represent classical mixing, although it has generally infinitely many terms. For continuous variables, if $\mathcal{H}$ is infinite-dimensional, the sum must be replaced by an integral.

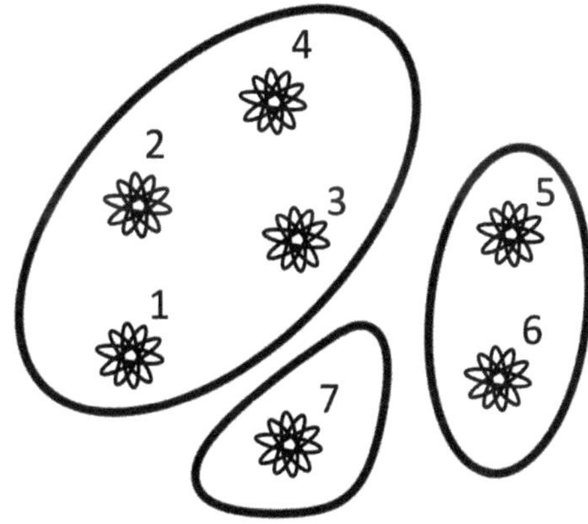

Fig. 4.4 Collection of $N = 7$ distinguishable atoms and the partition $\{\{1, 2, 3, 4\}, \{5, 6\}, \{7\}\}$. Atoms in the same partition add to a global GME state, i.e. they cannot be further separated without degradation (losing part of the state's description). The entanglement depth refers to the minimal number of GME parties needed to describe the state. In this case: $K = 4$

Note that if $K = 1$, then $\mathcal{P}_1 = [N]$ and $\hat{\rho}$ is *fully separable* (SEP), that is, not entangled. In general, entanglement among K parties is necessary to generate a K-producible state. From a resource point of view, generating multipartite entanglement is not for free. Therefore, one may be interested in the minimal K for which a decomposition as per Eq. (4.40) exists. Such value is the so-called *entanglement depth* and quantifies the minimal number of genuine multipartite entangled parties required to describe the state [61, 189]. In particular, if $K = N$, then we say that the state is genuinely-multipartite entangled (GME). It is clear from the construction that the set of states with entanglement depth K, $\mathcal{R}_K$, is convex and $\mathcal{R}_k \subset \mathcal{R}_{k+1}$. Moreover, under local operations, the entanglement depth can only be reduced. which illustrates the many levels of depth in which a many-body state can be entangled, from fully separable to genuinely N-body entangled. As compared with the previous *sections*, this classification is coarse: two states with the same entanglement depth may be locally inequivalent. However, such operations can not increase the value of K, which makes it a valid entanglement measure.

4.4.1 Useful (and Useless) Many-Body Entanglement

As explained in previous *sections*, quantum entanglement is a resource. However, most of the quantum states are useless, in the sense that their power can not be exploited. Here, we will focus on some features of two classes of many-body states which have stood out for their physical relevance and usefulness in current applications: matrix product states and symmetric states.

4.4.1.1 Matrix Product States

Many-Body States Are Tensors We interpret the coefficients of the many-body state Eq. (4.38) as an N-tensor $\Psi : [d] \times [d] \times \ldots \times [d] \to \mathbb{C}$ such that $(s_1, s_2, .., s_N) \mapsto \Psi_{s_1 s_2 .. s_N}$. Notice how for separable Ψ, the tensor factorizes (compare with Eq. (4.23)),

$$\Psi_{s_1 s_2 \ldots s_N} = \psi_{s_1} \psi_{s_2} \ldots \psi_{s_N} , \tag{4.41}$$

and only $O(dN)$ coefficients are needed to be specified. Such a case is realized in noninteracting systems. As soon as interaction is switched on, entanglement is generated and the number of necessary parameters grows exponentially $O\left(d^N\right)$. However, the interaction mechanisms realized in Nature are very particular. Which entanglement patterns are produced? Can they be efficiently handled?

The interactions are encoded in a Hamiltonian $\hat{H}$. Typically, its physical realizations possess the following basic properties [190]:

1. **Locality**: Fundamental interactions happen locally,[25] that is, within compact neighborhoods $\{X\}$, with support independent from N, of a topological space L. Such geometry structures the Hamiltonian as:

$$\hat{H}_{\text{loc.}} = \sum_{X \subset L} \hat{H}_X . \tag{4.42}$$

2. **Gap:** In the thermodynamic limit $N \to \infty$, the energy difference between the first excitation and the ground state is finite.

Here, we will be interested in the properties in equilibrium, i.e., ground states (GSs). One of the main results of the *section* is that the entanglement content of such states remains scalable:

Area Law of Entanglement
Let $|GS\rangle$ be a ground state of a gapped local Hamiltonian $\hat{H}_{\text{loc.}}$. The entanglement entropy S_X with respect to a neighborhood X scales with the area of the boundary of X, $|\partial X|$.

Notice that this is not what one usually expects for random pure many-body states. These would scale much more dramatically, with the volume $|X|$, $S_X \sim (N \log d - 1)/2$, i.e. S_X is extensive [191,192]. The entropy growth with the system's size is first analyzed from field theory approaches [193], also in the context of black-hole physics [194]. Within quantum information, the 1D area law was rigorously established first for Gaussian models [195], and subsequently, under more general assumptions by Hastings [196]. For such Hastings bound was improved in

[25] Notice how e.g. the Coulomb law of electrostatics is nonlocal. However, it is an effective theory. The fundamental interaction consists in giving and receiving virtual photons, as a local process.

Refs. [197, 198]. Another worth mentioning theorem is that exponential-decaying correlations imply area law [199]. The extension of the previous results to higher dimensions is an open problem and a matter of intensive ongoing research [200]. Thus far, the problem is addressed for Gaussian models [201] and quasi-free fermionic and bosonic lattices [202]. Unexpectedly, the entropy scaling in such systems depends on the statistics (bosonic or fermionic).

The conditions imposed by the area law are quite stringent and only fulfilled by a set of states restricted to a *corner* of the Hilbert space. In 1D those can be efficiently approximated by matrix-product states [203]:

Matrix Product States Matrix product states (MPS) can be defined as a factorization in matrices (2-tensors) [cf. Eq. (4.41)],

$$\Psi^{(D)}_{s_1 s_2 .. s_N} = \sum_{\{\alpha\in[D]\}} \psi^{\alpha_1}_{s_1} \psi^{\alpha_1\alpha_2}_{s_2} \dots \psi^{\alpha_{N-1}}_{s_N} = \vec{\psi}^{\,T}_{s_1} \psi_{s_2} \dots \vec{\psi}_{s_N} . \tag{4.43}$$

Here, the scaling is $O\left(D^2 N d\right)$. We refer to the size of such matrices, D, as the *bond dimension*. For $D = 1$, we recover the separable case Eq. (4.41). Note that any many-body state Ψ can be factorized in this way by repeated singular value decompositions (SVD) for a sufficiently large bond dimension (see Fig. 4.5). In the same figure, we introduce the diagrammatic notation which is very useful to visualize tensor operations [204, 205].

From the picture above (Fig. 4.5), we learn that the bond dimension in a link is lower-bounded by the Schmidt rank with the respective bipartite cut. Consequently, for generic states, one has $D = d^{\lfloor N/2 \rfloor} \sim \exp(N)$. As the bipartite entanglement entropy is upper-bounded by the Schmidt rank:

> Area law states in one dimension are well-approximated by matrix product states.

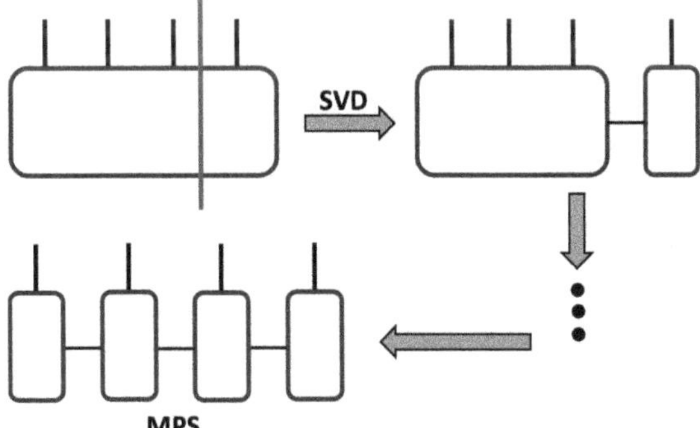

Fig. 4.5 In the diagrammatic notation an N-tensor is represented as an N-legged box. Each leg represents a vector space of dimension d. By iterating SVD steps, one can factorize the tensor as an MPS

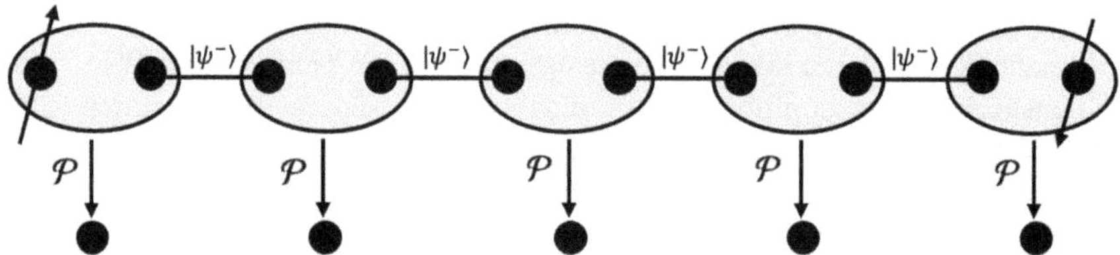

Fig. 4.6 The GS is understood as an MPS by deploying spin singlets $|\psi_-\rangle$ between bonds and projecting, $\mathcal{P}$, adjacent sites to the physical space

Specifically, if we want to approximate the ground state within an error ϵ per spin, $|||\text{GS}\rangle - |\Psi^{(D)}\rangle|| \leq \epsilon/N$, a truncation of the bond dimension at $D \sim \text{poly}(\text{N}, \epsilon)$ is sufficient [206] (cf. the exponential scaling for generic states).

MPS must be regarded as a variational ansatzes. There are more or less efficient algorithms to find an approximation of the $|\text{GS}\rangle$ by optimizing the set of the many small tensors $\{\psi\}$. One of the most widely used is the density-matrix renormalization group (DMRG) [207, 208]. However, there are some Hamiltonians, whose ground state is exactly an MPS of constant bond dimension. Below, we give an example of it:

Example: Affleck-Kennedy-Lieb-Tasaki (AKLT) Model
We place N spin-1 in an open chain interacting according to the Hamiltonian [209]:

$$\hat{H}_{\text{AKLT}} = \sum_{i=1}^{N-1} \left[\hat{\mathbf{s}}_i \cdot \hat{\mathbf{s}}_{i+1} + \frac{1}{3} (\hat{\mathbf{s}}_i \cdot \hat{\mathbf{s}}_{i+1})^2 \right] , \tag{4.44}$$

where $\mathbf{s} = (\hat{s}_x, \hat{s}_y, \hat{s}_z)$ is the vector of spin-1 matrices.

Here, we show how a bond dimension of $D = 2$ is sufficient to represent the ground state of $\hat{H}_{\text{AKLT}}$. In the following, we offer an explicit construction of it:

Valence Bond Picture We embed the spin-1 local degrees of freedom into two auxiliary spin-1/2. Then, we deploy $N - 1$ spin singlets $|\psi^-\rangle = (|\uparrow\downarrow\rangle - |\downarrow\uparrow\rangle)/\sqrt{2}$ between bonds. The physical spin-1 is then recovered as it should be; after projection, $\mathcal{P}$, onto the symmetric sector of two adjacent spins. As sketched in Fig. 4.6, such construction corresponds to an MPS of bond dimension $D = 2$: the-so-called valence bond state $|\text{VBS}\rangle$ [210] and constitutes a GS of $\hat{H}_{\text{AKLT}}$.

More generally, one can replace the singlet with any GME state in dimension D and the symmetric projection with some linear map $\mathcal{P} : \mathbb{C}^D \otimes \mathbb{C}^D \to \mathbb{C}^d$. In fact, such construction is instrumental to implementing gauge

(continued)

invariance (e.g. for high-energy physics simulators [211]) and to generalizing MPS to higher dimensions through the so-called projected entangled pair states (PEPS) [212].

Haldane Phase and Entanglement Spectrum The VBS is a finite-size precursor of the Haldane phase, which, among others, is characterized by the presence of topological edge states (see Fig. 4.6). In the thermodynamic limit, the entanglement spectrum [i.e., the eigenvalues of the (half chain) reduced density matrix] of the $|\mathrm{VBS}\rangle$ are exactly doubly-degenerate. As proven in Ref. [213], in spin-1 chains, such degeneracy can be lifted only if (1) there is a quantum phase transition, (2) certain symmetries of the Hamiltonian are spontaneously broken. Consequently, such feature constitutes a robust signature of the Haldane phase, and more generally, it results in a tool to classify symmetry-protected topological phases in 1D, which was completed in Ref. [214].

Gapless States and Criticality Gap closing implies divergence of the correlation length[26] in the system and may entail strong macroscopic consequences, like a critical point of a quantum phase transition. On the other hand, area-law-like theorems only provide sufficient conditions. The reverse is not true; e.g., there exist gapless Hamiltonians for which the GS entanglement content remains scalable. In particular, for some specific 1D systems, the bipartite entanglement entropy of a cut of length L scales as $S_L \sim \log(L/a)$, where a is the lattice spacing [215].[27] Moreover, some efforts focused on the scaling of such quantity with the gap close to the critical point [216].

Quenches In this *subsection* we have seen how MPS correctly reproduce the ground state properties of 1D local gapped Hamiltonians. One may ask whether MPS are useful for non-equilibrium states, such as quenches. This is generally not the case as entanglement increases rapidly with time, thus generating the volume law [217, 218]. However, the system may thermalize, and actually, quantum entanglement may no longer be necessary as correlations can be explained classically through mixing [219, 220].

[26] i.e., even though interactions are local, particles far apart become correlated.

[27] Note how it still scales exponentially better than a generic pure state.

Table 4.2 Typical scaling of the bipartite entanglement entropy of region X (versus the rest) for various classes of states of physical interest

Family of states	Scaling of S_X	Observations
Generic pure states	$\|X\|$	Curse of dimensionality
GS of local, gapped Hamiltonians	$\|\partial X\|$	Proven for 1D system, approximable by MPS
Gapless states and criticality	$\|\partial X\| \log \|X\|$	Only for 1D specific cases
Quenches	$\|X\|$	For long times, however system may thermalize

To close the subject, we summarize in Table 4.2 the different scaling laws outlined in this *subsection*:

4.4.1.2 Symmetric States and Sufficient Separability Criteria

Symmetries and invariance are of utmost importance in physics. They not only are instrumental in simplifying many problems but also fundamental to defining elementary interactions. Here we focus on symmetric states, i.e., those invariant under all permutation matrices.[28] The total symmetric sector of N qudits, is spanned by the set [221]:

$$\left\{ |S_{\mathbf{k}}\rangle = \binom{N}{\mathbf{k}} \sum_{\pi \in \mathfrak{S}_N} \pi \left(\bigotimes_{a=0}^{d-1} |a\rangle^{\otimes k_a} \right) \right\}_{\mathbf{k} \vdash N}, \tag{4.45}$$

where $\mathbf{k} = (k_0, k_1, .., k_{d-1})$ is a partition of N, $(\mathbf{k} \vdash N)$, i.e. a set of positive integers adding to N and the sum in the superposition is over unique permutations. The prefactor is the multinomial coefficient $\binom{N}{\mathbf{k}} = N!/\prod_{a=0}^{d-1} k_a!$.[29] The dimension of this subspace grows polynomially in N, with $\binom{N+d-1}{d-1} \sim O((N+d-1)^{(d-1)}/(d-1)!)$, which allows to process significant system sizes for low d.[30]

[28] Do not confuse *symmetric* states by *permutation-invariant* states. The latter refers to when the symmetry condition is relaxed to recover the same state up to a global phase, $\hat{T}|\Psi\rangle = e^{i\phi}|\Psi\rangle$, where T is a permutation matrix.

[29] However, the state is unnormalized, we may define the dual vectors $|\tilde{S}_k\rangle = \binom{N}{\mathbf{k}}^{-1}|S_k\rangle$ such that $\left\langle \tilde{S}_k \middle| \tilde{S}_{k'} \right\rangle = \delta_{kk'}$.

[30] For $d = 2$, we will ease the notation and denote by $|S_k\rangle$, $|S_{(k,N-k)}\rangle$.

The symmetry constraints open the possibility for a scalable characterization of entanglement, leading to many unexpected results regarding the sufficiency of entanglement criteria. For instance [222, 223],

Let $\hat{\rho}_{\text{DS}}$ be a diagonal symmetric N-qubit state (N even), i.e. of the form $\hat{\rho}_{\text{DS}} = \sum_{k=0}^{N} p_k |S_k\rangle\langle S_k|$, with $\{p_k\}$ a probability distribution.

Then, $\hat{\rho}_{\text{DS}}$ is separable if and only if $\hat{\rho}_{\text{DS}}$ is PPT with respect to the largest bipartition $N/2 : N/2$.

Check Ref. [224] for a generalization to qudits.

Since the state is (in particular) permutation invariant, the PT does not depend on the specific parties considered but just on the total number of them. Moreover, the condition can be easily verified as $\hat{\rho}_{\text{DS}}$ is PPT with respect to the bipartition $n : N - n$ iff the pair of Hankel matrices,

$$\hat{M}_0(n) = \begin{pmatrix} p_0 & \cdots & p_n \\ \cdots & \cdots & \cdots \\ p_n & \cdots & p_{2n} \end{pmatrix}, \quad \hat{M}_1(n) = \begin{pmatrix} p_1 & \cdots & p_{n+1} \\ \cdots & \cdots & \cdots \\ p_{n+1} & \cdots & p_{2n+1} \end{pmatrix}, \tag{4.46}$$

are positive semidefinite (PSD). The structure of such matrices also reveals that PPT with n^* implies PPT of all bipartitions with $n \leq n^*$ as they appear as principal minors.

We will complete the exploration by providing a sketch of the proof of PPT sufficiency:

Proof [222] By definition $\hat{\rho}_{\text{DS}}$ is separable if for *all* entanglement witnesses $\hat{W}$, $\text{Tr}(\hat{W}\hat{\rho}_{\text{DS}}) \geq 0$. On the other hand, any entanglement witness detecting such states can be parametrized as a convex combination of:

$$\hat{T} = \sum_{0 \leq i,j \leq N/2} a_i a_j \left|\tilde{S}_{i+j}\right\rangle\left\langle\tilde{S}_{i+j}\right|, \quad \hat{R} = \sum_{0 \leq i,j \leq (N-1)/2} b_i b_j \left|\tilde{S}_{i+j+1}\right\rangle\left\langle\tilde{S}_{i+j+1}\right|,$$

where $\mathbf{a} := \{a_i\}, \mathbf{b} := \{b_i\}$ are real coefficients. By applying them to our state, we obtain $\text{Tr}(\hat{T}\hat{\rho}_{\text{DS}}) = \mathbf{a}^T \hat{M}_0 \mathbf{a}$, $\text{Tr}(\hat{R}\hat{\rho}_{\text{DS}}) = \mathbf{b}^T \hat{M}_1 \mathbf{b}$.[31] Such quantities are non-negative for all $\mathbf{a}, \mathbf{b}$, if and only if $\hat{M}_{0,1}$ are PSD. Finally, as per the previous observation, PSD of the Hankel matrices $\hat{M}_{0,1}$ implies PPT of $\hat{\rho}_{\text{DS}}$ with respect to $N/2 : N/2$.

[31] Again, easing the notation $M_{0,1} := M_{0,1}(N/2)$.

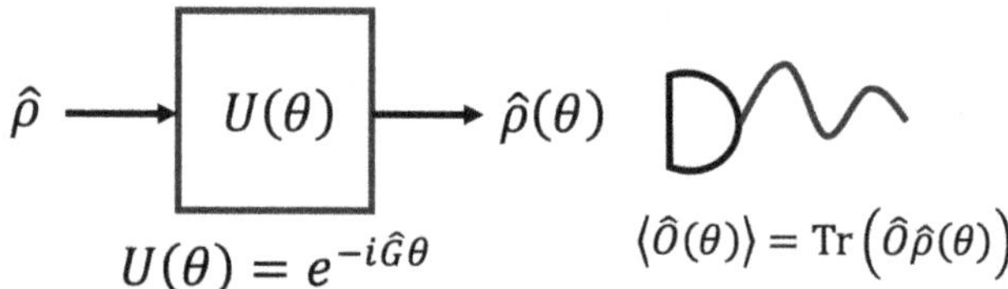

Fig. 4.7 A parameter θ is encoded in a quantum state $\hat{\rho}(\theta)$ by linear interferometry, i.e. as a phase shift in a unitary manner: $\hat{\rho}(\theta) = e^{-i\theta\hat{G}}\hat{\rho}e^{i\theta\hat{G}}$. After the encoding, the state is measured against an observable O

Permutation Symmetry and Spin For qubits, ($d = 2$) symmetric states are closely related to spatial rotations. Its associated group invariant is the total angular momentum (or spin), $\mathbf{S}^2 = S(S+1)$, where S is the maximal spin projection. In such a setting, symmetric states have well-defined collective spin, $S = S_{\max} = Ns$, with $s = (d-1)/2 = 1/2$ the local spin, resulting from the coherent collective participation of the N subsystems. The basis Eq. (4.45) is then constructed via coarse-graining the manifold of total maximal spin by identifying vectors that can be related by a permutation among the N subsystems (i.e., with the same spin projection). In higher dimensions $d > 2$, symmetric states span different total spin sectors, from the macroscopic, giant spin $S = S_{\max}$ to the singlet $S = S_{\min} = 0$. This latter case corresponds to rotation invariant states.

Symmetric States in Nature The prototypical realization of symmetric states is the spinor Bose-Einstein condensate (e.g. of sodium or rubidium atoms) under the single-mode approximation [225]. Similar collective models are realized under the denomination of Dicke models e.g. in cavity quantum-electrodynamics [226] and in nuclear physics (as Lipkin shell models [227]). These systems may display transitions between quantum chaotic and regular dynamics, which can also be explored from a Bell nonlocality perspective [228]. Symmetric selection rules are a standard way to scale quantum entanglement to the macroscopic regime. Hence, creating giant entangled states with depths of many thousands of atoms [189, 229]. As we show in the following, entanglement depth is a precious resource for sensing applications [230]:

Quantum-Enhanced Metrology
Here, the goal is to infer the value of an unknown parameter θ (e.g., a magnetic field [231], a space/time interval [232], temperature [233–235]). To this end, we employ a linear interferometer in order to encode the parameter as a phase in a quantum state via a unitary generated by $\hat{G}$, $\hat{\rho}(\theta) = e^{-i\theta\hat{G}}\hat{\rho}e^{i\theta\hat{G}}$ (see Fig. 4.7). Finally, we measure an observable O.

(continued)

After $\nu \to \infty$ number of shots, one can compute how the expectation value evolves in θ, $\langle \hat{O}(\theta) \rangle$ and infer the value of θ from it. The precision of such a task is quantified by the variance of the estimator, $(\Delta\theta)^2_{\text{est.}}$, which is bounded by the chain of inequalities [39, 236]:

$$\nu^2(\Delta\theta)^2_{\text{est.}} \geq \frac{(\Delta\hat{O})^2_{\hat{\rho}}}{\langle i[\hat{G}, \hat{O}]\rangle^2_{\hat{\rho}}} \geq \min_{\hat{O}} \frac{(\Delta\hat{O})^2_{\hat{\rho}}}{\langle i[\hat{G}, \hat{O}]\rangle^2_{\hat{\rho}}} := \frac{1}{F_Q[\hat{\rho}, \hat{G}]} . \tag{4.47}$$

The ultimate bound F_Q is the so-called quantum Fisher information (QFI) [237]. As a central property, the QFI is a convex function of the state $\hat{\rho}$. States of high quantum Fisher information are used to enhance measurements mainly in quantum optics [238], for gravitational wave detectors [232] or magnetometry with BECs [231]. Consider that we encode the phase collectively in N-qubits via a generator with structure $\hat{S}_z = \sum_{i\in[N]} \hat{\sigma}_{z,i}/2$, $\hat{\sigma}_z = |\uparrow\rangle\langle\uparrow| - |\downarrow\rangle\langle\downarrow|$. For such a case, using the convexity property, for any K-producible state $\hat{\rho}_K$ [239, 240]:

$$F_Q[\hat{\rho}_K, \hat{S}_z] \leq NK . \tag{4.48}$$

For $K = 1$, the bound (4.48) is saturated by coherent spins states such as $[(|\uparrow\rangle + |\downarrow\rangle)/\sqrt{2}]^{\otimes N}$ while for $K = N$ (a.k.a. the Heisenberg limit), the maximal sensitivity is achieved by the GHZ state $(|\uparrow\rangle^{\otimes N} + |\downarrow\rangle^{\otimes N})/\sqrt{2}$.

The QFI is a highly nonlinear function of the states, which makes it challenging to infer experimentally. In Ref. [241] machine-learning techniques [97] were developed to infer such quantity in a realistic scenario. Likewise, Ref. [45] puts forward theoretical tools to certify the metrological resource content of the state, as quantified by the QFI, if only partial information is available.

From inequality (4.48), we learn:

Increasing multipartite entanglement K and/or the system size N is necessary to improve the precision in phase estimation tasks.

Combining with Eq. (4.47) we build an *entanglement depth witness*. Consider sensing collective rotations around the z-axis, $\hat{G} = \hat{S}_z$, and probing the spin projection in an orthogonal direction, $\hat{O} = \hat{S}_x$. Then, one is able to recover the

Wineland spin squeezing criterion [242, 243]:[32]

$$\frac{\langle \hat{S}_x^2 \rangle_{\hat{\rho}}}{\langle \hat{S}_y \rangle_{\hat{\rho}}^2} < \frac{1}{NK} \Longrightarrow \hat{\rho} \text{ has at least entanglement depth } K+1, \tag{4.49}$$

or stronger versions of it (check Refs. [39,45]). Precisely, the next section is devoted to the detection of quantum entanglement.

4.4.2 Scalable Entanglement Certification

As we have seen, quantum entanglement is omnipresent. However, in the macroscopic scale, such feature is generally hidden in inaccessible higher-order correlations, and the world appears classical to us. With this, we highlight the importance of entanglement detection and the way we probe our systems. For instance, a state could be highly entangled; however, if we only measure commuting observables, the resulting correlations will always have a classical explanation. The same is true if only measure a correlator $\hat{O}_1 = \otimes_{i\in[N]}\hat{o}_{1,i}$. In such case, all quantum-feasible correlations $\langle \hat{O}_1 \rangle$ are also reproducible classically—with a separable state. However, if we combine two product observables $\hat{W} = k_1\hat{O}_1 + k_2\hat{O}_2$, as shown in Fig. 4.8 (cf. Fig. 4.1), then potentially some values of $\langle \hat{W} \rangle$ achievable with the quantum theory are not simulable classically. In the present subsection, we want to explore how to certify such fact in many-body systems, where already in-depth reviews exists, Ref. [244, 245].

To do so, we bound $\langle \hat{W} \rangle$ over separable states, i.e. find $\beta = \min_{\hat{\rho}\in\mathcal{R}_1} \mathrm{Tr}(\hat{W}\hat{\rho})$.[33] By convexity of the set of correlations accessible by separable states, $\langle \hat{W} \rangle_{\hat{\rho}} - \beta < 0$ implies $\hat{\rho}$ is entangled, i.e. entanglement is necessary to produce such statistics.

The Scalability Issue Note that even in many-body systems, few correlators $\{\hat{O}_a\}_a := \hat{\mathbf{O}}$ may be sufficient to detect entanglement, without the need for a full tomographic reconstruction of the state (which otherwise quickly becomes unfeasible with the system size). From a theoretical side, a challenge is to compute the bound β (see Fig. 4.8). In order to overcome such difficulty, one may relax the problem and instead aim to quantify weaker bounds β', which are not tight but sufficient to detect entangled certain families of physically relevant many-body states.

[32] We align the mean spin $\langle \hat{\mathbf{S}} \rangle$ along the y-axis such that $(\Delta \hat{S}_x)^2 = \langle \hat{S}_x^2 \rangle$.

[33] Note that by convexity of $\mathcal{R}_1$, optimization over pure states $\hat{\rho} = |\Psi\rangle\langle\Psi|$ is sufficient.

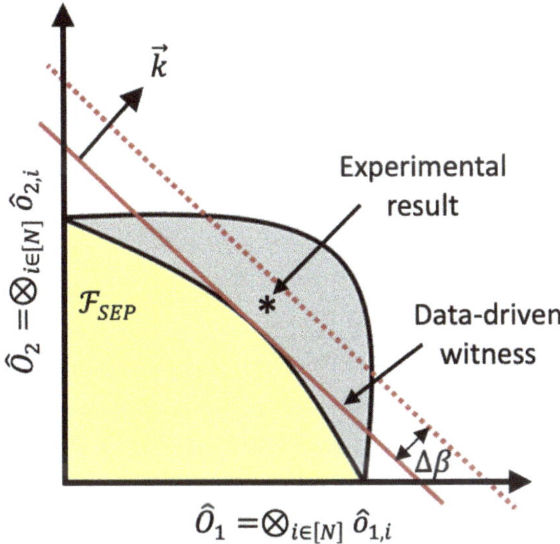

Fig. 4.8 In yellow $\mathcal{F}_{\text{SEP}}$, region of expectation values (correlations) compatible with a separable state. Such set is demarcated by (tight) entanglement witnesses (e.g. solid red line) characterized by the coefficients **k**. The relaxation of the bound β to $\beta' < \beta$ leads to a weaker witness (dashed red line), which may not be robust enough to detect some experimental data with genuine quantum origin (star). However, β' can be less challenging to compute than β. Finally, the gray set contains all correlations reproducible by the quantum framework—but not classically recoverable

The Data-Driven Approach A further question that one may ask is the optimal witness (e.g. the best values of $\mathbf{k} = (k_1, k_2)$ of the previous discussion) to certify a given set of few experimentally-inferred[34] mean values $\langle \hat{\mathbf{O}} \rangle$, hereinafter called *data*. Refs. [40, 45, 50, 246, 247] provide generic tools to solve such problem in the many-body scenario for a particular class of witnesses.

Finally, from the technological aspect, entanglement is a pivotal resource, which is exploited in emergent quantum-enhanced applications. For this reason, the certification of such a feature is a necessary first step to assess the presumed advantage such devices may provide. The remainder of the subsection will be devoted to the detection of many-body quantum entanglement and Bell correlation.

4.4.2.1 Detecting Many-Body Entanglement in Rotation-Invariant States

We start by giving an example of scalable entanglement witness by taking advantage of symmetries:

[34] i.e. after partial (incomplete) tomography.

Example: Entanglement Witness Tailored to the Many-Body Spin Singlet
Consider a system of N spin-s particles, from which we only can infer the expectation value of the total spin $\langle \hat{\mathbf{S}}^2 \rangle$. What can we say about the entanglement of the underlying quantum state? We can bound the expectation value over separable states Eq. (4.40) $K = 1$:

$$\langle \hat{\mathbf{S}}^2 \rangle_{\hat{\rho}_1} = \langle \left(\sum_{i\in[N]} \mathbf{s}_i \right)^2 \rangle_{\hat{\rho}_1} = \underbrace{\sum_{i\in[N]} \langle \mathbf{s}_i^2 \rangle_{\hat{\rho}_1}}_{=Ns(s+1)} + \underbrace{\sum_{i\neq j\in[N]} \langle \mathbf{s}_i \cdot \mathbf{s}_j \rangle_{\hat{\rho}_1}}_{\sum_\psi p_\psi \sum_{i\neq j\in[N]} \langle \mathbf{s} \rangle_{\psi_i} \cdot \langle \mathbf{s} \rangle_{\psi_j}} = \tag{4.50}$$

$$= Ns(s+1) + \underbrace{\sum_\psi p_\psi \langle \sum_{i\in[N]} \mathbf{s}_i \rangle_\psi \cdot \langle \sum_{j\in[N]} \mathbf{s}_j \rangle_\psi}_{\geq 0} - \underbrace{\sum_{i\in[N]} \sum_\psi \langle \mathbf{s}_i \rangle_{\psi_i}^2}_{\leq Ns^2} \geq \tag{4.51}$$

$$\geq Ns \tag{4.52}$$

Hence, for any N-partite separable state,

$$\langle \hat{\mathbf{S}}^2 \rangle \geq Ns \,. \tag{4.53}$$

The witness Eq. (4.53) is maximally violated by the many-body spin singlet $\mathbf{S}^2 = 0$. Note that it does not detect metrologically useful entanglement, i.e., sensitivity on rotations, as there always exists a rotation invariant state compatible with $\langle \hat{\mathbf{S}}^2 \rangle$ [cf. Eq. (4.49)].

Using the above example, one can ask about sufficiency. For instance, how can we characterize the set of mean spin $\langle \hat{\mathbf{S}} \rangle$ and its fluctuations (i.e., second moments) $\langle \hat{\mathbf{S}} \hat{\mathbf{S}}^T \rangle$ that are compatible with a separable state? Remarkably, for arbitrary N-partite spin-1/2 systems ($d = 2$), all entanglement that can be detected with such data can be summarized in only eight inequalities [49]. Such result was generalized to arbitrary spin [49, 248]. In Ref. [50], we formalize a data-driven approach to detect entanglement in spin ensembles with second moments of arbitrary collective observables (beyond spin projections). In such cases, we exploit Zeeman population measurements to unveil new witnesses tailored to relevant states prepared in spinor BEC experiments.

4.4.2.2 Bell Correlation

Bell correlations are one of the strongest tests of non-classicality. Here, in the so-called device-independent paradigm, we abandon any detailed assumption on the systems, e.g. if it can be described with quantum physics, the implementation of the measurements, etc. [249,250]. The minimal description is encapsulated in only three numbers (N, J, M). As before, N is the number of parties in which we distribute a resource (e.g., a quantum state), J is the number of measurements settings each party is able to choose, and M is the number of measurement outcomes. After several rounds of collecting results, the parties can infer statistics, e.g. correlations in the form of joint expectation values.[35] We say that a correlation is classical if there exists a local hidden variable (LHV) model reproducing it. In other words, if it factorizes up to a *hidden* statistical mixing:[36]

$$a, b \in [J], i, j \in [M]\,, \quad \langle \mathfrak{m}_{a,i} \mathfrak{m}_{a,j} \rangle_{\mathrm{LHV}} = \sum_{\lambda} p_{\lambda} \langle \mathfrak{m}_{a,i} \rangle_{\lambda} \langle \mathfrak{m}_{b,j} \rangle_{\lambda}\,, \tag{4.54}$$

where $\{p_{\lambda}\}$ is a probability distribution over the LHV λ. If for a given set of correlations (data), such probability does not exist, we denote them (Bell) nonlocal.[37]

It is surprising how such a simple description is sufficient to derive many nontrivial results. For instance, if quantum formalism is assumed, it is not difficult to show that all separable states cannot produce correlations beyond LHV. But the converse is not true, there are entangled states that do not violate any factorization of the form Eq. (4.54) [141]. The set of LHV by construction forms a convex set. Consequently, nonlocality is signaled by a violation of a witness (a.k.a. Bell inequality) as in Fig. 4.8. From this discussion, we conclude:

Violating Bell inequalities allows one to certify the preparation of entangled states from minimal assumptions—in a device-independent manner.

Perhaps, more importantly, there exist (entangled) states which *do* violate a Bell inequality [80]. With this fact, quantum formalism is justified as it derives predictions that cannot be classically reproduced.

Bell Nonlocality Versus Bell Correlation In a standard Bell scenario, the inference of the necessary correlations to violate a multipartite Bell inequality quickly becomes intractable as the number of parties N increases. Yet, for those that are permutationally invariant (PI), under some assumptions, it is not necessary to perform individual measurements as the very same correlations can be inferred

[35] Parties need to choose their settings for each round in a statistically independent way. This can be ensured by space-like separation. The lack of such a requirement makes LHV unfalsifiable.

[36] Note that we deliberately removed the hats $\hat{\mathfrak{m}}$ from the measurements $\mathfrak{m}$ as quantum mechanics is not assumed. The expectation values should be understood in the purely statistical sense.

[37] i.e., they require nonlocal hidden variables, beyond Eq. (4.54).

via moments of collective observables. For example, consider two parties and the quantum PI correlation $C_{ab} = 2(\langle \hat{a} \otimes \hat{b} \rangle + \langle \hat{b} \otimes \hat{a} \rangle)$. The inference of such expectation values seem to require individual addressing. However, it can also be expressed as $C_{ab} = \langle \hat{M}_a \hat{M}_b + \text{h.c.} \rangle - \langle \hat{M}_{ab} \rangle$, where $\{\hat{M}_a = \hat{a} \otimes \mathbb{I} + \mathbb{I} \otimes \hat{a},\ \hat{M}_b = \hat{b} \otimes \mathbb{I} + \mathbb{I} \otimes \hat{b},\ \hat{M}_{ab} = (\hat{a}\hat{b} + \hat{b}\hat{a}) \otimes \mathbb{I} + \mathbb{I} \otimes (\hat{a}\hat{b} + \hat{b}\hat{a})\}$ are collective observables. Once the single-particle observables are specified, e.g. Pauli matrices $\{\hat{a} = \hat{\sigma}_x, \hat{b} = \hat{\sigma}_y\}$, then $\hat{M}_{ab}$ corresponds to a new observable (in this case $\hat{M}_{ab} = 0$). The two-particle term can be symmetrized $\langle \hat{M}_a \hat{M}_b + \text{h.c.} \rangle = [\langle (\hat{M}_a + \hat{M}_b)^2 \rangle - \langle (\hat{M}_a - \hat{M}_b)^2 \rangle]/2$. Thus, they may be evaluated by measuring $\hat{M}_a \pm \hat{M}_b$, corresponding to the collective spin in $x \pm y$ orientation.

The latter strategy is scalable in N. This is the natural measurement approach for quantum many-body systems, specifically cold atomic ensembles and Bose-Einstein condensates (BECs) [225]. However, in such systems, in addition, one can no longer guarantee space-like separation between parties.

Therefore, as we, (1) do not require closing loopholes, (2) demand the validity of quantum mechanics, (3) need a correct calibration of the settings e.g. spin orientation, we can no longer talk about Bell nonlocality. Instead, we use the term *Bell correlations* to refer to the violation of multipartite Bell inequalities in a device-dependent way, from witnesses typically involving collective observables. From the assertions above, it is clear that Bell correlations are weaker than Bell nonlocality, but they are still instrumental to certify quantum entanglement. The detection of Bell correlations in many-body systems through collective measurements has been of great success in experiments with atomic ensembles [251, 252] and in the context of quantum simulators [253–258]. Beyond collective measurements, Bell correlation can also be detected from local data in 1D translation-invariant systems and other geometries [259–262]. Finally, it has been shown that as the number of particles grow, so does the violation of some Bell inequalities, like MABK [263–265] and WWWŻB [266–268] ones.

In order to derive the Bell inequalities, we need to bound functionals over LHV models, which might be challenging in the multipartite scenario. Below we outline a strategy to do so by interpreting the optimization as a statistical physics problem [246]:

Many-Body Bell Inequalities: A Statistical Physics Approach

Suppose we have a generic multipartite Bell inequality based on up to two-body correlation functions:

$$\mathcal{B} = \sum_{i \in [N]} \sum_{a \in [K]} k_{ai} \langle \mathfrak{m}_{a,i} \rangle + \sum_{i \neq j \in [N]} \sum_{a,b \in [K]} q_{ai,bj} \langle \mathfrak{m}_{a,i} \mathfrak{m}_{b,j} \rangle \,, \tag{4.55}$$

(continued)

where $\{k, q\}$ are real coefficients defining the Bell inequality. The goal is to find the classical bound, i.e. $\beta_{\mathrm{LHV}} = \min_{\mathrm{LHV}} \mathcal{B}$.

Local Deterministic Strategies According to Fine's theorem [269], for this task it is sufficient to consider extremal LHV models, i.e., deterministic strategies $\langle \mathfrak{m}_{a,i} \rangle = \sigma_{a,i} \in [M]$ (otherwise, any randomness is absorbed into the probability $\{p_\lambda\}$) and $\langle \mathfrak{m}_{a,i} \mathfrak{m}_{b,j} \rangle = \langle \mathfrak{m}_{a,i} \rangle \langle \mathfrak{m}_{b,j} \rangle$. Then, the problem can be reinterpreted as to find the ground energy of an Ising-like Hamiltonian, $\beta_{\mathrm{LHV}} = \min_{\{\sigma \in [M]\}} H$:

$$H = \sum_{i \in [N]} \sum_{a \in [K]} k_{ai} \sigma_{a,i} + \sum_{i \neq j \in [N]} \sum_{a,b \in [K]} q_{ai,bj} \sigma_{a,i} \sigma_{b,j} \,, \tag{4.56}$$

which can be solved with standard statistical-mechanical techniques such as simulated annealing, Monte Carlo, etc.

Within the data-driven paradigm, the problem is not only to find the classical bound, but the value of the coefficients $\{k, q\}$ which (hopefully) detects the given data $\{\langle \mathfrak{m}_{a,i} \rangle, \langle \mathfrak{m}_{a,i} \mathfrak{m}_{b,j} \rangle\}$ Bell nonlocal. As shown in Ref. [246], such a problem can be phrased as a convex optimization task with a well-defined solution.

As usual, in order to make the problem scalable, symmetries can be exploited. For instance, if one considers PI, that is $k_{ai} = k_a$, $q_{ai,bj} = k_{ab}$, the coarse-grained Hamiltonian Eq. (4.56) will only depend on the total number of parties which display outcome $r \in [M]$ when $a \in [J]$ is measured.

In the Bell scenario, the nature of the measurements remains unspecified. In a concrete physical situation, therefore, one has to optimize the settings to observe the maximal violation. This problem is, in general, highly non-convex and challenging in the multipartite regime. Even so, by exploiting symmetries, the first scalable Bell inequality derived in Ref. [270] led to the following Bell correlation witness [251]:

$$\frac{4\langle \hat{S}_x^2 \rangle}{N} < \frac{1}{2} \left[1 - \sqrt{1 - \left(\frac{2\langle \hat{S}_y \rangle}{N} \right)^2} \right] \Longrightarrow \text{Bell correlation} \,. \tag{4.57}$$

Importantly, Ineq. (4.57) is maximally violated by spin-squeezed states [cf. Eq. (4.49)], but it is more demanding than the Wineland criterion Eq. (4.49). In Ref. [247] Ineq. (4.57) has been generalized to arbitrary local spin with a novel data-driven method which avoids the issue above. Similar witness are obtained for three-level multipartite Bell inequalities [271, 272].

We conclude the *section* by highlighting that results like Eqs. (4.49), (4.53), and (4.57), which are based on the same data, connect, on the same footing, three relevant quantum resources: metrology, entanglement and Bell nonlocality [273]. In particular, such data is inferred from experimentally accessible observables, e.g. those routinely exploited in ultracold atom platforms. Thus, eliciting the study of useful connections between quantum resources in realistic scenarios.

4.5 Open Problems

Here, we put forth a non-exhaustive list of open problems that were highlighted within the scope of this chapter. We hope that this discussion will rekindle the spark in the readers and the quantum information community to revisit these problems and address them in the coming years.

1. *nPT bound entanglement*: All PPT entangled states are bound entangled [27]. Is it also the other way around? In other words, are there bound entangled states that are not PPT?
2. *Classification of AME states*: Two AME states can be nonetheless LU-inequivalent [138]. The problem is to classify AME states in LU-inequivalent and/or other entanglement classes of interest.
3. *Sufficient conditions for multipartite LOCC convertibility*: As we have seen, Nielsen's theorem [163] provides necessary and sufficient conditions for LOCC conversion in bipartite systems. In the multipartite setting, there are proposals to relax the conditions of LOCC to local separable (SEP) operations [164], by considering stabilizing symmetries [274]. While previous generalizations provide only necessary conditions, the quest for sufficient criteria of LOCC existence, beyond trivial symmetries, is open [275].
4. *Unified theory of multipartite entanglement*: To this date, there does not exist a unified theory of multipartite entanglement despite the numerous efforts over the past three decades. This problem is intrinsically related to the mathematical problem of tensor classification [276].
5. *Higher dimensional area laws*: Rigorous proof of area law theorem/theorems in higher dimensions and suitability of PEPS in approximating higher dimensional ground states. states. Significant results are established for Gaussian models such as harmonic lattices [201] and quasi-free fermionic and bosonic systems [202]. It would be beneficial to provide proofs under more general assumptions.
6. *Entanglement versus thermalization in many-body systems:* Under time evolution, entanglement generally increases rapidly. Such a scaling restricts the usefulness of many-body methods, like MPS, to short times. However, in typical scenarios, few-body correlations eventually thermalize and the necessary quantum entanglement to explain such statistics may remain nonetheless scalable. The task is to develop efficient standard tools to address thermalization via MPS or other low-entanglement ansatzes.

7. *The separability problem in symmetric states*: As discussed in Sect. 4.4, there are no PPT entangled symmetric diagonal qubit states, which implies that the separability problem (in such subspace) is solved. Here, the task is to extend these results to qudit systems beyond the restricted set of Ref. [224], the bipartite case [77] or even diagonal states.
8. *Metrological resources from a quantum information viewpoint*: Here, we highlight two questions: (1) Does Bell nonlocality provide any advantage in the metrological task that is not explained by quantum entanglement, cf. Eq. (4.48)? (2) As proved in Ref. [277], there exist PPT entangled states that nonetheless are metrologically useful. Which PnCP maps detect metrologically useful entanglement?

For a more detailed and elaborate list of open problems in quantum theory in general, please also look at Ref. [119], and the following links.[38]

Acknowledgments We thank Karol Życzkowski, Anna Sanpera, Irénée Frérot, and Flavio Baccari for fruitful discussions and their valuable feedback. A.K.S. acknowledges support from the European Union's Horizon 2020 Research and Innovation Programme under the Marie Skłodowska-Curie Grant Agreement No. 847517. ICFO group acknowledges support from: ERC AdG NOQIA; MCIN/AEI (PGC2018-0910.13039/501100011033, CEX2019-000910-S/10.13039/501100011033, Plan National FIDEUA PID2019-106901GB-I00, Plan National STAMEENA PID2022-139099NB-I00 project funded by MCIN/AEI/10.13039/501100011033 and by the "European Union NextGenerationEU/PRTR" (PRTR-C17.I1), FPI); QUANTERA MAQS PCI2019-111828-2); QUANTERA DYNAMITE PCI2022-132919 (QuantERA II Programme co-funded by European Union's Horizon 2020 program under Grant Agreement No 101017733), Ministry of Economic Affairs and Digital Transformation of the Spanish Government through the QUANTUM ENIA project call—Quantum Spain project, and by the European Union through the Recovery, Transformation, and Resilience Plan—NextGenerationEU within the framework of the Digital Spain 2026 Agenda; Fundació Cellex; Fundació Mir-Puig; Generalitat de Catalunya (European Social Fund FEDER and CERCA program, AGAUR Grant No. 2021 SGR 01452, QuantumCAT U16-011424, co-funded by ERDF Operational Program of Catalonia 2014-2020); Barcelona Supercomputing Center MareNostrum (FI-2023-1-0013); EU Quantum Flagship (PASQuanS2.1, 101113690); EU Horizon 2020 FET-OPEN OPTOlogic (Grant No 899794); EU Horizon Europe Program (Grant Agreement 101080086 - NeQST), ICFO Internal "QuantumGaudi" project; European Union's Horizon 2020 program under the Marie Skłodowska-Curie grant agreement No 847648; "La Caixa" Junior Leaders fellowships, "La Caixa" Foundation (ID 100010434): CF/BQ/PR23/11980043. Views and opinions expressed are, however, those of the author(s) only and do not necessarily reflect those of the European Union, European Commission, European Climate, Infrastructure and Environment Executive Agency (CINEA), or any other granting authority. Neither the European Union nor any granting authority can be held responsible for them.

[38] "https://oqp.iqoqi.oeaw.ac.at/open-quantum-problems", "https://giedke.dipc.org/Benasque/futqi2023.pdf".

References

1. L. Amico, R. Fazio, A. Osterloh, V. Vedral, Rev. Mod. Phys. **80**, 517 (2008). https://journals.aps.org/rmp/abstract/10.1103/RevModPhys.80.517 226
2. R. Horodecki, P. Horodecki, M. Horodecki, K. Horodecki, Rev. Mod. Phys. **81**, 865 (2009). https://doi.org/10.1103/RevModPhys.81.865. https://link.aps.org/doi/10.1103/RevModPhys.81.865 242, 245
3. N. Laflorencie, Phys. Rep. **646**, 1–59 (2016). https://doi.org/10.1016/j.physrep.2016.06.008
4. I. Bengtsson, K. Zyczkowski. A brief introduction to multipartite entanglement (2016). DOI 10.48550/ARXIV.1612.07747. URL https://arxiv.org/abs/1612.07747
5. M. Walter, D. Gross, J. Eisert, Quantum Inf., 293–330 (2016). https://doi.org/10.1002/9783527805785.ch14 226, 247, 249, 250, 251
6. J. Tura, A.B. Sainz, T. Graß, R. Augusiak, A. Acín, M. Lewenstein, *Proceedings of the International School of Physics "Enrico Fermi", "Quantum Matter at Ultralow Temperatures"*, vol. Course 191 (IOS/SIF, Amsterdam/Bologna, 2016), p. 505. https://doi.org/10.3254/978-1-61499-694-1-505 226
7. S. Das, T. Chanda, M. Lewenstein, A. Sanpera, A. Sen De, U. Sen. The separability versus entanglement problem (2016). https://doi.org/10.1002/9783527805785.ch8
8. M. Lewenstein, A. Sanpera, V. Ahufinger, *Ultracold Atoms in Optical Lattices: Simulating Quantum Many-body Systems* (Oxford University Press, 2012). https://doi.org/10.1093/acprof:oso/9780199573127.001.0001 226
9. C.W. Helstrom, J. Stat. Phys. **1**(2), 231 (1969). https://doi.org/10.1007/BF01007479 227
10. M.A. Nielsen, I.L. Chuang, *Quantum Computation and Quantum Information: 10th Anniversary Edition* (Cambridge University Press, 2012). https://doi.org/10.1017/CBO9780511976667 227, 234
11. A. Peres (ed.), *Quantum Theory: Concepts and Methods* (Springer Netherlands, 2002). https://doi.org/10.1007/0-306-47120-5 228
12. M. Born, Z. Phys. **37**, 863 (1926). https://doi.org/10.1007/BF01397477 228
13. W. Heisenberg, Z. Phys. **43**, 172 (1927). https://doi.org/10.1007/BF01397280 228
14. H. Robertson, Phys. Rev. **34**, 163 (1929). https://doi.org/10.1103/PhysRev.34.163 228
15. H.A. Carteret, D.R. Terno, K. Życzkowski, Phys. Rev. A **77** (2008). https://doi.org/10.1103/physreva.77.042113 229
16. I. Bengtsson, K. Życzkowski, *Geometry of Quantum States: An Introduction to Quantum Entanglement* (Cambridge University Press, 2006). https://doi.org/10.1017/CBO9780511535048 230, 242
17. J. Watrous, *The Theory of Quantum Information* (Cambridge University Press, 2018). https://doi.org/10.1017/9781316848142 230
18. M. Choi, Linear Algebra Appl. **10**, 285 (1975). https://doi.org/10.1016/0024-3795(75)90075-0 230
19. P. Horodecki, Phys. Lett. A **232**, 333 (1997). https://doi.org/10.1016/s0375-9601(97)00416-7 231
20. H. Wiseman, J. Vaccaro, Phys. Rev. Lett. **91** (2003). https://doi.org/10.1103/PhysRevLett.91.097902 231
21. M. Dowling, A. Doherty, H. Wiseman, Phys. Rev. A **73** (2006). https://doi.org/10.1103/PhysRevA.73.052323 231
22. L. Gurvits, *Proc. of the 35th ACM symp. on Theory of comp.* (2003), pp. 10–19 233
23. M. Plenio, S. Virmani, Quantum Inf. Comput. **7**(1 & 2), 1 (2007). https://doi.org/10.26421/QIC7.1-2-1 234, 242
24. G. Tóth, T. Moroder, O. Gühne, Phys. Rev. Lett. **114** (2015). https://doi.org/10.1103/PhysRevLett.114.160501 235
25. M. Horodecki, P. Horodecki, R. Horodecki, Phys. Lett. A **223**, 1 (1996). https://doi.org/10.1016/S0375-9601(96)00706-2 235, 236
26. A. Peres, Phys. Rev. Lett. **77**, 1413 (1996). https://doi.org/10.1103/physrevlett.77.1413 235

27. M. Horodecki, P. Horodecki, R. Horodecki, Phys. Rev. Lett. **80**, 5239 (1998). https://doi.org/10.1103/physrevlett.80.5239 235, 242, 272
28. M. Lewenstein, B. Kraus, J. Cirac, P. Horodecki, Phys. Rev. A **62** (2000). https://doi.org/10.1103/PhysRevA.62.052310 237, 238, 242
29. D. Bruß, J. Cirac, P. Horodecki, F. Hulpke, B. Kraus, M. Lewenstein, A. Sanpera, J. Mod. Opt. **49**, 1399 (2002). https://doi.org/10.1080/09500340110105975 235
30. K. Chen, L.A. Wu, Quantum Inf. Comput. **3**, 193 (2003) 235
31. O. Rudolph, Quantum Inf. Process. **4**, 219 (2005). https://doi.org/10.1007/s11128-005-5664-1 235
32. P. Horodecki, Phys. Rev. A **68** (2003). https://doi.org/10.1103/PhysRevA.68.052101 235, 237
33. D. Abanin, E. Demler, Phys. Rev. Lett. **109** (2012). https://doi.org/10.1103/PhysRevLett.109.020504 235
34. J. Shang, O. Gühne, Phys. Rev. Lett. **120** (2018). https://doi.org/10.1103/PhysRevLett.120.050506 235
35. P. Pandya, O. Sakarya, M. Wieśniak, Phys. Rev. A **102** (2020). https://doi.org/10.1103/PhysRevA.102.012409
36. M. Wieśniak, P. Pandya, O. Sakarya, B. Woloncewicz, Quantum Rep. **2**, 49 (2020). https://doi.org/10.3390/quantum2010004 235
37. M. Lewenstein, A. Sanpera, Phys. Rev. Lett. **80**, 2261 (1998). https://doi.org/10.1103/PhysRevLett.80.2261 235, 253
38. A. Rico, F. Huber, Entanglement detection with trace polynomials (2023). https://arxiv.org/abs/2303.07761 235
39. M. Gessner, A. Smerzi, L. Pezzè, Phys. Rev. Lett. **122**, 090503 (2019). https://doi.org/10.1103/PhysRevLett.122.090503. https://link.aps.org/doi/10.1103/PhysRevLett.122.090503 235, 265, 266
40. I. Frérot, F. Baccari, A. Acín, PRX Quantum **3** (2022). https://doi.org/10.1103/PRXQuantum.3.010342 267
41. I. Frérot, T. Roscilde, Phys. Rev. Lett. **127** (2021). https://doi.org/10.1103/PhysRevLett.127.040401 235
42. P. Hauke, M. Heyl, L. Tagliacozzo, P. Zoller, Nat. Phys. **12**, 778 (2016). https://doi.org/10.1038/nphys3700 235
43. G. Mathew, S. Silva, A. Jain, A. Mohan, D. Adroja, V. Sakai, C. Tomy, A. Banerjee, R. Goreti, V.N. Aswathi, R. Singh, D. Jaiswal-Nagar, Phys. Rev. Res. **2** (2020). https://doi.org/10.1103/PhysRevResearch.2.043329
44. A. Scheie, P. Laurell, A. Samarakoon, B. Lake, S. Nagler, G. Granroth, S. Okamoto, G. Alvarez, D. Tennant, Phys. Rev. B **103** (2021). https://doi.org/10.1103/PhysRevB.103.224434
45. G. Müller-Rigat, A.K. Srivastava, S. Kurdziałek, G. Rajchel-Mieldzioć, M. Lewenstein, I. Frérot, Quantum **7**, 1152 (2023). https://doi.org/10.22331/q-2023-10-24-1152 235, 265, 266, 267
46. A. Elben, R. Kueng, H. Huang, R. van Bijnen, C. Kokail, M. Dalmonte, P. Calabrese, B. Kraus, J. Preskill, P. Zoller, B. Vermersch, Phys. Rev. Lett. **125** (2020). https://doi.org/10.1103/PhysRevLett.125.200501 235
47. B. Vermersch, M. Ljubotina, J.I. Cirac, P. Zoller, M. Serbyn, L. Piroli. Many-body entropies and entanglement from polynomially-many local measurements (2023). https://doi.org/10.48550/ARXIV.2311.08108. https://arxiv.org/abs/2311.08108 235
48. G. Tóth, C. Knapp, O. Gühne, H. Briegel, Phys. Rev. A **79** (2009). https://doi.org/10.1103/PhysRevA.79.042334 235
49. G. Vitagliano, I. Apellaniz, I.L. Egusquiza, G. Tóth, Phys. Rev. A **89**, 032307 (2014). https://doi.org/10.1103/PhysRevA.89.032307. https://link.aps.org/doi/10.1103/PhysRevA.89.032307 268
50. G. Müller-Rigat, M. Lewenstein, I. Frérot, Quantum **6**, 887 (2022). https://doi.org/10.22331/q-2022-12-29-887 235, 267, 268

51. Č. Brukner, V. Vedral, A. Zeilinger, Phys. Rev. A **73** (2006). https://doi.org/10.1103/PhysRevA.73.012110 235, 236
52. J. Estève, C. Gross, A. Weller, S. Giovanazzi, M. Oberthaler, Nature **455**, 1216 (2008). https://doi.org/10.1038/nature07332
53. F. Haas, J. Volz, R. Gehr, J. Reichel, J. Estève, Science **344**, 180 (2014). https://doi.org/10.1126/science.1248905
54. T. Lanting, A. Przybysz, A. Smirnov, F. Spedalieri, M. Amin, A. Berkley, R. Harris, F. Altomare, S. Boixo, P. Bunyk, N. Dickson, C. Enderud, J. Hilton, E. Hoskinson, M. Johnson, E. Ladizinsky, N. Ladizinsky, R. Neufeld, T. Oh, I. Perminov, C. Rich, M. Thom, E. Tolkacheva, S. Uchaikin, A. Wilson, G. Rose, Phys. Rev. X **4** (2014). https://doi.org/10.1103/PhysRevX.4.021041 235
55. M. Nielsen, Preprint. arxiv quant-ph/0011036 (1998). https://arxiv.org/abs/0011036 236
56. M. Wieśniak, V. Vedral, Č. Brukner, New J. Phys. **7**, 258 (2005). https://doi.org/10.1088/1367-2630/7/1/258 236
57. M. Wieśniak, V. Vedral, Č. Brukner, Phys. Rev. B **78** (2008). https://doi.org/10.1103/PhysRevB.78.064108 236
58. M. Dowling, A. Doherty, S. Bartlett, Phys. Rev. A **70** (2004). https://doi.org/10.1103/PhysRevA.70.062113 236
59. G. Tóth, Phys. Rev. A **71** (2005). https://doi.org/10.1103/PhysRevA.71.010301
60. J. Tura, G. De las Cuevas, R. Augusiak, M. Lewenstein, A. Acín, J. Cirac, Phys. Rev. X **7** (2017). https://doi.org/10.1103/PhysRevX.7.021005 236
61. O. Gühne, G. Tóth, H.J. Briegel, New J. Phys. **7**, 229 (2005). https://doi.org/10.1088/1367-2630/7/1/229 236, 257
62. O. Gühne, G. Tóth, Phys. Rep. **474**, 1 (2009). https://doi.org/10.1016/j.physrep.2009.02.004 236, 254
63. A.O. Pittenger, M.H. Rubin, Linear Algebra Appl. **346**(1–3), 47 (2002). https://doi.org/10.1016/S0024-3795(01)00524-9 237
64. C. Spengler, M. Huber, S. Brierley, T. Adaktylos, B. Hiesmayr, Phys. Rev. A **86** (2012). https://doi.org/10.1103/PhysRevA.86.022311 237
65. D. Chruściński, G. Sarbicki, F. Wudarski, Phys. Rev. A **97** (2018). https://doi.org/10.1103/PhysRevA.97.032318 237
66. J. Bae, B. Hiesmayr, D. McNulty, New J. Phys. **21**, 013012 (2019). https://doi.org/10.1088/1367-2630/aaf8cf
67. B. Hiesmayr, D. McNulty, S. Baek, S. Singha Roy, J. Bae, D. Chruściński, New J. Phys. **23**, 093018 (2021). https://doi.org/10.1088/1367-2630/ac20ea
68. J. Bavaresco, N. Herrera Valencia, C. Klöckl, M. Pivoluska, P. Erker, N. Friis, M. Malik, M. Huber, Nat. Phys. **14**, 1032 (2018). https://doi.org/10.1038/s41567-018-0203-z 237
69. F. Huber, I. Klep, V. Magron, J. Volčič, Commun. Math. Phys. **396**, 1051 (2022). https://doi.org/10.1007/s00220-022-04485-9 237
70. O. Gühne, M. Mechler, G. Tóth, P. Adam, Phys. Rev. A **74** (2006). https://doi.org/10.1103/PhysRevA.74.010301 237
71. J.B. Lasserre, Sufficient conditions for a real polynomial to be a sum of squares (2006). https://arxiv.org/abs/math/0612358 238
72. S. Woronowicz, Rep. Math. Phys. **10**(2), 165 (1976). https://doi.org/10.1016/0034-4877(76)90038-0 238
73. D. Hilbert, Math. Ann. **32**(3), 342 (1888). https://doi.org/10.1007/bf01443605 238
74. T. Motzkin, *Inequalities* (Proc. Sympos. Wright-Patterson Air Force Base, Ohio, 1965) (1967), pp. 205–224 238
75. M.M. Wolf, Quantum channels and operations – guided tour (Lecture notes) (2012). https://mediatum.ub.tum.de/doc/1701036/document.pdf 238
76. M. Lewenstein, R. Augusiak, D. Chruściński, S. Rana, J. Samsonowicz, Phys. Rev. A **93**(4) (2016). https://doi.org/10.1103/PhysRevA.93.042335 239

77. J. Tura, A. Aloy, R. Quesada, M. Lewenstein, A. Sanpera, Quantum **2**, 45 (2018). https://doi.org/10.22331/q-2018-01-12-45 239, 273
78. A. Einstein, B. Podolsky, N. Rosen, Phys. Rev. **47**(10), 777 (1935). https://doi.org/10.1103/PhysRev.47.777 240
79. N. Bohr, Phys. Rev. **48**(8), 696 (1935). https://doi.org/10.1103/PhysRev.48.696 240
80. J.S. Bell, Phys. Phys. Fiz. **1**, 195 (1964) 240, 269
81. J. Clauser, M. Horne, A. Shimony, R. Holt, Phys. Rev. Lett. **23**, 880 (1969). https://doi.org/10.1103/PhysRevLett.23.880 240
82. N.P.O.A. 2024, The Nobel prize in physics 2022 (2022). https://www.nobelprize.org/prizes/physics/2022/summary/ 241
83. S. Pironio, A. Acín, S. Massar, A. de la Giroday, D. Matsukevich, P. Maunz, S. Olmschenk, D. Hayes, L. Luo, T. Manning, C. Monroe, Nature **464**, 1021 (2010). https://doi.org/10.1038/nature09008 241
84. B. Terhal, Phys. Lett. A **271**, 319 (2000). https://doi.org/10.1016/S0375-9601(00)00401-1 241
85. S. Wiesner, ACM SIGACT News **15**, 78 (1983). https://doi.org/10.1145/1008908.1008920 241
86. A. Ekert, Phys. Rev. Lett. **67**, 661 (1991). https://doi.org/10.1103/PhysRevLett.67.661 241
87. D. Jonathan, M. Plenio, Phys. Rev. Lett. **83**, 3566 (1999). https://doi.org/10.1103/PhysRevLett.83.3566 241
88. C. Bennett, G. Brassard, C. Crépeau, R. Jozsa, A. Peres, W. Wootters, Phys. Rev. Lett. **70**, 1895 (1993). https://doi.org/10.1103/PhysRevLett.70.1895 241
89. R. Cleve, D. Gottesman, H. Lo, Phys. Rev. Lett. **83**, 648 (1999). https://doi.org/10.1103/PhysRevLett.83.648 241
90. Y.B. Sheng, L. Zhou, G.L. Long, Sci. Bull. **67**(4), 367–374 (2022). https://doi.org/10.1016/j.scib.2021.11.002 242
91. H. Briegel, W. Dür, J. Cirac, P. Zoller, Phys. Rev. Lett. **81**, 5932 (1998). https://doi.org/10.1103/PhysRevLett.81.5932 242
92. C. Mora, H. Briegel, Phys. Rev. Lett. **95** (2005). https://doi.org/10.1103/PhysRevLett.95.200503 242
93. M. Schleier-Smith, I. Leroux, V. Vuletić, Phys. Rev. Lett. **104** (2010). https://doi.org/10.1103/PhysRevLett.104.073604 242
94. C. Gross, T. Zibold, E. Nicklas, J. Estève, M. Oberthaler, Nature **464**, 1165 (2010). https://doi.org/10.1038/nature08919
95. M. Riedel, P. Böhi, Y. Li, T. Hänsch, A. Sinatra, P. Treutlein, Nature **464**, 1170 (2010). https://doi.org/10.1038/nature08988 242
96. Y.B. Sheng, L. Zhou, Sci. Bull. **62**(14), 1025–1029 (2017). https://doi.org/10.1016/j.scib.2017.06.007 242
97. A. Dawid, J. Arnold, B. Requena, A. Gresch, M. Płodzień, K. Donatella, K.A. Nicoli, P. Stornati, R. Koch, M. Büttner, R. Okuła, G. Muñoz-Gil, R.A. Vargas-Hernández, A. Cervera-Lierta, J. Carrasquilla, V. Dunjko, M. Gabrié, P. Huembeli, E. van Nieuwenburg, F. Vicentini, L. Wang, S.J. Wetzel, G. Carleo, E. Greplová, R. Krems, F. Marquardt, M. Tomza, M. Lewenstein, A. Dauphin, Modern applications of machine learning in quantum sciences (2022). https://doi.org/10.48550/ARXIV.2204.04198. https://arxiv.org/abs/2204.04198 242, 265
98. C. Bennett, G. Brassard, S. Popescu, B. Schumacher, J. Smolin, W. Wootters, Phys. Rev. Lett. **76**, 722 (1996). https://doi.org/10.1103/PhysRevLett.76.722 242
99. C. Bennett, D. DiVincenzo, J. Smolin, W. Wootters, Phys. Rev. A **54**, 3824 (1996). https://doi.org/10.1103/PhysRevA.54.3824 248
100. D. Deutsch, A. Ekert, R. Jozsa, C. Macchiavello, S. Popescu, A. Sanpera, Phys. Rev. Lett. **77**, 2818 (1996). https://doi.org/10.1103/PhysRevLett.77.2818 242
101. J. Pan, C. Simon, Č. Brukner, A. Zeilinger, Nature **410**, 1067 (2001). https://doi.org/10.1038/35074041 242

102. Y. Sheng, F. Deng, H. Zhou, Phys. Rev. A **77** (2008). https://doi.org/10.1103/PhysRevA.77.042308
103. M. Zwerger, H. Briegel, W. Dür, Phys. Rev. Lett. **110** (2013). https://doi.org/10.1103/PhysRevLett.110.260503
104. M. Zwerger, H. Briegel, W. Dür, Phys. Rev. A **90** (2014). https://doi.org/10.1103/PhysRevA.90.012314
105. P. Yan, L. Zhou, W. Zhong, Y. Sheng, Front. Phys. **17** (2021). https://doi.org/10.1007/s11467-021-1103-8
106. H. Qin, M. Du, X. Li, W. Zhong, L. Zhou, Y. Sheng, Efficient multipartite entanglement purification with non-identical states (2023). https://doi.org/10.48550/ARXIV.2311.10250. https://arxiv.org/abs/2311.10250 242
107. P. Kwiat, S. Barraza-Lopez, A. Stefanov, N. Gisin, Nature **409**, 1014 (2001). https://doi.org/10.1038/35059017 242
108. J. Pan, S. Gasparoni, R. Ursin, G. Weihs, A. Zeilinger, Nature **423**, 417 (2003). https://doi.org/10.1038/nature01623
109. Z. Zhao, T. Yang, Y. Chen, A. Zhang, J. Pan, Phys. Rev. Lett. **90** (2003). https://doi.org/10.1103/PhysRevLett.90.207901
110. T. Yamamoto, M. Koashi, Ş.K. Özdemir, N. Imoto, Nature **421**, 343 (2003). https://doi.org/10.1038/nature01358
111. P. Walther, K. Resch, Č. Brukner, A. Steinberg, J. Pan, A. Zeilinger, Phys. Rev. Lett. **94** (2005). https://doi.org/10.1103/PhysRevLett.94.040504
112. R. Reichle, D. Leibfried, E. Knill, J. Britton, R. Blakestad, J. Jost, C. Langer, R. Ozeri, S. Seidelin, D. Wineland, Nature **443**, 838 (2006). https://doi.org/10.1038/nature05146
113. N. Kalb, A. Reiserer, P. Humphreys, J. Bakermans, S. Kamerling, N. Nickerson, S. Benjamin, D. Twitchen, M. Markham, R. Hanson, Science **356**, 928 (2017). https://doi.org/10.1126/science.aan0070
114. X. Hu, C. Huang, Y. Sheng, L. Zhou, B. Liu, Y. Guo, C. Zhang, W. Xing, Y. Huang, C. Li, G. Guo, Phys. Rev. Lett. **126** (2021). https://doi.org/10.1103/PhysRevLett.126.010503
115. S. Ecker, P. Sohr, L. Bulla, M. Huber, M. Bohmann, R. Ursin, Phys. Rev. Lett. **127** (2021). https://doi.org/10.1103/PhysRevLett.127.040506
116. H. Yan, Y. Zhong, H. Chang, A. Bienfait, M. Chou, C. Conner, É. Dumur, J. Grebel, R. Povey, A. Cleland, Phys. Rev. Lett. **128** (2022). https://doi.org/10.1103/PhysRevLett.128.080504 242
117. M. Murao, M. Plenio, S. Popescu, V. Vedral, P. Knight, Phys. Rev. A **57**, R4075 (1998). https://doi.org/10.1103/PhysRevA.57.R4075 242
118. Y. Sheng, F. Deng, B. Zhao, T. Wang, H. Zhou, Eur. Phys. J. D **55**, 235 (2009). https://doi.org/10.1140/epjd/e2009-00237-y 242
119. P. Horodecki, L. Rudnicki, K. Życzkowski, PRX Quantum **3** (2022). https://doi.org/10.1103/prxquantum.3.010101 242, 273
120. P. Horodecki, M. Horodecki, R. Horodecki, Phys. Rev. Lett. **82**, 1056 (1999). https://doi.org/10.1103/physrevlett.82.1056 242
121. K.F. Pál, G. Tóth, E. Bene, T. Vértesi, Phys. Rev. Res. **3** (2021). https://doi.org/10.1103/physrevresearch.3.023101 242
122. T. Moroder, O. Gittsovich, M. Huber, O. Gühne, Phys. Rev. Lett. **113** (2014). https://doi.org/10.1103/PhysRevLett.113.050404 242
123. K. Horodecki, M. Horodecki, P. Horodecki, J. Oppenheim, Phys. Rev. Lett. **94** (2005). https://doi.org/10.1103/PhysRevLett.94.160502 242
124. K. Horodecki, L. Pankowski, M. Horodecki, P. Horodecki, IEEE Trans. Inf. Theory **54**, 2621 (2008). https://doi.org/10.1109/TIT.2008.921709 242
125. L. Masanes, Phys. Rev. Lett. **96** (2006). https://doi.org/10.1103/PhysRevLett.96.150501 242
126. O. Rudolph, Phys. Rev. A **67** (2003). https://doi.org/10.1103/PhysRevA.67.032312 242
127. B. Hiesmayr, W. Löffler, New J. Phys. **15**, 083036 (2013). https://doi.org/10.1088/1367-2630/15/8/083036 242

128. J. Bae, A. Bera, D. Chruściński, B. Hiesmayr, D. McNulty, J. Phys. A Math. Theor. **55**, 505303 (2022). https://doi.org/10.1088/1751-8121/acaa16 242
129. G. Vidal, J. Mod. Opt. **47**(2–3), 355 (2000). https://www.tandfonline.com/doi/abs/10.1080/09500340008244048 242
130. C.H. Bennett, S. Popescu, D. Rohrlich, J.A. Smolin, A.V. Thapliyal, Phys. Rev. A **63**, 012307 (2000). https://doi.org/10.1103/PhysRevA.63.012307. https://link.aps.org/doi/10.1103/PhysRevA.63.012307 246
131. D. Bruß, J. Math. Phys. **43**(9), 4237 (2002). https://doi.org/10.1063/1.1494474
132. R. Augusiak, F.M. Cucchietti, M. Lewenstein, *Many-Body Physics from a Quantum Information Perspective* (Springer Berlin Heidelberg, Berlin, Heidelberg, 2012), pp. 245–294. https://doi.org/10.1007/978-3-642-10449-7_6
133. R. Bertlmann, N. Friis, *Modern Quantum Theory: From Quantum Mechanics to Entanglement and Quantum Information* (Oxford University Press, Oxford, 2023). https://doi.org/10.1093/oso/9780199683338.001.0001 242
134. C.H. Bennett, D.P. DiVincenzo, T. Mor, P.W. Shor, J.A. Smolin, B.M. Terhal, Phys. Rev. Lett. **82**(26), 5385 (1999). https://doi.org/10.1103/PhysRevLett.82.5385 243, 245, 253, 254
135. W. Helwig, W. Cui, J.I. Latorre, A. Riera, H. Lo, Phys. Rev. A **86**(5) (2012). https://doi.org/10.1103/PhysRevA.86.052335 244
136. D. Goyeneche, D. Alsina, J.I. Latorre, A. Riera, K. Życzkowski, Phys. Rev. A **92**(3) (2015). https://doi.org/10.1103/PhysRevA.92.032316 244
137. F. Pastawski, B. Yoshida, D. Harlow, J. Preskill, J. High Energy Phys. **2015**(6) (2015). https://doi.org/10.1007/JHEP06(2015)149 244
138. S.A. Rather, N. Ramadas, V. Kodiyalam, A. Lakshminarayan, Phys. Rev. A **108**(3) (2023). https://doi.org/10.1103/PhysRevA.108.032412 244, 272
139. F. Huber, C. Eltschka, J. Siewert, O. Gühne, J. Phys. A Math. Theor. **51**(17), 175301 (2018). https://doi.org/10.1088/1751-8121/aaade5 244
140. S.A. Rather, A. Burchardt, W. Bruzda, G. Rajchel-Mieldzioć, A. Lakshminarayan, K. Życzkowski, Phys. Rev. Lett. **128**(8) (2022). https://doi.org/10.1103/PhysRevLett.128.080507 244
141. R.F. Werner, Phys. Rev. A **40**, 4277 (1989). https://link.aps.org/doi/10.1103/PhysRevA.40.4277 245, 269
142. W. Dür, J.I. Cirac, Phys. Rev. A **61**(4) (2000). https://doi.org/10.1103/PhysRevA.61.042314 245, 250, 253
143. E. Chitambar, G. Gour, Rev. Mod. Phys. **91**(2) (2019). https://doi.org/10.1103/RevModPhys.91.025001 245
144. P. Contreras-Tejada, C. Palazuelos, J.I. de Vicente, Phys. Rev. Lett. **122**(12) (2019). https://doi.org/10.1103/PhysRevLett.122.120503
145. K. Schwaiger, D. Sauerwein, M. Cuquet, J. de Vicente, B. Kraus, Phys. Rev. Lett. **115** (2015). https://doi.org/10.1103/PhysRevLett.115.150502 245
146. K. Kuroiwa, R. Takagi, G. Adesso, H. Yamasaki, Every quantum helps: Operational advantage of quantum resources beyond convexity. Phys. Rev. Lett. **132** (2024). https://doi.org/10.1103/PhysRevLett.132.150201 246
147. R. Salazar, J. Czartowski, R.R. Rodríguez, G. Rajchel-Mieldzioć, P. Horodecki, K. Życzkowski, Quantum resource theories beyond convexity. arXiv (2024). https://arxiv.org/abs/2405.05785 246
148. N. Linden, S. Popescu, S. Popescu, Fortschr. Phys. **46**(4–5), 567 (1998). https://doi.org/10.1002/(SICI)1521-3978(199806)46:4/5<567::AID-PROP567>3.0.CO;2-H 247
149. H.A. Carteret, A. Higuchi, A. Sudbery, J. Math. Phys. **41**(12), 7932 (2000). https://doi.org/10.1063/1.1319516 247
150. A. Peres, Phys. Lett. A **202**(1), 16 (1995). https://doi.org/10.1016/0375-9601(95)00315-T 247
151. A.V. Thapliyal, Phys. Rev. A **59**(5), 3336 (1999). https://doi.org/10.1103/PhysRevA.59.3336 247

152. P. Zanardi, C. Zalka, L. Faoro, Phys. Rev. A **62** (2000). https://doi.org/10.1103/PhysRevA.62.030301 247
153. J. Eisert, Phys. Rev. Lett. **127** (2021). https://doi.org/10.1103/PhysRevLett.127.020501 247
154. T. Linowski, G. Rajchel-Mieldzioć, K. Życzkowski, J. Phys. A Math. Theor. **53**, 125303 (2020). https://doi.org/10.1088/1751-8121/ab749a 247
155. L. Clarisse, S. Ghosh, S. Severini, A. Sudbery, Phys. Lett. A **365**, 400 (2007). https://doi.org/10.1016/j.physleta.2007.02.001 247
156. N. Linden, J. Smolin, A. Winter, Phys. Rev. Lett. **103** (2009). https://doi.org/10.1103/PhysRevLett.103.030501 247
157. E. Chitambar, D. Leung, L. Mančinska, M. Ozols, A. Winter, Commun. Math. Phys. **328**(1), 303 (2014). https://doi.org/10.1007/s00220-014-1953-9 248, 249
158. C. Bennett, D. DiVincenzo, C. Fuchs, T. Mor, E. Rains, P. Shor, J. Smolin, W. Wootters, Phys. Rev. A **59**, 1070 (1999). https://doi.org/10.1103/PhysRevA.59.1070 248
159. M.J. Donald, M. Horodecki, O. Rudolph, J. Math. Phys. **43**(9), 4252 (2002). https://doi.org/10.1063/1.1495917 248
160. C.H. Bennett, H.J. Bernstein, S. Popescu, B. Schumacher, Phys. Rev. A **53**(4), 2046 (1996). https://doi.org/10.1103/PhysRevA.53.2046 248
161. V. Vedral, M.B. Plenio, M.A. Rippin, P.L. Knight, Phys. Rev. Lett. **78**(12), 2275 (1997). https://doi.org/10.1103/PhysRevLett.78.2275 248
162. E. Chitambar, Phys. Rev. Lett. **107**(19) (2011). https://doi.org/10.1103/PhysRevLett.107.190502 248
163. M.A. Nielsen, Phys. Rev. Lett. **83**(2), 436 (1999). https://doi.org/10.1103/PhysRevLett.83.436 249, 272
164. G. Gour, N.R. Wallach, New J. Phys. **13**(7), 073013 (2011). https://doi.org/10.1088/1367-2630/13/7/073013 249, 272
165. W. Dür, G. Vidal, J.I. Cirac, Phys. Rev. A **62**(6) (2000). https://doi.org/10.1103/PhysRevA.62.062314 249, 250, 251, 252, 253
166. F. Verstraete, J. Dehaene, B. De Moor, H. Verschelde, Phys. Rev. A **65**(5) (2002). https://doi.org/10.1103/PhysRevA.65.052112 249, 254
167. A. Acín, A. Andrianov, E. Jané, R. Tarrach, J. Phys. A Math. Gen. **34**(35), 6725 (2001). https://doi.org/10.1088/0305-4470/34/35/301 250, 251, 252, 253
168. A. Acín, D. Bruß, M. Lewenstein, A. Sanpera, Phys. Rev. Lett. **87**(4) (2001). https://doi.org/10.1103/PhysRevLett.87.040401 250, 253, 254
169. I.M. Gelfand, M.M. Kapranov, A.V. Zelevinsky, *Discriminants, Resultants, and Multidimensional Determinants* (Birkhäuser Boston, 1994). https://doi.org/10.1007/978-0-8176-4771-1 251
170. A. Miyake, Phys. Rev. A **67**(1) (2003). https://doi.org/10.1103/PhysRevA.67.012108 251
171. V. Coffman, J. Kundu, W.K. Wootters, Phys. Rev. A **61**(5) (2000). https://doi.org/10.1103/PhysRevA.61.052306 251, 252
172. A. Osterloh, J. Siewert, Int. J. Quantum Inf. **04**(03), 531 (2006). https://doi.org/10.1142/S0219749906001980 251
173. Y. Makhlin, Quantum Inf. Process. **1**(4), 243 (2002). https://doi.org/10.1023/A:1022144002391 251
174. D.M. Greenberger, M.A. Horne, A. Zeilinger, *Going Beyond Bell's Theorem* (Springer Netherlands, 1989), pp. 69–72. https://doi.org/10.1007/978-94-017-0849-4_10 252
175. P. Vrana, M. Christandl, J. Math. Phys. **56**(2) (2015). https://doi.org/10.1063/1.4908106 252
176. M.B. Plenio, V. Vedral, J. Phys. A Math. Gen. **34**(35), 6997 (2001). https://doi.org/10.1088/0305-4470/34/35/325 253
177. J. Eisert, H.J. Briegel, Phys. Rev. A **64**(2) (2001). https://doi.org/10.1103/PhysRevA.64.022306 253
178. A. Sanpera, D. Bruß, M. Lewenstein, Phys. Rev. A **63**(5) (2001). https://doi.org/10.1103/PhysRevA.63.050301 253
179. M. Płodzień, J. Chwedeńczuk, M. Lewenstein, G. Rajchel-Mieldzioć, Entanglement classification and *non-k*-separability certification via Greenberger-Horne-Zeilinger-class fidelity. arXiv (2024). https://arxiv.org/abs/2406.10662 253

180. M. Walter, B. Doran, D. Gross, M. Christandl, Science **340**, 1205 (2013). https://doi.org/10.1126/science.1232957 253
181. A. Sawicki, M. Walter, M. Kuś, J. Phys. A Math. Theor. **46**, 055304 (2013). https://doi.org/10.1088/1751-8113/46/5/055304
182. A. Sawicki, M. Oszmaniec, M. Kuś, Phys. Rev. A **86** (2012). https://doi.org/10.1103/PhysRevA.86.040304
183. A. Sawicki, M. Oszmaniec, M. Kuś, Rev. Math. Phys. **26**, 1450004 (2014). https://doi.org/10.1142/S0129055X14500044
184. T. Maciążek, A. Sawicki, J. Phys. A Math. Theor. **51**, 07LT01 (2018). https://doi.org/10.1088/1751-8121/aaa4d7
185. M. Enríquez, F. Delgado, K. Życzkowski, Entropy **20**, 745 (2018). https://doi.org/10.3390/e20100745 253
186. M.H. Anderson, J.R. Ensher, M.R.. Matthews, C.E. Wieman, E.A. Cornell, Science **269**, 198 (1995). https://doi.org/10.1126/science.269.5221.198 255
187. B. Zeng, X. Chen, D.L. Zhou, X.G. Wen, *Quantum Information Meets Quantum Matter: From Quantum Entanglement to Topological Phases of Many-Body Systems* (Springer New York, 2019). https://doi.org/10.1007/978-1-4939-9084-9 255
188. S. Sachdev, *Quantum Phases of Matter* (Cambridge University Press/Harvard University, Massachusetts, 2023). https://doi.org/10.1017/9781009212717 255
189. B. Lücke, J. Peise, G. Vitagliano, J. Arlt, L. Santos, G. Tóth, C. Klempt, Phys. Rev. Lett. **112** (2014). https://doi.org/10.1103/physrevlett.112.155304 257, 264
190. J. Eisert, M. Cramer, M.B. Plenio, Rev. Mod. Phys. **82**(1), 277 (2010). https://doi.org/10.1103/RevModPhys.82.277 258
191. D.N. Page, Phys. Rev. Lett. **71**, 1291 (1993). https://doi.org/10.1103/physrevlett.71.1291 258
192. E. Bianchi, L. Hackl, M. Kieburg, M. Rigol, L. Vidmar, PRX Quantum **3**(3) (2022). https://doi.org/10.1103/PRXQuantum.3.030201 258
193. M. Srednicki, Entropy and area. Phys. Rev. Lett. **71**(5), 666–669 (1993). https://doi.org/10.1103/PhysRevLett.71.666 258
194. L. Bombelli, R.K. Koul, J. Lee, R.D. Sorkin, Quantum source of entropy for black holes. Phys. Rev. D **34**(2), 373–383 (1986). https://doi.org/10.1103/PhysRevD.34.373 258
195. K. Audenaert, J. Eisert, M.B. Plenio, R.F. Werner, Entanglement properties of the harmonic chain. Phys. Rev. A **66**(4) (2002). https://doi.org/10.1103/PhysRevA.66.042327 258
196. M.B. Hastings, J. Stat. Mech. Theory Exp. **2007**, P08024 (2007). https://doi.org/10.1088/1742-5468/2007/08/p08024 258
197. I. Arad, Z. Landau, U. Vazirani, Phys. Rev. B **85** (2012). https://doi.org/10.1103/PhysRevB.85.195145 259
198. I. Arad, A. Kitaev, Z. Landau, U. Vazirani, An area law and sub-exponential algorithm for 1d systems (2013). https://arxiv.org/abs/1301.1162 259
199. F.G.S.L. Brandão, M. Horodecki, Commun. Math. Phys. **333**, 761 (2014). https://doi.org/10.1007/s00220-014-2213-8 259
200. A. Anshu, I. Arad, D. Gosset, *Proceedings of the 54th Annual ACM SIGACT Symposium on Theory of Computing* (ACM, 2022), STOC '22. https://doi.org/10.1145/3519935.3519962 259
201. M.B. Plenio, J. Eisert, J. Dreißig, M. Cramer, Entropy, entanglement, and area: analytical results for harmonic lattice systems. Phys. Rev. Lett. **94**(6). https://doi.org/10.1103/PhysRevLett.94.060503 259, 272
202. M. Cramer, J. Eisert, M.B. Plenio, J. Dreißig, Phys. Rev. A **73** (2006). https://doi.org/10.1103/PhysRevA.73.012309 259, 272
203. D. Perez-Garcia, F. Verstraete, M.M. Wolf, J.I. Cirac, arXiv preprint quant-ph/0608197 (2006). https://arxiv.org/abs/quant-ph/0608197 259
204. G. Vidal, Phys. Rev. Lett. **91** (2003). https://doi.org/10.1103/PhysRevLett.91.147902 259
205. G. Vidal, Phys. Rev. Lett. **93**, 040502 (2004). https://doi.org/10.1103/PhysRevLett.93.040502. https://link.aps.org/doi/10.1103/PhysRevLett.93.040502 259

206. F. Verstraete, J.I. Cirac, Phys. Rev. B **73**(9) (2006). https://doi.org/10.1103/physrevb.73.094423 260
207. S.R. White, Phys. Rev. Lett. **69**(19), 2863 (1992). https://doi.org/10.1103/physrevlett.69.2863 260
208. U. Schollwöck, Rev. Mod. Phys. **77**(1), 259 (2005). https://doi.org/10.1103/revmodphys.77.259 260
209. I. Affleck, T. Kennedy, E.H. Lieb, H. Tasaki, Phys. Rev. Lett. **59**, 799 (1987). https://link.aps.org/doi/10.1103/PhysRevLett.59.799 260
210. M. Fannes, B. Nachtergaele, R.F. Werner, J. Phys. A Math. Gen. **24**(4), L185 (1991). https://doi.org/10.1088/0305-4470/24/4/005 260
211. L. Tagliacozzo, A. Celi, M. Lewenstein, Phys. Rev. X **4**(4) (2014). https://doi.org/10.1103/physrevx.4.041024 261
212. F. Verstraete, J.I. Cirac, Renormalization algorithms for quantum-many body systems in two and higher dimensions (2004). https://doi.org/10.48550/ARXIV.COND-MAT/0407066. https://arxiv.org/abs/cond-mat/0407066 261
213. F. Pollmann, A.M. Turner, E. Berg, M. Oshikawa, Phys. Rev. B **81**(6) (2010). https://doi.org/10.1103/PhysRevB.81.064439 261
214. X. Chen, Z.C. Gu, X.G. Wen, Phys. Rev. B **84**(23) (2011). https://doi.org/10.1103/PhysRevB.84.235128 261
215. P. Calabrese, J. Cardy, J. Stat. Mech. Theory Exp. **2004**(06), P06002 (2004). https://doi.org/10.1088/1742-5468/2004/06/P06002 261
216. D. Gottesman, M.B. Hastings, New J. Phys. **12**, 025002 (2010). https://doi.org/10.1088/1367-2630/12/2/025002 261
217. P. Calabrese, J. Cardy, J. Stat. Mech. Theory Exp. **2005**(04), P04010 (2005). https://doi.org/10.1088/1742-5468/2005/04/P04010 261
218. N. Schuch, M.M. Wolf, K.G.H. Vollbrecht, J.I. Cirac, New J. Phys. **10**(3), 033032 (2008). https://doi.org/10.1088/1367-2630/10/3/033032 261
219. E. Leviatan, F. Pollmann, J.H. Bardarson, D.A. Huse, E. Altman, Quantum thermalization dynamics with matrix-product states (2017). https://doi.org/10.48550/ARXIV.1702.08894. https://arxiv.org/abs/1702.08894 261
220. M. Frías-Pérez, L. Tagliacozzo, M.C. Bañuls, Converting long-range entanglement into mixture: tensor-network approach to local equilibration (2023). https://arxiv.org/abs/2308.04291 261
221. A. Aloy, M. Fadel, J. Tura, New J. Phys. **23**(3), 033026 (2021). https://doi.org/10.1088/1367-2630/abe15e 262
222. N. Yu, Phys. Rev. A **94**(6) (2016). https://doi.org/10.1103/PhysRevA.94.060101 263
223. R. Quesada, S. Rana, A. Sanpera, Phys. Rev. A **95**(4) (2017). https://doi.org/10.1103/PhysRevA.95.042128 263
224. A. Rutkowski, M. Banacki, M. Marciniak, Phys. Rev. A **99**(2) (2019). https://doi.org/10.1103/PhysRevA.99.022309 263, 273
225. Y. Kawaguchi, M. Ueda, Phys. Rep. **520**(5), 253 (2012). https://doi.org/10.1016/j.physrep.2012.07.005 264, 270
226. R.H. Dicke, Phys. Rev. **93**, 99 (1954). https://doi.org/10.1103/PhysRev.93.99. https://link.aps.org/doi/10.1103/PhysRev.93.99 264
227. H.J. Lipkin, Phys. Rev. **110**, 1395 (1958). https://doi.org/10.1103/PhysRev.110.1395. https://link.aps.org/doi/10.1103/PhysRev.110.1395 264
228. A. Aloy, G. Müller-Rigat, M. Lewenstein, J. Tura, M. Fadel, Bell inequalities as a tool to probe quantum chaos. arXiv (2024). https://arxiv.org/abs/2406.11791 264
229. Y.Q. Zou, L.N. Wu, Q. Liu, X.Y. Luo, S.F. Guo, J.H. Cao, M.K. Tey, L. You, Proc. Natl. Acad. Sci. **115**(25), 6381 (2018). https://doi.org/10.1073/pnas.1715105115 264
230. L. Pezzè, A. Smerzi, M.K. Oberthaler, R. Schmied, P. Treutlein, Rev. Mod. Phys. **90**, 035005 (2018). https://link.aps.org/doi/10.1103/RevModPhys.90.035005 264
231. R.J. Sewell, M. Koschorreck, M. Napolitano, B. Dubost, N. Behbood, M.W. Mitchell, Phys. Rev. Lett. **109**, 253605 (2012). https://doi.org/10.1103/PhysRevLett.109.253605. https://link.aps.org/doi/10.1103/PhysRevLett.109.253605 264, 265

232. M. Tse, H. Yu, N. Kijbunchoo, A. Fernandez-Galiana, P. Dupej, L. Barsotti, C.D. Blair, D.D. Brown, S.E. Dwyer, A. Effler, M. Evans, P. Fritschel, V.V. Frolov, A.C. Green, G.L. Mansell, F. Matichard, N. Mavalvala, D.E. McClelland, L. McCuller, T. McRae, J. Miller, A. Mullavey, E. Oelker, I.Y. Phinney, D. Sigg, B.J.J. Slagmolen, T. Vo, R.L. Ward, C. Whittle, R. Abbott, C. Adams, R.X. Adhikari, A. Ananyeva, S. Appert, K. Arai, J.S. Areeda, Y. Asali, S.M. Aston, C. Austin, A.M. Baer, M. Ball, S.W. Ballmer, S. Banagiri, D. Barker, J. Bartlett, B.K. Berger, J. Betzwieser, D. Bhattacharjee, G. Billingsley, S. Biscans, R.M. Blair, N. Bode, P. Booker, R. Bork, A. Bramley, A.F. Brooks, A. Buikema, C. Cahillane, K.C. Cannon, X. Chen, A.A. Ciobanu, F. Clara, S.J. Cooper, K.R. Corley, S.T. Countryman, P.B. Covas, D.C. Coyne, L.E.H. Datrier, D. Davis, C. Di Fronzo, J.C. Driggers, T. Etzel, T.M. Evans, J. Feicht, P. Fulda, M. Fyffe, J.A. Giaime, K.D. Giardina, P. Godwin, E. Goetz, S. Gras, C. Gray, R. Gray, A. Gupta, E.K. Gustafson, R. Gustafson, J. Hanks, J. Hanson, T. Hardwick, R.K. Hasskew, M.C. Heintze, A.F. Helmling-Cornell, N.A. Holland, J.D. Jones, S. Kandhasamy, S. Karki, M. Kasprzack, K. Kawabe, P.J. King, J.S. Kissel, R. Kumar, M. Landry, B.B. Lane, B. Lantz, M. Laxen, Y.K. Lecoeuche, J. Leviton, J. Liu, M. Lormand, A.P. Lundgren, R. Macas, M. MacInnis, D.M. Macleod, S. Márka, Z. Márka, D.V. Martynov, K. Mason, T.J. Massinger, R. McCarthy, S. McCormick, J. McIver, G. Mendell, K. Merfeld, E.L. Merilh, F. Meylahn, T. Mistry, R. Mittleman, G. Moreno, C.M. Mow-Lowry, S. Mozzon, T.J.N. Nelson, P. Nguyen, L.K. Nuttall, J. Oberling, R.J. Oram, B. O'Reilly, C. Osthelder, D.J. Ottaway, H. Overmier, J.R. Palamos, W. Parker, E. Payne, A. Pele, C.J. Perez, M. Pirello, H. Radkins, K.E. Ramirez, J.W. Richardson, K. Riles, N.A. Robertson, J.G. Rollins, C.L. Romel, J.H. Romie, M.P. Ross, K. Ryan, T. Sadecki, E.J. Sanchez, L.E. Sanchez, T.R. Saravanan, R.L. Savage, D. Schaetzl, R. Schnabel, R.M.S. Schofield, E. Schwartz, D. Sellers, T.J. Shaffer, J.R. Smith, S. Soni, B. Sorazu, A.P. Spencer, K.A. Strain, L. Sun, M.J. Szczepańczyk, M. Thomas, P. Thomas, K.A. Thorne, K. Toland, C.I. Torrie, G. Traylor, A.L. Urban, G. Vajente, G. Valdes, D.C. Vander-Hyde, P.J. Veitch, K. Venkateswara, G. Venugopalan, A.D. Viets, C. Vorvick, M. Wade, J. Warner, B. Weaver, R. Weiss, B. Willke, C.C. Wipf, L. Xiao, H. Yamamoto, M.J. Yap, H. Yu, L. Zhang, M.E. Zucker, J. Zweizig, Phys. Rev. Lett. **123**, 231107 (2019). https://link.aps.org/doi/10.1103/PhysRevLett.123.231107 264, 265
233. T.M. Stace, Phys. Rev. A **82**(1) (2010). https://doi.org/10.1103/PhysRevA.82.011611 264
234. L.A. Correa, M. Mehboudi, G. Adesso, A. Sanpera, Phys. Rev. Lett. **114**(22) (2015). https://doi.org/10.1103/PhysRevLett.114.220405
235. A.K. Srivastava, U. Bhattacharya, M. Lewenstein, M. Płodzień. Topological quantum thermometry (2023). https://arxiv.org/abs/2311.14524 264
236. F. Fröwis, R. Schmied, N. Gisin, Phys. Rev. A **92**, 012102 (2015). https://link.aps.org/doi/10.1103/PhysRevA.92.012102 265
237. S.L. Braunstein, C.M. Caves, Phys. Rev. Lett. **72**, 3439 (1994). https://link.aps.org/doi/10.1103/PhysRevLett.72.3439 265
238. T. Martos, M. Lewenstein, G. Rajchel-Mieldzioć, P. Stammer. Metrological robustness of high photon number optical cat states (2023). https://doi.org/10.48550/ARXIV.2311.01371. https://arxiv.org/abs/2311.01371 265
239. P. Hyllus, W. Laskowski, R. Krischek, C. Schwemmer, W. Wieczorek, H. Weinfurter, L. Pezzé, A. Smerzi, Phys. Rev. A **85**, 022321 (2012). https://doi.org/10.1103/PhysRevA.85.022321. https://link.aps.org/doi/10.1103/PhysRevA.85.022321 265
240. G. Tóth, Phys. Rev. A **85**, 022322 (2012). https://doi.org/10.1103/PhysRevA.85.022322. https://link.aps.org/doi/10.1103/PhysRevA.85.022322 265
241. A. Palmieri, G. Müller-Rigat, A. Srivastava, M. Lewenstein, G. Rajchel-Mieldzioć, M. Płodzień, Enhancing quantum state tomography via resource-efficient attention-based neural networks (2023). https://doi.org/10.48550/ARXIV.2309.10616. https://arxiv.org/abs/2309.10616 265
242. D.J. Wineland, J.J. Bollinger, W.M. Itano, D.J. Heinzen, Phys. Rev. A **50**, 67 (1994). https://doi.org/10.1103/PhysRevA.50.67. https://link.aps.org/doi/10.1103/PhysRevA.50.67 266
243. L.M. Duan, J.I. Cirac, P. Zoller, Phys. Rev. A **65**, 033619 (2002). https://doi.org/10.1103/PhysRevA.65.033619. https://link.aps.org/doi/10.1103/PhysRevA.65.033619 266

244. I. Frérot, M. Fadel, M. Lewenstein, Rep. Prog. Phys. **86**(11), 114001 (2023). https://doi.org/10.1088/1361-6633/acf8d7 266
245. N. Friis, G. Vitagliano, M. Malik, M. Huber, Nat. Rev. Phys. **1**(1), 72–87 (2018). https://doi.org/10.1038/s42254-018-0003-5 266
246. I. Frérot, T. Roscilde, Phys. Rev. Lett. **126**, 140504 (2021). https://doi.org/10.1103/PhysRevLett.126.140504. https://link.aps.org/doi/10.1103/PhysRevLett.126.140504 267, 270, 271
247. G. Müller-Rigat, A. Aloy, M. Lewenstein, I. Frérot, PRX Quantum **2**, 030329 (2021). https://doi.org/10.1103/PRXQuantum.2.030329. https://link.aps.org/doi/10.1103/PRXQuantum.2.030329 267, 271
248. G. Vitagliano, P. Hyllus, I.n.L. Egusquiza, G. Tóth, Phys. Rev. Lett. **107**, 240502 (2011). https://doi.org/10.1103/PhysRevLett.107.240502. https://link.aps.org/doi/10.1103/PhysRevLett.107.240502 268
249. V. Scarani, Preprint. arXiv:1303.3081 (2013). https://doi.org/10.48550/ARXIV.1303.3081. https://arxiv.org/abs/1303.3081 269
250. N. Brunner, D. Cavalcanti, S. Pironio, V. Scarani, S. Wehner, Rev. Mod. Phys. **86**, 419 (2014). https://doi.org/10.1103/RevModPhys.86.419. https://link.aps.org/doi/10.1103/RevModPhys.86.419 269
251. R. Schmied, J.D. Bancal, B. Allard, M. Fadel, V. Scarani, P. Treutlein, N. Sangouard, Science **352**(6284), 441 (2016). https://doi.org/10.1126/science.aad8665 270, 271
252. N.J. Engelsen, R. Krishnakumar, O. Hosten, M.A. Kasevich, Phys. Rev. Lett. **118**, 140401 (2017). https://doi.org/10.1103/PhysRevLett.118.140401. https://link.aps.org/doi/10.1103/PhysRevLett.118.140401 270
253. M. Płodzień, M. Kościelski, E. Witkowska, A. Sinatra, Phys. Rev. A **102** (2020). https://doi.org/10.1103/PhysRevA.102.013328 270
254. M. Płodzień, M. Lewenstein, E. Witkowska, J. Chwedeńczuk, Phys. Rev. Lett. **129**(25) (2022). https://doi.org/10.1103/PhysRevLett.129.250402
255. M. Płodzień, T. Wasak, E. Witkowska, M. Lewenstein, J. Chwedeńczuk, Generation of scalable many-body bell correlations in spin chains with short-range two-body interactions (2023). https://doi.org/10.48550/ARXIV.2306.06173. https://arxiv.org/abs/2306.06173
256. T. Hernández Yanes, M. Płodzień, M. Mackoit Sinkevičienė, G. Žlabys, G. Juzeliūnas, E. Witkowska, Phys. Rev. Lett. **129** (2022). https://doi.org/10.1103/PhysRevLett.129.090403
257. T. Hernández Yanes, G. Žlabys, M. Płodzień, D. Burba, M. Sinkevičienė, E. Witkowska, G. Juzeliūnas, Phys. Rev. B **108** (2023). https://doi.org/10.1103/PhysRevB.108.104301
258. M. Dziurawiec, T. Hernández Yanes, M. Płodzień, M. Gajda, M. Lewenstein, E. Witkowska, Phys. Rev. A **107** (2023). https://doi.org/10.1103/PhysRevA.107.013311 270
259. Z. Wang, S. Singh, M. Navascués, Phys. Rev. Lett. **118**(23) (2017). https://doi.org/10.1103/PhysRevLett.118.230401 270
260. M. Navascués, S. Singh, A. Acín, Phys. Rev. X **10**(2) (2020). https://doi.org/10.1103/PhysRevX.10.021064
261. M. Navascués, F. Baccari, A. Acín, Quantum **5**, 589 (2021). https://doi.org/10.22331/q-2021-11-25-589
262. M. Płodzień, J. Chwedeńczuk , M. Lewenstein, Inherent quantum resources in the stationary spin chains. arXiv (2024). https://arxiv.org/abs/2405.16974 270
263. N. Mermin, Phys. Rev. Lett. **65**, 1838 (1990). https://doi.org/10.1103/PhysRevLett.65.1838 270
264. M. Ardehali, Phys. Rev. A **46**, 5375 (1992). https://doi.org/10.1103/PhysRevA.46.5375
265. A. Belinskii, D. Klyshko, Phys. Uspekhi **36**, 653 (1993). https://doi.org/10.1070/PU1993v036n08ABEH002299 270
266. R. Werner, M. Wolf, Phys. Rev. A **64** (2001). https://doi.org/10.1103/PhysRevA.64.032112 270
267. H. Weinfurter, M. Żukowski, Phys. Rev. A **64** (2001). https://doi.org/10.1103/PhysRevA.64.010102

268. M. Żukowski, Č. Brukner, Phys. Rev. Lett. **88** (2002). https://doi.org/10.1103/PhysRevLett.88.210401 270
269. A. Fine, Phys. Rev. Lett. **48**, 291 (1982). https://doi.org/10.1103/PhysRevLett.48.291. https://link.aps.org/doi/10.1103/PhysRevLett.48.291 271
270. J. Tura, R. Augusiak, A.B. Sainz, T. Vértesi, M. Lewenstein, A. Acín, Science **344**(6189), 1256 (2014). https://doi.org/10.1126/science.1247715 271
271. G. Müller-Rigat, A. Aloy, M. Lewenstein, M. Fadel, J. Tura, Three-outcome multipartite Bell inequalities: applications to dimension witnessing and spin-nematic squeezing in many-body systems. arXiv (2024). https://arxiv.org/abs/2406.12823 271
272. A. Aloy, G. Müller-Rigat, J. Tura, M. Fadel, Deriving three-outcome permutationally invariant Bell inequalities. arxiv (2024). https://arxiv.org/abs/2406.11792 271
273. A. Niezgoda, J. Chwedeńczuk, Many-body nonlocality as a resource for quantum-enhanced metrology. Phys. Rev. Lett. **126**(21) (2021). https://doi.org/10.1103/PhysRevLett.126.210506 272
274. D. Sauerwein, N.R. Wallach, G. Gour, B. Kraus, Phys. Rev. X **8**(3) (2018). https://doi.org/10.1103/PhysRevX.8.031020 272
275. N.K.H. Li, C. Spee, M. Hebenstreit, J.I. de Vicente, B. Kraus, Identifying families of multipartite states with non-trivial local entanglement transformations (2023). https://doi.org/10.48550/ARXIV.2302.03139. https://arxiv.org/abs/2302.03139 272
276. W. Bruzda, S. Friedland, K. Życzkowski, Linear Multilinear Algebra, 1–64 (2023). https://doi.org/10.1080/03081087.2023.2211717 272
277. G. Tóth, T. Vértesi, Phys. Rev. Lett. **120** (2018). https://doi.org/10.1103/PhysRevLett.120.020506 273

www.ingramcontent.com/pod-product-compliance
Ingram Content Group UK Ltd.
Pitfield, Milton Keynes, MK11 3LW, UK
UKHW021833270726
14058UKWH00001B/130

* 9 7 8 3 0 3 1 5 5 6 5 6 2 *